Worldwide Perspectives on Geographical Indications

Emilie Vandecandelaere
Delphine Marie-Vivien • Erik Thévenod-Mottet
Maria Bouhaddane • Valérie Pieprzownik
Florence Tartanac • Ida Puzone
Editors

Worldwide Perspectives on Geographical Indications

Crossed views between researchers, policy makers and practitioners

 Springer

Editors

Emilie Vandecandelaere
Food and Agriculture Organization
Rome, Italy

Delphine Marie-Vivien
CIRAD
Montpellier, France

Erik Thévenod-Mottet
IPI
Bern, Switzerland

Maria Bouhaddane
CIRAD
Montpellier, France

Valérie Pieprzownik
Food and Agriculture Organization
Rome, Italy

Florence Tartanac
Food and Agriculture Organization
Rome, Italy

Ida Puzone
oriGIn
Genève, Switzerland

ISBN 978-3-031-71640-9   ISBN 978-3-031-71641-6   (eBook)
https://doi.org/10.1007/978-3-031-71641-6

This work was supported by Food and Agricultural Organisation

The original submitted manuscript has been translated into English. The translation was done using artificial intelligence. A subsequent revision was performed by the author(s) to further refine the work and to ensure that the translation is appropriate concerning content and scientific correctness. It may, however, read stylistically different from a conventional translation.

The opinions expressed in this publication are those of the authors/editors and do not necessarily reflect the views of the FAO: Food and Agriculture Organization of the United Nations, its Board of Directors, or the countries they represent.

This Springer imprint is published by the registered company Springer Nature Switzerland AG
The registered company address is: Gewerbestrasse 11, 6330 Cham, Switzerland

If disposing of this product, please recycle the paper.

Acknowledgement

This book is the result of many rich contributions prepared by different authors to provide crossed views on geographical indications development in the world. All the authors (see list page 168) are all thanked for their great contributions on various topics, their specific knowledge and their interest in sharing their experience or research, as well as for their patience in addressing the editor committee requests.

Contents

About the Authors[1]

Neşe Altıntaş graduated from the Faculty of Agriculture at Gaziosmanpaşa University as a top student. She earned her Master's degree at Gaziosmanpaşa University. She earned a doctor's degree at the Graduate School of Natural and Applied Sciences, Ankara University, in 2015 with a GPA of 4.00/4.00. The thesis she prepared was awarded as the "Thesis with the Highest Publication Score" among the theses prepared in the department that year. She started her career at the Ministry of Agriculture and Forestry in 1993. She has been working as an Expert in Agriculture and Forestry, and Geographical Indications Coordinator.

Fadhila Bacha is a PhD student at the Doctoral School of Economics and Management Sciences (EDEG), Agro Institute of Montpellier and Montpellier Interdisciplinary Center on Sustainable agri-food systems—Social and Nutritional Sciences research, research unit (UMR MoISA). Her thesis is focused on the role of collective action and stakeholder strategies in improving the sustainability and competitiveness of the GI value chain through the case study GI *Deglet Nour of Tolga* (Algeria).

Giovanni Belletti is a Full Professor in Agricultural Economics and Policy at the Department of Economics and Management, University of Firenze. His main research areas in economics concern quality in the agri-food systems with particular reference to the valorisation of typical products and the protection of geographical indications, value chains and alternative food networks, food and rural development governance and policies.

Claire Bernard-Mongin works on supporting environmental transitions in territories. A political scientist by training, she has specialised in the management of natural resources in an international context. Holding a doctorate in management sciences, she is particularly interested in the construction of collective strategies with

---

[1] The order follows the first appearance of the names of the authors in the book.

an agro-environmental purpose. She contributes to establishing and equipping spaces for dialogue and joint action between practitioners, knowledge producers and public policies. She has worked at the International Centre for Advanced Mediterranean Agronomic Studies, then at the Ministry of Agriculture and of Food in France, before joining the International Centre for Agricultural Research for Development in 2020, where she mobilises labels and signs of quality and origin in support of agroecological and climate transitions for territories.

Caroline Blot is the Head of the Wines, Ciders and Spirits division at INAO, and in charge of the technical and regulatory expertise of recognition and modification files of specifications and of the technical and regulatory monitoring of AOP, PGI and IG sectors in the wine, cider and spirits sectors.

Maria Bouhaddane is a Researcher at Cirad (French Agricultural Research Centre for International Development). She holds a PhD in Management Science from Clermont Auvergne University and specialises in Food Economics and Marketing. Her research centres around sustainable and origin-linked labelling strategies, in particular geographical indications, and their impact on consumer behaviour.

Tifenn Corre is an Engineer in Statistics, Econometrics and Databases within the Rural Development Observatory (US-ODR) unit of INRAE. She manages data related to agricultural employment and aid from the second pillar of the common agricultural policy, integrated into the information system of the ODR's agricultural systems and policies. She uses this information system for her technical assistance mission in monitoring and evaluating rural development programmes and for conducting statistical analyses such as impact assessments of public policies or typologies of territories.

Patricia Covarrubia a Reader in Law at The University of Buckingham, is an IP consultant at the Latin America IPR SME Helpdesk. She holds degrees from Venezuela, Southampton, and Brunel University London. She has taught at several institutions, including Brunel University and BPP, School of Law, London. Dr Covarrubia manages the IPTango blog, a top-rated IP weblog. Her research spans geographical indications, compulsory licences in pharmaceuticals, fashion and traditional knowledge protection. She has received numerous awards and grants, published extensively and is an active researcher.

Maurizio Crupi holds a joint doctorate from Alicante and Maastricht Universities, with a doctoral thesis funded by the European Commission under the Marie Skłodowska-Curie Action. Maurizio is *colaborador honorifico* of the Commercial Law Department of the University of Alicante, where he lectures on intellectual property law. Maurizio is currently working as a legal assistant at the Boards of Appeal of the European Union Intellectual Property Office (EUIPO).

Sertaç Dokuzlu graduated from Ankara University, Faculty of Agriculture, Department of Dairy Technology in 1992. She started working as a Research Assistant at Bursa Uludag University Agricultural Faculty Department of Agricultural Economics in 1994. She received the title of Professor in 2000 in the field of Agricultural Management. Prof. Dokuzlu has been working on geographical indications and carrying out national and international researches and projects since 1995

Hart N. Feuer is an Associate Professor of Food Studies at Kyoto University, Japan. His work has centred around the valorisation of traditional cuisines through inter-generational knowledge transfer, heritage food certifications and sociological approaches to nutrition. His work focuses mostly on East Asia, with recent topics including wild edible plant cuisine (jungle food), food literacy and agri-food geographical indications.

Gilles Flutet is the Head of the Territories and Delimitation Service at INAO, and in charge of defining the production areas of IGs and, in this capacity, leads a reflection with his teams on the adaptation of production areas to climate change

Fatiha Fort is a Full Professor at the Institut Agro de Montpellier and conducts her research activities within the UMR MOISA (Joint Research Unit on Markets, Organizations, Institutions and Actors' Strategies). Her research is at the interface between strategy and marketing applied to the territory. She has been interested in the marketing of local products, products under labels and sustainable consumption. More recently, her work has focused on place branding and territorial marketing.

Stéphane Fournier is a Professor in Economics at the Institut Agro Montpellier and a Member of the Joint Research Unit "Innovation". He has been working on local innovation processes and product qualification and their interactions with territorial development for 25 years, with a particular interest in Geographical Indications. He has conducted research on this subject mainly in France, West Africa (Senegal, Benin and Cameroon) and Southeast Asia (Indonesia and Vietnam).

Juan J. Ferrero-García is a biologist. He began his professional career at the Doñana Biological Station (C.S.I.C.). In 1997, he joined the Administration of the *Junta de Extremadura*, where he has worked since then in different units of the regional administration (Extremadura Agri-food Laboratory, Plant Health Service and Nature Conservation Service). Since 2015, he has been the Head of Section of Designations of Origin of the Agricultural and Food Quality Service. In addition, he is the author or co-author of more than 30 academic and outreach publications.

Jacques Gautier National Inspector of the INAO, ensures technical monitoring and relations with research and development organisations, particularly on issues related to climate change and the environment, Vice-President of the expert group "Sustainable Development and Climate Change" of the OIV.

Vicente Gimeno Beviá is an Assistant Professor of Commercial Law at the University of Alicante and an Academic Coordinator in the Trademarks and Designs Module of the Master in Intellectual Property and Digital Innovation, Magister Lvcentinvs. He is the author of 2 books and more than 20 papers in reputable journals and collective books about different topics including intellectual property and corporate law. He has participated in relevant public research projects funded by European and national authorities. He has been a Visiting Researcher at Max-Planck-Institut for Comparative and International Private Law in Hamburg and at the University of Miami.

Xin Gu is an Associated Professor at the Intellectual Property Development and Research Center of CNIPA. He holds a PHD in Law. He investigated the protection status of geographical indications in China, the economic contribution of geographical indication products, and the legislation of specialised laws on geographical indications.

Paola Guerrero Andreu is an Intellectual Property Lawyer. She holds a Master's in Intellectual Property, Lvcentinvs Magister (ML/LLM) from the University of Alicante. She is the Coordinator of the Seal of Origin Program at the National Institute of Industrial Property, (INAPI) Chile.

Fatima El Hadad-Gauthier is a Lecturer-Researcher at the Mediterranean Agronomic Institute of Montpellier (CIHEAM-IAMM). She holds a PhD in Economics from the University of Montpellier-I (France). She is a Member of the MoISA (Montpellier Interdisciplinary Center on Sustainable agri-food systems—Social and Nutritional Sciences) research unit. Her research focuses on the sustainability of food value chain: organisational models, governance and actors' strategies.

Gero Laurenz Höhn earned his PhD in Business Economics from KU Leuven in 2023, where he currently serves as a Research Associate. Since his Master's studies, he has been intrigued by the promises and constraints of terroir, aiming to demystify its implications for traditional wines and foods. Specialising in food economics and business, his research primarily assesses critically the sustainability potential of geographical indications, with a focus on aspects such as prices, nutrition, animal welfare and consumer preferences. Besides his research, he is a Lecturer at IMC Krems, teaching on international agriculture and trade policies.

Martijn Huysmans is an Associate Professor at the School of Economics, Utrecht University, where he teaches in the Politics, Philosophy, and Economics (PPE) bachelor programme. His research focuses on the Political Economy of the EU, with Geographical Indications as his main specialty. He obtained his PhD at KU Leuven in 2018 and was a Visiting Fellow at LUISS University Rome in 2022.

Özden İlhan graduated from the Department of Food Engineering, Hacettepe University, in 2011. She holds a PhD in Food Science from Ankara University, where she specialised in classification of geographical origin of agricultural products. She has been working as an Intellectual Property Examiner at the Turkish Patent and Trademark Office since 2016 under the Geographical Indications Department. She is currently working as a Fellow at Lisbon Registry for the International Protection of Appellations of Origin (AOs) and Geographical Indications (GIs) under the World Intellectual Property Organization (WIPO).

Pape Tahirou Kanouté is the Executive Director of the ETDS (Economy Territories and Development Services) study office, now an NGO, which he founded in 2012. Pape Tahirou Kanoute is an agro-economist engineer specialised in organisation and quality in agricultural and agri-food supply chains and in economy and governance of territories. He has been working for 15 years on territorial development and geographical indications in West Africa, particularly in Senegal. He coordinates the project for the registration of the GI *Casamance madd* supported by the FAO, WIPO and the GI Facility (CIRAD/AFD). He was a member of the working group responsible for preparing the establishment of the National Committee of GIs in Senegal (CNIG). He is co-opted as an expert technical support to this committee.

Junko Kimura is a Professor of Marketing at the Faculty of Business Administration, Hosei University in Japan. She was a Visiting Professor at the University of Ca'Foscari, Venezia. She holds a PhD in Commerce from Kobe University, and an MA in Communication from the State University of New York, College at Brockport. Her research focuses on Rural Development, Agri-food Marketing, SDGs and Geographical Indications (GI). She consults the Ministry of Agriculture, Forestry and Fisheries in Japan for registration of GI applicants, and the Ministry of Finance for Geographical Indication alcohol.

Johann Kirsten is the Director of the Bureau for Economic Research since 1 August 2016. Before this appointment, he was the Head of the Department of Agricultural Economics, Extension and Rural Development in the Faculty of Natural and Agricultural Sciences at the University of Pretoria for 20 years. His research activities cover a number of themes but remain focused on critical aspects of agricultural policy in South Africa. In addition, recent topics include a research programme evaluating the economic options for establishing geographical indications and related certification schemes in South African agriculture and food markets. The establishment and commercialisation of South Africa's first non-wine geographical indication, namely "Karoo lamb", is a direct result of this research.

Andrea Marescotti is an Associate Professor of Agricultural Economics at the Department of Economics and Management, University of Florence (Italy). His main research interests range between theoretical and empirical studies on farmers' marketing strategies; socio-economic analysis and evaluation of geographical indications protection; short supply-chains analysis.

Delphine Marie-Vivien joined CIRAD in 1999, a French public institute dedicated to agricultural research for developing countries. She was a Visiting Researcher at NLSIU, Bangalore, India, from 2005 to 2008. She defended her PhD thesis on "The law of GI in India compared to French, EU and International laws", Paris University. She was, from 2012 to 2018 in Vietnam, working on GIs for AFD, EUIPO, FAO, UNIDO, Swiss IPI in Asia, South America and Africa by elaborating GI laws, examination manual, specifications, control systems and producer's association. She organises the international training on GIs InterGI. She teaches IP and GI law, writes scientific papers and supervises PhD students.

Julia Martín-Cerrato is a graduate of Veterinary Science, specialising in Bromatology and Nutrition. Her professional activity has always been linked with the Administration of the *Junta de Extremadura*. She began in 1999 as a Public Health Veterinarian in slaughterhouses and health centres, later joining the Department of Agriculture and Livestock, where she has worked in various sections related to the Agricultural and Food Quality and Agricultural Production. She was the Head of the Agricultural and Food Quality Service until the end of 2023.

Laurent Mayoux Is Deputy of the INAO's territories and delimitation service, a specialist in vineyards, the Vice-President of the expert group "Genetic Resources and Vine Selection" of the OIV (International Organisation of Vine and Wine) and leads the reflection on the adaptation of production areas to climate change.

Samir Messaili was born in Constantine, Algeria, and studied agronomy at the Agricultural Institute of Guelma. He is an agricultural advisor (1st Degree) and was a civil servant in the administration of Agriculture between 1988 and 2023. He has been a founding member and vice-president of the IMESSENDA association for the promotion and protection of the *Bouhezza cheese* denomination since 2016 and was responsible for the labelling project of *Bouhezza cheese* from 2018 to 2023.

Sylvette Monier-Dilhan a research officer at INRAE, has conducted research in the economics of quality: consumer behaviour and organisation of sectors. For her applied research, she has mobilised and processed numerous individual data files from surveys or administrative records. Her exchanges with institutions, data provider partners and economic actors have allowed her to feed her research issues and contribute at the national level to the creation of the Economic Observatory of SIQO.

Latha R. Nair currently chairs the Trademark, Copyright and Geographical Indications (GIs) practice at K&S Partners, India. A passionate advocate of GI protection, she has extensive experience of over 25 years in advising and protecting some well-known domestic and foreign GIs. She has also published widely on the topic of protection of GIs, including a co-authored book in 2005 titled *Geographical Indications: A search for identity*. All her publications are available on her firm's website www.knspartners.com. She is currently a principal editor of the searchable online guide on GIs, Certification Marks and Collective Marks published by the International Trademark Association (INTA).

Orachos Napasintuwong is an Associate Professor at the Department of Agricultural and Resource Economics, Kasetsart University. Her teaching and research areas are economics of biotechnology, consumer preferences, economic analysis of technology adoption and diffusion, seed industry, agricultural value chain, food policy and rice economy in Southeast Asia. She has published several international articles including consumer preferences for geographical indication rice and sustainability of geographical rice, among others, and serves as Associate Editor of *Asian Journal of Applied Economics*, *Asian Journal of Agriculture and Development* and *Agro Ekonomi*. She received her PhD in Food and Resource Economics from the University of Florida.

Daniël van Noord obtained his Bachelor's degree in Economics at Utrecht University with a thesis on Geographical Indications and quality. In 2022, he obtained a Research Master's in History and a Master's in Strategic Management from the Rotterdam School of Management. He currently works as a Software Engineer in Utrecht.

Alexandra Ognov is the Head of the Agricultural and Agri-food Products division of the INAO, and in charge of the technical and regulatory expertise of the files for recognition and modification of specifications and of the technical and regulatory monitoring of the AOP, PGI and TSG sectors of the agri-food industry.

Chitra Parayil is a Professor in the Department of Agricultural Economics at the College of Agriculture, Vellanikkara, Thrissur, Kerala. She has 22 years of teaching experience across all degree levels; Bachelor's, Master's and PhD. She has nine peer-reviewed journal articles and completed three externally aided research projects funded by ICSSR, Centre for Development Studies and the State Planning Board and two State Plan projects funded by the Kerala Agricultural University (KAU). She is currently working on a project funded by the KAU state plan on developing a model supply chain for smallholder vegetable farmers in Kerala.

Claire Philippoteaux is a Political Scientist from the Institute of Political Studies—Sciences Po, from Bordeaux in France, with a Master's degree in Public and Political Communication from the same institute. She has been working from Colombia for 18 years, in fundraising, management and implementation of international cooperation projects. Since 2013, she coordinates the Colombian-Swiss intellectual property project—COLIPRI, with a focus on Geographical Indications. She has worked on the topic of GIs and in general the signs of collective vocation with resources from the Swiss, American government and United Nations agencies in Colombia and in Sri Lanka. She organises training spaces on DOs for Colombia and Latin America in general.

Barbara Pick is an International Consultant for the Food and Agriculture Organization of the United Nations (FAO) and the World Intellectual Property Organization (WIPO) and a Research Associate with the French Agricultural Research Centre for International Development (CIRAD). Her interests include the issues related to intellectual property and development, including geographical indications and farmers' rights in the context of agricultural development.

Valérie Pieprzownik works as an expert in Geographical Indications at the Food and Agriculture Organization of the United Nations (FAO) in the Food and Nutrition Division since 2023. She is an agricultural and environmental engineer with a Master's degree in Law. She comes from the French Ministry of Agriculture where she dealt with the whole economy of the wine sector as the Head of the Wine and Other Beverages Office and where she was also for many years the Head of the Quality Office, being in charge of all the official modes of valorisation of products including GIs and organic farming.

Yudy Paola Pineda Suarez is an Industrial Engineer from the Universidad Distrital Francisco José de Caldas in Bogotá (Colombia), with a specialisation in Hygiene, Safety and Health at Work from the same university. She has taken courses on Trademarks and Industrial Property with the National University of Colombia and the Superintendency of Industry and Commerce of Colombia. She was the Legal Representative of the Federation of Entrepreneurs of the Productive Chain of Bocadillo Veleño—FedeVeleños, who manages the Denomination of Origin "Bocadillo Veleño". She is currently the Administrator of the Bocadillo Producer La Selección SAS, located in Moniquirá, Boyacá, Colombia, and continues to be part of the Board of Directors of FedeVeleños.

Kerry Purcell is a Senior Lecturer of Law at the University of Buckingham and brings a blend of skills and knowledge to the field of international governance. Her research and publications primarily focus on ethical considerations such as respecting cultural representation within international governance and medical law. She has also worked in healthcare advocating for patient autonomy. Dr Purcell further sits on fitness to practice medicine panels and conducts investigations herself in terms of determining fitness to study/ practice medicine. Additionally, Dr Purcell is a specialist book reviewer for Oxford University Press.

Ida Puzone is the Operations and Member Relations Manager at oriGIn, the global alliance of GIs. In this role, she works with the Managing Director to implement the organisation's strategies and main activities, oversee specific projects and manage relationships with members. Ida began her career at an Italian law firm. Following this, after a period at the international group "Pirelli Cavi e Sistemi Spa", she moved to the Office for Harmonization in the Internal Market (OHIM, now EUIPO) in Alicante, Spain, where she focused on intellectual property. Ida holds a law degree from the University Federico II in Naples, Italy, and an LLM in Intellectual Property and Information Technology (Magister Lvcentinvs) from the University of Alicante.

A. M. Radhika currently serves as an Assistant Professor in the Department of Agricultural Economics, School of Agricultural Sciences, Amrita Vishwa Vidyapeetham, Coimbatore Campus. She has a doctoral degree in Agricultural Economics. Her contributions are mainly in the fields of intellectual property rights, Impact evaluation, traditional rice cultivation, agricultural marketing and so forth.

Julie Regolo is an Engineer in Statistics, Econometrics and Databases within the Rural Development Observatory (US-ODR) unit of INRAE. She is responsible for the development and animation of the Territorial Observatory of Signs of Identification of Quality and Origin (OT-SIQO). After obtaining her PhD in Development Economics and International Trade from the University of Geneva, she worked as an Economic Studies Officer at INAO for several years. She participates in studies and research articles in economics and econometrics on SIQO and on the impact of public policies.

Cyrille Rigolot is a Research scientist at France's National Research Institute for Agriculture, Food and Environment (INRAE) Clermont-Ferrand, Joint Research Unit (UMR) Territoires, France. He was a Visiting Scientist at the Research Institute for Humanity and Nature (RIHN) Kyoto and the Institute for Future Initiatives (IFI) Tokyo, Japan, from 2020 to 2022; a Postdoc at Commonwealth Scientific and Industrial Research Organisation (CSIRO) Brisbane, Australia, from 2014 to 2015; and a Research Engineer at INRAE Rennes, France, from 2007 to 2011. His main research interests include agroecology, livestock farming systems, sustainability transformations and transdisciplinarity.

Miranda Risang Ayu Palar is the head of Intellectual Property Centre on Regulation and Application Studies, Faculty of Law, Universitas Padjadjaran, Indonesia. She teaches aspects of Intellectual Property Law, especially Geographical Indication and Communal Intellectual Property. She was a Visiting Lecturer/Examiner in Australia (UTS), Japan (JIII/JIPII), Singapore (SMU), Thailand (Kasetsart Uni) and India (KLA). As a speaker/expert, she was also involved in international seminars, conferences, workshops, research and a number of multilateral negotiations (WIPO). Her concepts have been accommodated in the Indonesian Law on Trademarks and Geographical Indications 20/2016 and the Indonesian Government Regulation on Communal Intellectual Property 56/2022.

Suzana Romeiro Araújo is an Adjunct Professor at the Socio-Environmental and Water Resources Institute of the Federal Rural University of Amazonia (UFRA), Brazil at the graduate level and a Permanent Professor of the Postgraduate Program in Agronomy at UFRA. She is an Agronomist Engineer whose principal research interests are the investigation of soil quality indicators and the development of agricultural landscapes. She has a PhD in Soil Sciences from the University of São Paulo (USP).

Maimouna Sambou is a facilitator of producers organisations and support NGOs. She holds a first-cycle international diploma for the training of peasant leaders and managers. She has supported numerous management committees for rural development projects. She has actively invested in the process of setting up the *Casamance madd* GI, a topic she defended at an international agriculture conference in Rome in 2019. In November 2019, she was appointed Head of the Association for the Protection and Promotion of the Geographical Indication "Madd de Casamance" (APPIGMAC), bringing together the actors of the madd sector from the three regions of natural Casamance.

Luis F. Samper leads 4.0 Brands based in Bogota, Colombia, and acts as Sustainability Manager at oriGIn. He has over 20 years of experience in collective marketing, sustainability and intellectual property strategies for different products in Latin America, Asia and Africa. Before becoming a consultant, he led Café de Colombia's IP, promotional and communication strategies around the world, developing new brands including the Juan Valdez® Café brand architecture. He is frequently invited as a guest speaker in several global venues, including WIPO's biannual Geographical Indications Symposium. He is an Economist, has a Master's in Law (MML) from Uniandes, a Specialist in Intangible assets and valuation from Uniexternado in Colombia and has a Master's in Business Administration (MBA) from Columbia University, New York.

Kae Sekine is a Professor of Agricultural Economics at Graduate School of Economics, Aichi Gakuin University, Japan. She obtained her PhD in Economics from Kyoto University in 2011. She has professional experiences at INRA (2007-2010), Rikkyo University (2011-2014), CFS/HLPE (2012-2013) and FAO (2018-2019). She is a co-author/editor of Alessandro Bonanno, Kae Sekine, Hart N. Feuer (Eds.). 2019. Geographical Indications and Global Agri-Food: Development and Democratization, Routledge. Her research focuses on sustainable agri-food system and institutions that contribute to its construction, including geographical indications, world agricultural heritages, organic public procurement, among others.

Nazlı Şimşek graduated from the Department of Food Engineering, Hacettepe University, in 2004 with a graduation thesis on Cross-linking of Milk Protein Molecule. She has worked in a wide variety of areas in the Ministry of Agriculture and Forestry, in the Internal Audit Unit of the Agriculture and Rural Development Support Institution, IPARD Management Authority and Marketing Department since 2009, and she still continues to work in the geographical indications unit. Many projects are being carried out, including projects for producer systems, projects using TAIEX tools and geographical indications.

Fanta Sow is a facilitator of the association for the protection and promotion of the geographical indication madd from Casamance (APPIGMAC). Fanta Sow holds a degree in Secretarial and Commercial Assistance. She has also specialised in marketing and communication and in the transformation of local products with the ITA (Institute of Food Technology). Recruited in 2020 as a facilitator by APPIGMAC with the support of WIPO, she has put her experience and knowledge of the agri-food sector in Casamance at the service of the project to label the *Casamance madd* as a geographical indication.

Paulo de Tarso Anunciação de Melo has a Law degree from the University of Amazonia and a Master's degree in Intellectual Property and Technology Transfer for Innovation from the PROFNIT Graduate Program at the Federal Institute of Pará (IFPA, Brazil). He has an MBA from the Massachusetts Institute of Business— Global Business Management—ABRACOMEX. He is currently a mentor at the Guamá Science and Technology Park (Belém, Brazil) in the area of Intellectual Property, a member of the Technical Forum for Geographical Indication and Collective Trademarks of the State of Pará (BRAZIL), Secretary General of the Intellectual Property and Innovation Commission of the Brazilian Lawyers' Organization and Founding Partner of the law firm FERREIRA MELO BARROSO (FMB ADVOCACIA).

Florence Tartanac has been working at the Food and Agriculture Organization of the United Nations (FAO) since 2001, and she is Senior Officer in the Food and Nutrition Division at FAO Headquarters in Rome, Italy. She is the Team Leader of the Market Linkages and Value Chains team and her areas of expertise are the following: sustainable food value chain development; geographical indications; public food procurement; small and medium food enterprises development. Before FAO, she worked 10 years in Guatemala, for the French Cooperation, the Institute of Nutrition for Central America and Panama (INCAP) and the United Nations Industrial Development Organization (UNIDO). She also worked 5 years in the FAO regional office for Latin America and the Caribbean in Santiago, Chile. As academic background, she is a Food Engineer and has a PhD in Economical Geography from Paris University.

Erik Thévenod-Mottet is an Advisor for geographical indications at the Swiss Federal Institute of Intellectual Property. He has previously taken part in a number of European scientific research projects on GIs and worked for a GI certification body and for wine organisations in Switzerland. His work focuses both on legal issues relating to GIs and on the ethnobiological aspects of GIs.

Emilie Vandecandelaere is an agricultural and food economist, working at FAO since 2007 as Food Quality and Geographical Indications Specialist, with areas of expertise in marketing, territorial markets, sustainable food systems, nutrition and investments. Building on technical assistance in Latin America, Africa, Eastern Europe and Asia, she developed various publications to support concrete actions and benefits. Before joining FAO in 2007, Emilie was a civil servant for four years at the French Ministry of agriculture, General Directorate on Food, working on labelling and nutrition standards, and the national food policy.

Massimo Vittori is the Head of oriGIn—the global alliance of Geographical Indications, and is in charge of the organisation's planning and management. Having previously worked for several UN development agencies, he has a 20-year experience in intellectual property rights, trade laws, the protection and promotion of geographical indications, sustainability issues related to geographical indications and cooperation activities involving public and private stakeholders. He holds a Master of Laws (LLM) in Intellectual Property from the University of Turin and a Master's degree (DEA) in International Law from the Geneva Graduate Institute of International and Development Studies. Massimo regularly delivers seminars and lectures on intellectual property, geographical indications and trade-related issues in several European universities.

Hui Xu is an Associated Professor at the China Center for International Economic Exchange, after an experience as Research Fellow in the intellectual Property Development and Research Center of CNIPA for 15 years. He holds a PhD in Economics. He investigated the protection status of geographical indications in China, the economic contribution of geographical indication products and the legislation of specialised laws on geographical indications.

Xuan Yang is an Associated Professor at the Intellectual Property Development and Research Center of CNIPA. He holds a PhD in Management. He investigated the protection status of geographical indications in China, the economic contribution of geographical indication products and the legislation of specialised laws on geographical indications.

Andrea Zappalaglio is a Lecturer in Intellectual Property Law at the School of Law of the University of Sheffield (U.K.). Prior to that, he worked as a Senior Research Fellow at Max Planck Institute for Innovation and Competition (Munich, Germany), where he led the research team on the Law of Geographical Indications.

Chapter 1
Introduction—A Worldwide Perspective of Geographical Indications in a Time of Changes: Crossed Views Between Researchers, Policy Makers and Practitioners

Emilie Vandecandelaere, Delphine Marie-Vivien, Erik Thévenod-Mottet, Maria Bouhaddane, Valérie Pieprzownik, Florence Tartanac, and Ida Puzone

Abbreviations

ASEAN	Association of Southeast Asian Nations
PDO	Protected Denomination of Origin
PGI	Protected Geographical Indications
SDG	Sustainable Development Goals
TSG	Traditional Specialty Guaranteed
TRIPS	Trade Related Aspects of Intellectual Property rights agreement

Geographical indications (GIs) are signs used to designate products having a specific geographical origin and possessing qualities or a reputation that are due to that origin. Actually, the world's food and artisanal heritage encompasses a multitude of products linked to their origin that rely on the knowledge, skills, practices and traditions developed collectively by local producers over time and transmitted across generations. GIs are an intellectual property right granting exclusive right to use the

E. Vandecandelaere (✉) · V. Pieprzownik · F. Tartanac
Food and Agriculture Organization, Rome, Italy
e-mail: emilie.vandecandelaere@fao.org

D. Marie-Vivien · M. Bouhaddane
International Centre for Agricultural Research for Development (CIRAD), Montpellier, France

E. Thévenod-Mottet
Swiss Federal Institute of Intellectual Property (IPI), Bern, Switzerland

I. Puzone
oriGIn – the global alliance of Geographical Indications, Geneva, Switzerland

© The Author(s) 2025
E. Vandecandelaere et al. (eds.), *Worldwide Perspectives on Geographical Indications*, https://doi.org/10.1007/978-3-031-71641-6_1

1

name that can particularly benefit local producers, and particularly smallholders, and be therefore a tool to preserve and promote origin.

While the commercial success of origin products is very ancient since Antiquity, laws governing the registration and protection of the names of the origin products as Geographical Indications have mushroomed around the world in a very diverse manner, translating the variety of meanings and objectives attached to them. These range from market-based approaches targeting unfair competition practices, to approaches directed at non-market objectives, including territorial development, and preservation of cultural heritage and natural resources, driven by public policies inspired by success stories in context of long-standing GI development.

Considering the plurality of actors and interests involved in the public and private sectors, researchers in all parts of the world have explored the connections between GIs and their many dimensions. These include, among others: the nature of the links between the products' qualities and their geographical origin and how it is translated into specifications; the role of public authorities; the collective organization of producers; quality control systems; governance and enforcement mechanisms; the recognition of the specific know-how of local producers, including women and indigenous and local communities; international and national protection, including the type of legal protection; marketing issues; biodiversity conservation; preservation of environment and cultural heritage; sustainable development; food heritage and healthy diets when related to food products; and tourism.

In this context, in 2022, to give space and time for sharing knowledge and experiences on these topics among researchers, policies makers and practitioners, the Food and Agriculture Organizations of the UN and the French agricultural research and cooperation organization (Cirad) in collaboration with the Swiss Institute for Intellectual Property and oriGIn, have organized in Montpellier the conference "*WORLDWIDE PERSPECTIVES ON GEOGRAPHICAL INDICA-TIONS: An international conference for researchers, policy makers and practitioners*". The conference gathered more than 200 researchers, public authorities, producers and their collective organizations, public authorities and international organizations, coming from 47 countries representing all the regions of the world: Europe, Asia/Oceania, Africa, South and North America, who showed the importance of GI and of its multitude facets, through new perspectives, approaches and practices that have led to an increasingly complex, heterogeneous, dense and evolving picture of GIs.

The core objective of this book is to reflect the variety and density of the topics raised during the conference, regarding the nature of GIs and their development in various contexts of the world. This analysis is reflected in xx chapters organized in four parts:

- Part I explores the unique nature of GIs as intellectual property right,
- Part II provides an overview of GI protection systems and the role of public actors;
- Part III looks at the relation between GIs and territorial development; and
- Part IV showcases the sustainability of GIs.

The originality of this book, disseminating the outcomes of the Conference, resides in its multi-actors perspective. Practitioners describe their experience in implementing GIs in the field and public authorities their roles in regulating and supporting GIs at international, national and product level, both providing important insights and specific approaches to enrich the landscape of GI trends. Academics analysis through empirical and theoretical research GI strength, weaknesses and challenges, in different contexts, with a comparative approach, to put light on certain phenomenon. Crossing stakeholders views in each part offers a dynamic and multifaceted understanding of the topics and possible evolutions, as described in details below.

1.1 Part I: The Unique Nature of GIs as Intellectual Property Right

More than for other intellectual property rights, the legal history of GIs is characterized by a variety of concepts, a diversity of legal instruments and overlaps with other areas of law. Over the last decades, this history has been particularly rich in developments, at both national and international level. Looking at international harmonization, we are still in the middle of the road. It is not surprising, therefore, that legal issues relating to GIs and to nature of this unique concept are abundant and topical, with this part providing relevant and diverse overviews.

For cultural reasons, the concept of GI has emerged mainly in Southern Europe, first under the concept of appellation of origin and eventually the concept of geographical indication, the latter encompassing the former. The European Union (EU) common legal framework has evolved since its first introduction in 1992, with the last regulation recently adopted in April 2024, EU Reg. 2024/1143 merging GIs protection for wine, spirit drinks and agricultural products, as well as traditional specialties guaranteed and optional quality terms for agricultural products, together with for the first time a legal framework for the protection of GIs for craft and industrial products, the EU Regulation 2023/2411. The respective international references are the 1958 Lisbon Agreement for the Protection of Appellations of Origin and their International Registration (Lisbon Agreement), and the 1994 Trade-Related Aspects of Intellectual Property Rights Agreement (TRIPS Agreement) including a chapter on GIs. Maintaining this differentiation between AO and GI, as in the Geneva Act of the Lisbon Agreement of 2015, or choosing to recognise only one category, opens up fields of investigation that are still relatively unexplored about the nature of the link to the origin codified in the specifications.

Looking at the EU, from the perspective of consumer expectations and choice criteria, in Chap. 2, Martijn Huysmans, Daniël van Noord and Gero Laurenz Höhn point certain paradoxes within the vast corpus of PDOs and PGIs registered in the EU. Looking beyond a few specific PDO/PGIs questioning cases, characterized by the considerable extension of the geographical area, the extended area of origin of the raw material for some PDOs or the choice of the name of the GI referring to a place which is not within the defined geographical area (examples of Gouda and Stilton), the authors look at the 'black box' of the specifications. Behind the official symbols and logos that signal to consumers that they belong to the large family of GIs, the uniqueness of each GI, embedded in its specifications, poses a major challenge for consumers. Indeed, while belonging to the same GI family, not all GIs have the same meaning for consumers, world-famous names are different from 'small' GIs, as well as local scale GIs compared to international scale GIs.

The study by Tifenn Corre, Sylvette Monier-Dilhan and Julie Regolo on the combination of PDO and organic certification in the context of the French cheese market in Chap. 3 takes a similar view of the significance of GIs from consumer expectations and preferences. While, generally speaking, the two quality labels are complementary and do not conflict, the situation varies greatly depending on the level of awareness of the PDO in question. In short, there is certainly a GI family, but its members have contrasting fates, which makes it difficult to generalise. One might think that the distinction between PDOs and PGIs would at least show more uniform siblings. But this is not so obvious.

Indeed, in Chap. 4, Maurizio Crupi examines the difference between PDO and PGI, looking at amendments in specifications for processed meat products. This study raises a number of interesting points, on the evolving balance between tradition and innovation, or on divergent national trends towards stricter requirements or greater flexibility regarding the link to the origin. It does not, however, provide universal conclusions, due to the inherent limitations of the corpus under consideration and, here too, to the casuistry that each GI justifies as soon as looking into the details.

While the EU legal framework on quality signs includes since 1992 another legal, tool, the Traditional Specialty Guaranteed (TSG), it is still a blind spot that Vincente Gimeno Beviá's reveals in Chap. 5. While the TSG is not considered to be a GI, being a quality standard with no link to a specific geographical origin, the chapter shows that some TSGs are clearly on the fringes of GIs, either because certain elements of their specifications de facto imply a geographical origin, or because the name in question evokes a precise geographical origin. This is the case for the *Jámon serrano* TSG, which is the subject of this chapter. The discussions underway within the industry to decide whether or not to move from TSG to PGI shed new light on the issues and meanings of the GI concept.

The degree of sophistication of the GI "system" in the EU has evolved considerably, influenced by the context of the internationalization of the GI concept in other parts of the world, especially since the implementation of the TRIPS Agreement with a marked acceleration over the past decade, providing fertile ground for reconsidering the fundamentals of the legal conceptualization of GIs.

The experiences of South America, analyzed in Chap. 6 by Patricia Covarrubia and Kerry Purcell, and of Algeria, presented by Fadhila Bacha and Fatima El Hadad-Gauthier in Chap. 7, show the extent to which consistency between the legal and institutional framework and the motivations of producers is crucial for GI success. Does the success of GIs in Europe require a 'cultural' rather than a technological transfer? Can the recommendations put forward by the authors encourage the development of GIs in the interests of producers? These two chapters shed a harsh light on the limits of adopting GI legislation when the elements needed to bring it to life are lacking, such as the motivation of value chain actors to protect the reputation of their product.

In spite of a wide number of countries with GI legal and institutional framework, with now a broad consensus on the benefits that GIs can bring, in particular with many examples in developing countries, the risk of GIs existing only on paper, also called sleeping GIs is a real challenge.

Unexpectedly, silent GIs also exist in Europe, as described by Andrea Zappalaglio, Giovanni Belletti, and Andrea Marescotti's Chap. 8. Using an original methodology, the authors open up prospects for future work that could help to refine, through an analysis of failures, the conditions necessary for the success of a GI. Another "blind spot" in the GI ecosystem that needs to be addressed!

Finally, Latha R. Nair's contribution in Chap. 9 takes us into the virtual world, where the protection of GIs against frauds is becoming an important issue. The author shows that GIs are still struggling to gain a legitimate status similar to that enjoyed by trademarks in front of registered internet domain names comprising a GI name. The new EU Regulation 2024/1143 on PDOs and PGIs now explicitly addresses this issue.

In conclusion, the chapters of this part illustrate the many issues currently at stake in the conceptualization of GIs: a major work-in-progress that is in full swing, requiring dialogue between diverse and complementary experiences and perspectives.

1.2 Part II: GI Institutional Systems

In Part II the book explores the importance of the legal and institutional framework set up by public authorities in optimizing GIs objectives. GI public policies and institutional frameworks were primarily focused on ensuring the protection of the Intellectual Property right against frauds and infringement, as required by the TRIPS agreement, but are increasingly integrating other public policies, with a strong support to stakeholders in the application for GI registration and management. Among the legal approaches for GI protection, the *sui generis* system and the certification/collective trademark have emerged as the two main legal instruments to recognize and preserve the name, typicality, quality and reputation of products linked to their origin.

In two decades, a multitude of new GIs has been recognized in many different countries in all regions of the world, according to either a *sui generis* or a trademark system or both. For example, in the Chap. 10, the authors Hui Xu, Xuan Yang, and Xin Gu, describe the complex situation in China, a country with now 3 GI systems, two under the IP umbrella, sui generis GI and trademarks, and one under the Ministry of Agriculture umbrella, deriving from a progressive establishment of the institutional system. The chapter describes positive impacts of GIs on the local economy and is rich of lessons from country experiencing gradual establishment of their GI legal and institutional framework in order to address national context and evolving needs.

The Chap. 11 retraces the rich institutional experience since the initial legal framework established in 1995 until its final form of 2017. The authors Özden İlhan, Neşe Altıntaş, Nazlı Şimşek, and Sertaç Dokuzlu describe the different roles of the public institutions and their coordination, especially between the GI competent authority, the Turkish Patent and Trademark Office and the Ministry of Agriculture and Forestry, to maximize GI success. The chapter illustrate what are the key success factors through two emblematic GI cases *Gemlik Zeytini* (black table olive of Gemlik) *and Bursa Siyah İnciri* (Bursa Black Fig).

Chile represents another fruitful country experience to illustrate the role of public actors in promoting GI as an efficient a tool for rural development, as described in the Chap. 12. Paola Guerrero Andreu explains the importance of government programs for the promotion and protection of products of origin, and the need for an active role of the government in accompanying producers and artisans in the registration and use of their GI.

Establishing a legal framework for GIs may not be easy in countries where the concept is new and require perseverance from GI value chain actors and a strong dialogue between public and private actors to clarify the procedures. An example is the case of South Africa described by Johann Kirsten in the Chap. 13, where the registration phase of *Karoo Lamb* GI has required efforts to overcome the initial confusion of the evolving legal framework and its understanding by the value chain actors.

Another issue for newly established framework is the lack of coordination between public institutions and different areas of laws such as IP law dealing with GIs and food safety regulations. Claire Philippoteaux and Yudy Paola Pineda Suarez in Chap. 14, describe the issue of registered GIs for food products not complying with pre-existing food safety product standard. The authors show the consequences on control and certification and highlight the need of coordination between institutions as crucial for GI system efficiency.

The question of controls and certification includes controls in the market, which is a crucial aspect for GI efficiency, and seems to remain difficult in any country, either with long or recent experience in GI protection. In Chap. 15, Barbara Pick describes a very relevant comparison between France and Vietnam on how frauds are tackled, providing important lessons for other countries.

Finally, another important characteristic of the evolution of the GI institutional frameworks is the need of harmonization of national legislations of various countries, primarily at regional level. In the Chap. 16, Miranda Risang Ayu Palar exposes the situation of the Association of Southeast Asian Nations (ASEAN) where member countries have different legal means to protect GI but where through cooperation, the Intellectual Property (IP) offices look forward to aligning the way to demonstrate the link to origin which is reflected in the GI specification.

In conclusion, all chapters of this part, whoever the authors are, either public authority representatives, researchers or value chain actors, converge in showing the dynamic capacity of public actors to identify weaknesses and improve the GI scheme accordingly, strengthening institutional coordination and expanding the functions of GIs from policies protecting consumers and producers' interests to underpinning territorial development policies.

1.3 Part III: GIs and Territorial Development

The potential of GIs as a development tool is well-documented in the literature. Indeed, by anchoring a production in a specific territory and promoting its origin-based quality, GIs secure employment and livelihood opportunities in rural areas while also preserving cultural heritage, ancestral know-how and local natural resources, thus contributing to the economic, socio-cultural and environmental sustainability of local stakeholders.

Drawing from case studies of GIs both in developed and developing countries, this part explores potential impacts of GIs, in terms of territorial development. In the Chap. 17, Juan J. Ferrero-García and Julia Martín-Cerrato illustrates how PDOs and PGIs can make a significant contribution to the sustainable development of disadvantaged regions, as Extremadura region in Spain is, and where GIs form an integral part of the rural development policy strongly supported by the Regional Government through a wide range of actions; from funding operating expenses to carrying out direct promotional activities.

Territorial development and biodiversity are crucial issues for a GI product collected fresh from the forest by traditional pickers and then processed into sirop and other processed product by different stakeholders. Stéphane Fournier, Pape Tahirou Kanouté, Maimouna Sambou, and Fanta Sow show in their Chap. 18 on "*Madd de Casamance*", how a GI can meet a range of sustainability challenges in Senegal. A number of challenges had to be overcome in drawing up the specifications, in terms of the nature of the GI product (fresh and/or processed), the delimitation of the geographical area for processing, the control and traceability. The impact that registration of this GI could have on the sustainability of the sector is assessed ex ante.

In territories where GIs have been registered in more recent years, GIs are considered as a promising tool for reaching a balance between providing income to local communities of producers and environmental preservation. This is all the more valuable in territories, such as the Amazon in Brazil, where the rich biodiversity is under threat, as described by Paulo de Tarso Anunciação de Melo and Suzana Romeiro Araújo in the Chap. 19 on the GI transforming role in the Pará state.

While GIs hold considerable potential for territorial development, its fulfillment depends on stakeholders' way of operationalizing this tool. In some cases, GI implementation is motivated by governmental objectives of export expansion and economic growth, rather than the protection of local traditional know-how and cultural heritage preservation. For example in Japan, Hart N. Feuer and Fatiha Fort in their Chap. 20 describe how the GI authorities' top-down approach and utilitarian focus on boosting production have placed traditional products at a disadvantage in the case of the *Mikawa region*.

With a similar perspective, Kae Sekine, in her Chap. 21, depicts a first controversy case in Japan of alienation of legitimate stakeholders (i.e., traditional manufacturers) in favor of economic benefits of bigger modernized producers of *Hatcho Miso GI*. The trajectory of registration of *Hatcho Miso* GI has led to a nation-wide controversy, ultimately calling into question the GI system in Japan.

Furthermore, the level of state intervention in the registration and implementation of GIs impacts the degree to which producers can derive economic benefits. The Chap. 22 by authors Orachos Napasintuwong, Chitra Parayil, and A. M. Radhika, compares two cases of GIs on red rice in India and in Thailand and shows that GI premium price and export opportunities remain limited when the government assumes a large role in GI management, contrary to when the producers' group is more actively involved.

In conclusion of this part, the six chapters illustrate, well in different geographical and political contexts, how GI processes are contributing to territorial development, while they also highlight the current and future challenges that stakeholders must address in order to avoid pitfalls and fully leverage the benefits of GIs.

1.4   Part IV: GIs Processes and Sustainability

The Part IV aims at contributing to the debate on the potential roles and limits of GIs as a tool for sustainable development. An important milestone for sustainability at global level is the Agenda 2030 and its 17 Sustainable Development Goals (SDGs). In this perspective, a number of best GI practices can contribute to sustainable development and sustainable food systems. The cases presented in this part show how GI systems, embedded in their local context, offer an origin(al) way to contribute to this global agenda through a territorial approach.

First, the Chap. 23 by Junko Kimura and Cyrille Rigolot on the potential of GI to enhance Sustainable Development Goals, shows with GI Mishima potato as a case study in Japan that, from local stakeholders' point of view and through field observations, the GI system can contribute to at least nine SDGs at the production, transformation and commercialization stages, such as for example the employment of disabled people or nutritional education.

The perception of the GI value-chain actors is an interesting viewpoint to look at the GI impacts on sustainability, as presented in the Chap. 24 by Samir Messaili on the lessons learned from the *Bouhezza cheese* GI in Algeria through stakeholders' perceptions of the economic, social, environmental and cultural effects. The results of focus groups with producers suggest important impacts in terms of market value and volumes sold, reputation, increase of knowledge and environmental preservation.

Still, measuring the GI contribution to sustainability and SDGs is a complex matter, especially if sets of indicators, subjective and objective, need to be identified and informed for each pillar (economic, social and environmental). In the same time, it becomes crucial for GI producers themselves to be aware of their sustainability challenges to improve their GI system performances. In this perspective, the Chap. 25 by Emilie Vandecandelaere, Luis F. Samper, Florence Tartanac and Massimo Vittori, presents a practical methodology to empower GI organizations in defining their own sustainability roadmap though relevant alliances developed with relevant stakeholders involved in the consultation, building on the traditional three pillars, economic, social and environmental, plus the fourth governance pillar. The roadmap, built through a participative and bottom-up approach, allows identify their sustainability topic priorities, assess and monitor their contributions to these priorities and finally improve the GI systems performance accordingly and in a continuous and iterative manner.

Sustainability in production is also closely linked to the climate change challenges, impacting in particular the environmental and economic pillars. By modifying the natural conditions of production (temperature, humidity, etc.), climate change can undermine the causal relationship between product quality and the local area of production defined in the specifications, and jeopardize the possibility to maintain the GI product specificity in the long term. In the Chap. 26 Claire Bernard-Mongin, proposes to examine the strategic conditions for the emergence of a path of innovation for GI to adapt to climate change and maintain a strong link to origin thanks to a processual definition of the origin-linked quality rooted in the co-evolution of production practices with their environment.

In the Chap. 27, the authors Gilles Flutet, Caroline Blot, Jacques Gautier, Laurent Mayoux, and Alexandra Ognov confirm the need of GIs to adapt to climate change, in the case of France, by adjusting production rules while preserving the authenticity of their GI products and the promise made to consumers. The chapter describes the way the stakeholders from public and private sectors are collaborating, finding concrete solutions, including field and normative experimentations.

Looking specifically at the food sector, sustainability should be considered from a food system perspective, including nutrition and health. Food systems are now expected to re-align from just supplying food to providing, in sustainable manner, high-quality foods as part of healthy diets for all. Emilie Vandecandelaere and Florence Tartanac analyze the relations between GI foods and healthy diets in the Chap. 28. If many GI foods by nature are part of a healthy diets, because notably of their composition, their link to biodiversity and traditional diets, more evidence are needed. The authors advocate for more research and provide some recommendations to stakeholders.

Indeed, the link between GI food and sustainable consumption and consumer health is not so obvious for consumers, as demonstrated in the in the case Norway by Gun Roos and Virginie Amilien in the Chap. 29. Through an ethnographic approach in Norway fieldwork, the authors suggest that consumers lack knowledge of GI products and seldom associate GI products with sustainability and health in their everyday food practices and reflect on how GI products may develop their potential to contribute to a more diversified and healthy diet.

   The chapters of this part offered then a diversity of viewpoints on GI contributions to sustainability, from the SDGs to climate change, looking also at nutrition and healthy diets and highlighting the crucial roles of GI producers and alliances, the way GI processes can mobilize the three, and even four pillars of governance are rich and deserve to be explored further.

   In final conclusion, we wish that these four part, covering all aspects of GIs, getting into the last trends of GIs systems in all areas of the world, from a rich multi-actors perspective will enjoy the readers and give them food for thought.

Part I
Nature of GIs and Relations to Other Rights

Chapter 2
Do Geographical Indications Certify Origin and Quality?

Martijn Huysmans, Daniël van Noord, and Gero Laurenz Höhn

Abbreviations

EU	European Union
GI	Geographical Indication
PDO	Protected Designation of Origin
PGI	Protected Geographical Indications
SECPP	Search, experience, credence, Potemkin, and placebo

2.1 Introduction

The European Union (EU) has set up quality schemes to protect regional specialty foods such as Gouda Holland, Parma ham, and Champagne. Under these schemes the EU currently protects over 3500 geographical indications (GIs) for food and wine. In order to use a protected GI name, producers have to be located in the protected region and follow the product specification, which specifies the allowed ingredients and production methods. The premise of these GI systems is that the origin of a product has a direct influence on the quality of the good: the notion of terroir (Barham 2003). This link may arise due to geographical factors such as soil

M. Huysmans (✉)
School of Economics, Utrecht University, Utrecht, the Netherlands
e-mail: m.huysmans@uu.nl

D. van Noord
Utrecht, the Netherlands

G. L. Höhn
KU Leuven, Leuven, Belgium

and climate, as well as human factors such as ancestral know-how or *savoir-faire*. But does origin guarantee quality?

This article contributes to the literature on GIs and quality by separating out the role of GIs in certifying different attributes of quality. In addition, it models how different consumers may value those quality attributes differently. In particular, regional origin may be a direct quality attribute per se, as well as influence taste through terroir.

GIs are a contentious topic in international trade (Hughes 2006; Gangjee 2007; Rickard et al. 2015; Watson 2016; Hough 2016; Huysmans and Swinnen 2019; Beattie 2019; Huysmans 2022). In particular, the US sees them as a protectionist measure (Marette et al. 2008; Mancini et al. 2016). The US does not buy into the EU assertion that GIs are necessary to certify and protect high quality regional specialty products. In the context of trade this has led to a transatlantic "war on terroir" (Josling 2006). Clarifying the role of GIs in certifying quality, as this article attempts, is hence important.

2.2   The *Search, Experience, Credence, Potemkin, and Placebo* Attributes Model of Quality Applied to GIs

The standard model of quality in economics of information is the theory of *search, experience* and *credence* attributes. While search attributes such as the color of food are readily apparent (Jahn et al. 2005), experience attributes such as taste only become clear upon consumption. In practice, repeat buying may transform the overall taste of a type of food from an experience into a search attribute (Desquilbet and Monier-Dilhan 2015).

Credence attributes such as the absence of pesticide residue cannot be evaluated during normal use (Swinnen et al. 2015), and require costly tests for verification. Another example of a credence attribute of a food item is its nutritional value: it can be verified in the lab, but consumers will not experience it directly. Of course, nutritional information on packaging can transform nutritional value into a search attribute. Similarly, a GI logo might be a search attribute that stands for other quality attributes that are more difficult to verify by the consumer directly.

Tietzel and Weber (1991) extend this model by introducing *Potemkin*[1] attributes as somewhere in the middle between credence and *placebo* attributes. This leads to the *Search, experience, credence, Potemkin, and placebo* attributes (SECPP) model (van Noord 2019). While placebo attributes are nonexistent, Potemkin attributes are process attributes: real but not verifiable in the final product. Potemkin attributes can therefore not be verified after purchase (Tietzel and Weber 1991; Jahn et al. 2005).

---

[1] The name Potemkin comes from the Russian marshal Potemkin, who was charged with developing conquered territories in the Crimea in 1780. When Catherine II visited, Potemkin built some villages to fake the success of his program. Presumably, Catherine II cared not about the villages per se, but rather about the assumed underlying process of development in the region.

	Search	Experience	Credence	Potemkin	Placebo
General food Example	Color, Smell, Taste		Pesticides	Fairtrade	False health claims
Example of GI role	Rules on composition and process that affect taste		Origin per se	Sustainability Animal welfare	
				Authenticity	

Fig. 2.1 Quality attributes by increasing information asymmetry and verification cost

Typical examples of Potemkin attributes are fairtrade and climate-neutral. While there is a difference between a fairtrade and normal chocolate bar, this difference cannot be determined from the end product and audits of the production process are therefore required.

Figure 2.1, building on Jahn et al. (2005), gives an overview of the SECPP quality attributes. The attributes have been sorted by increasing verification cost from left to right. Examples of food quality attributes are given in the second row, while the third row illustrates how GIs may affect and certify these types of attributes. The next section provides the underlying details.

Search and Experience Attributes: Taste
In order to obtain a GI, a product specification needs to be established. However, while the link between geographical origin and quality has to be verbally described, there is no requirement for blind tasting or other independent verification in order for GI protection to be granted.[2] In practice, products' historical reputation seems key in GI applications (Zappalaglio 2021), rather than direct proof of a unique taste. This means that, at least in theory, the product specifications of some GI products may not be identifiable in blind tasting.

Unfortunately, no systematic empirical research has studied to what extent product specifications actually affect taste. Theory predicts that GIs should have minimum quality standards in order to be relevant for consumer decisions and hence valuable to producers (Winfree and McCluskey 2005).

In the EU, there are two main GI schemes: Protected Geographical Indications (PGIs) and the stricter Protected Designations of Origin (PDOs). For a PGI, only one production step needs to happen within the protected region. For a PDO, all production steps need to happen within the region. Both types of GIs have product specifications that go beyond specifying the origin.

Prima facie, some GI requirements seem very likely to affect taste. For instance, Mozzarella di Bufala Campana PDO can only be made using buffalo milk. Since this

---

[2]Some individual GIs do require tastings by qualified panels as part of their control plans, but EU regulations do not require blind tasting for GI applications.

is on the ingredient list, it can be considered a search attribute. When consumers are less aware of how composition links to taste, it remains an experience attribute. For instance, PDO Feta has to contain at least 70% sheep milk and at most 30% goat milk. Since most consumers do not know this but the requirement probably does affect the taste, this is arguably an experience attribute. GI certification can be beneficial to consumers by collapsing this experience attribute into a search attribute: by looking for the GI logo, consumers can look for products with a taste they like—without needing to know what causes this taste.

Credence Attributes: Origin Per Se

Most authors consider geographical origin per se (so apart from its potential influence on taste) to be a credence attribute (Desquilbet and Monier-Dilhan 2015; Li et al. 2017). Indeed, given that modern science can allow the determination of geographical origin from the end product, it is not in general a Potemkin attribute.

While all GIs have clear rules on the origin of raw materials, the rules may be more flexible than consumers think. This section lists some examples of the limitations of the GI-origin nexus.

First, the link between the region and the product may be less strict than sometimes assumed. In spite of the fact that a PDO normally requires all production steps to take place in the PDO area, the meat for Parma ham PDO may come from a number of regions adjacent to the province of Parma (Gangjee 2017).[3] PDO Parma ham is not a unique case, as other famous Italian PDO hams from San Daniele, Modena or Carpegna source pig meat from much larger regions than the small "core" production area.[4] Thus, the more strictly regulated PDOs are not necessarily more strictly regional products than PGIs because, for example, the sourcing area for pig meat of PGI Bayonne ham from France is smaller than the ones of PDO Parma and San Daniele ham (Höhn et al. 2024).

Second, some GI names have no direct link to a region—although the product specification always has to specify the allowed production zone and the link between the area and product quality. As trade policy-makers from the US and other GI-skeptic countries like to point out, Feta is not a place (Beattie 2019). The name reportedly comes from the Italian "fetta" or "slice" (Gangjee 2007), and the production zone covers a very large part of the Greek territory; all of the mainland plus the former prefecture of Lesbos.

Third, the exact geographical areas may not actually be constant over time. This may limit the sharp identification of GI names with specific regions. The areas of several French cheese GIs have been amended significantly: in 2014, the area for Livarot was increased by 146%. Other areas were decreased: Roquefort by 90% in

---

[3] A related but different issue is that over 45% of GI names refer to geographical units that are smaller than the production area specified by the GI (Zappalaglio et al. 2022: 21).

[4] Food GIs were introduced in the EU by regulation 2081/92. Articles 2.4 and 2.7 provided a two-year period for PDOs to be registered with larger sourcing than production areas. Many Italian cheeses and processed meat PDOs were registered under this exception (Fino et al. 2021).

2008, Camembert de Normandie by 52% in 2013. A very controversial GI in this respect is Champagne. With land selling for 1 million euros per hectare inside of the GI area and about 2 thousand euros outside, a proposed expansion has high stakes (Stevenson 2008).

Finally, paradoxically, the production of some GIs is not allowed in the eponymous regions. In the cases of Gouda North-Holland and Stilton, the PDO cheeses may not actually be produced in the eponymous cities. The city of Gouda is not actually in the Dutch province of North-Holland, so Gouda North-Holland PDO cannot be produced in Gouda. Historically, the name derives from the Gouda cheese market, not from the production area. Similarly, Stilton cheese PDO cannot be produced in the village of Stilton (Cambridgeshire), which is outside of the production zone comprising the counties of Leicestershire, Derbyshire and Nottinghamshire (Rippon 2014).

Credence, Potemkin or Placebo Attribute: Authenticity

A related concept to origin is authenticity (Rippon 2014). However, the concept of authenticity seems so flexible that it may range from a credence to a placebo attribute. Concerning the product requirements, (Gangjee 2017: 19) notes: "tradition and heritage are malleable resources, being actively reconstructed during the drafting process". For instance, larger firms involved in drafting the product specification may seek to allow more industrial methods of production or a larger geographical production area. In more extreme cases, new GIs and traditions may be invented outright (Grandi 2018). For instance, the Pomodoro di Pachino PGI is based on a varietal of tomatoes created in 1989 by the Israeli seed company Hazera Genetics (Grandi 2018: 71–74).

Potemkin Attributes: Sustainability, Animal Welfare

Finally, some GIs also regulate Potemkin or process attributes such as sustainability. For instance, GIs may protect rural livelihoods (Chilla et al. 2020; Crescenzi et al. 2022). Jobs related to GIs cannot be delocalized, and farmers whose production depends on conserving local ecosystems will produce more sustainably (Food and Agriculture Organization of the United Nations 2019). However, as a process attribute this is not verifiable in the end product, and production audits are required.

Another Potemkin attribute is animal welfare. GIs may have a potential to contribute—at least partially—to more animal-friendly conditions. For example, the French PGI Gruyère already stipulates a minimum of 150 grazing days per year.

A final, more concrete example of a Potemkin attribute is that Parmigiano Reggiano PDO cheese can only be sliced and packaged in the region. Since the place of packaging cannot be reliably determined from the end product, this is clearly a Potemkin attribute. Of course, this rule may also help in preventing fraud.[5] Clearly, GIs need to be enforced in order to be informative and valuable (Zhao et al. 2014; Marie-Vivien and Biénabe 2017).

---

[5]For a more extensive discussion of slicing and packaging rules, see Zappalaglio et al. (2022: 18–19).

Summary: Which Quality Attributes Do GIs Certify?

To summarize, at the very minimum GIs certify origin per se, a credence attribute. The product requirements in most cases are likely to also affect taste, which is a search or experience attribute. Also, GIs have the potential to regulate Potemkin attributes like sustainability or animal welfare. In addition, they may guarantee the authenticity of a product, although this is a malleable concept.

For those consumers who care about attributes such as origin and authenticity, a GI label can help them easily find the product they want. The label transforms these attributes into a search attribute. A consumer who buys a certified Camembert de Normandie PDO knows better what to expect than if she buys a generic Camembert. Herein lies the value of a GI: if the system functions properly, it informs the customer of product quality attributes she cares about, before making a purchase. The next section provides a simple economic model on how consumers may value these attributes.

2.3   Consumers' Quality Equations

This section sets up a stylized economic model to capture consumers' quality equations. Assume quality q has three main components: taste t, origin per se o, and authenticity a.[6] This can be written as $q = f(q_t, q_o, q_a)$. The simplest functional form is a linear quality equation with parameters θ_i, summing to 1, giving the importance to the consumer of dimension i:

$$q = \theta_t q_t + \theta_o q_o + \theta_a q_a \tag{2.1}$$

Depending on how much a consumer values each quality component, the parameters θ_i will be larger or smaller for that consumer. For instance, a consumer who only cares about taste will have $\theta_t = 1$, $\theta_o = 0$, $\theta_a = 0$.

Now assume that taste is determined by origin o and production method p: $q_t = g(q_o, q_p)$. The simplest example would again be a linear function:

$$q_t = g(q_o, q_p) = \alpha q_o + (1 - \alpha) q_p \tag{2.2}$$

Where $\alpha \in [0, 1[$ captures the importance of origin vis-à-vis production methods. If terroir or origin plays no role in taste, $\alpha = 0$. It seems natural to assume that the importance or origin in taste is strictly below 1, because production methods should play at least some role in taste. Combining (2.1) and (2.2), we obtain

---

[6]This model could be extended, for instance to include an ethical or environmental sustainability dimension, which can also play a role in consumer choices (Aurier and Sirieix 2016). However, since EU GI regulations do not have binding rules on this for all GIs, we do not focus on this aspect in our model.

$$q = (\alpha\theta_t + \theta_o)q_o + (1 - \alpha)\theta_t q_p + \theta_a q_a \qquad (2.3)$$

This shows that origin can have an indirect effect via taste (with weight $\alpha\theta_t$), as well as a direct effect (with weight θ_o).

Disregarding prices, a consumer will prefer the GI product over a generic product if

$$(\alpha\theta_t + \theta_o)\left(q_o^{GI} - q_o^{Gen}\right) + (1 - \alpha)\theta_t\left(q_p^{GI} - q_p^{Gen}\right) + \theta_a\left(q_a^{GI} - q_a^{Gen}\right) > 0 \qquad (2.4)$$

Equation (2.4) neatly summarizes the importance of consumers' quality equations and of empirical parameters linking GIs to quality. As argued above, on the whole GIs always certify origin to some extent, so one can assume $q_o^{GI} > q_o^{Gen}$. They may also certify taste—through a combination of origin and production methods—and authenticity.

If consumers only care about taste, origin per se and authenticity have no role in their quality equation: $\theta_o = \theta_a = 0$ and $\theta_t = 1$. For such consumers (4) reduces to $\alpha\left(q_o^{GI} - q_o^{Gen}\right) + (1 - \alpha)\left(q_p^{GI} - q_p^{Gen}\right) > 0$. For them, GIs are only informative if they actually affect taste—either because there is a direct effect of terroir ($\alpha > 0$ and using $q_o^{GI} > q_o^{Gen}$) or because the GI product specification ensures superior production methods ($q_p^{GI} > q_p^{Gen}$).

2.4 Empirical Evidence on the Valuation of GI Quality Attributes by Consumers

While GI products tend to have higher prices, sales, and exports (Loureiro and McCluskey 2000; Chever et al. 2012; Curzi and Olper 2012; Agostino and Trivieri 2014; Sorgho and Larue 2014; AND-International 2019; Raimondi et al. 2020; Vandecandelaere et al. 2020; Curzi and Huysmans 2022), not much literature exists on which actual or perceived quality attributes of GIs drive this. A study that has looked into Hessian apple wine finds that the willingness to pay for this GI comes more from a willingness to support local producers rather than from considerations of gustatory quality, i.e. taste (Teuber 2011). However, other studies have found that the strongest driver of using GI labels in shopping decisions is that they signal better quality (Verbeke et al. 2012).

In a meta-study on fruits and vegetables, Moser et al. (2011) find that origin was a significant driver of consumer choice in only one out of 8 studies. In a study for Italian olive oil, Van Der Lans et al. (2001) find for some local consumers both a positive effect of origin per se and indirectly through perceived quality. When differentiating between European and "new world" consumers outside of Europe, origin is less relevant for new world consumers (Moser et al. 2011). This is likely to be part of the explanation why GIs are more popular in Europe than elsewhere: the one quality attribute that all GIs regulate is geographical origin per se, which is not valued equally by new world consumers.

2.5   Conclusion

This article shows how the SECPP model of quality can be used to analyze the European GI quality schemes for PGIs and PDOs. The SECPP model adds Potemkin and placebo attributes to the standard model of search, experience, and credence attributes of quality.

While the GI product specification may impose rules that affect composition and taste—search and experience attributes—this is not independently verified by blind tasting or otherwise.

The main requirement for EU GIs is a specific geographical origin, which is a credence attribute. However, the link between origin and GIs is not always as straightforward as one may assume. For instance, Stilton cheese cannot be made in the village of Stilton, the pigs for Parma ham may come from outside the Parma province, and Feta is not a region. Yet overall, GIs do certify a certain regional origin.

In addition to origin, product specifications may affect other credence and Potemkin attributes, such as whether it was produced in a sustainable way and where the good was packaged. Finally, there may also be placebo attributes involved, such as when industrially produced products are perceived as authentic and traditional just because they are GI certified.

Not all consumers value credence and Potemkin attributes. Hence for some consumers geographical origin per se can have a weight in their quality equation while for others it does not. The European GI schemes can only guarantee higher quality for consumers who indeed value the guaranteed origin and product specification of a GI. For other consumers strict GI regulations may do more harm than good. By prohibiting the use of GI names for generic alternatives, even accompanied by "like" or "style" and their true geographical origin, search costs are increased rather than lowered for such consumers. For instance, a recipe they are trying to cook may call for Feta or for Roquefort. Consumers looking for cheaper alternatives to the authentic GI products have to know that the terms for the broader category are "white cheese" or "blue cheese", and may need more time finding a suitable alternative on the shelves.

Research indicates that non-European or "new world" consumers tend to care less about origin. Hence full EU GI protection on the US market may be unnecessary as well as unrealistic. In order to forbid the unqualified use of its GI names in the US, the EU might consider accepting the use of GI names combined with their true origin, as in "Feta-style cheese from Wisconsin".

In conclusion, by separating out the different quality attributes that GIs may affect, we hope to have clarified the possibilities and limits for GIs to certify quality and appropriate the brand value of regional specialty foods. In addition, the SECPP framework provides some common ground for a more nuanced transatlantic debate on GIs, regional origin and quality. Future empirical consumer research could investigate in more detail how specific GIs affect the different SECPP quality attributes.

Acknowledgements   The authors would like to thank Jo Swinnen, Daniele Curzi, Justin Hughes, and colleagues at the Applied Economics Section of Utrecht University for their comments. For research assistance related to this project, Martijn thanks Bruno Coucke at KU Leuven, Kate Wilson at KU Leuven and Stanford University, and Rhodé Looije and Lukas Bikker at Utrecht University. Finally, we thank participants of the 2022 'Worldwide Perspectives on GIs' conference in Montpellier, the 'AgEconMeet 2022' in Göttingen, the 2023 workshop "Geographical Indications and Fair Food" at Utrecht University, and the 2023 workshop "Place Identity in an Evolving Space" at the City Law school, London.

References

Agostino M, Trivieri F (2014) Geographical indication and wine exports. An empirical investigation considering the major European producers. Food Policy 46:22–36

AND-International (2019) Study on economic value of EU quality schemes, geographical indications (GIs) and traditional specialities guaranteed (TSGs). Publications Office of the European Union, Directorate-General for Agriculture and Rural Development

Aurier P, Sirieix L (2016) Marketing de l'agroalimentaire. Environnement, stratégies et plans d'action. Dunod, Paris

Barham E (2003) Translating terroir. J Rural Stud 19:127–138

Beattie A (2019) EU trade negotiators find non-Greek 'feta' hard to swallow. Financial Times

Chever T, Renault C, Renault S, Romieu V (2012) Value of production of agricultural products and foodstuffs, wines, aromatised wines and spirits protected by a geographical indication (GI). AND-International final report to the European Commission, TENDER N° AGRI–2011–EVAL–04 1–85

Chilla T, Fink B, Balling R et al (2020) The EU food label 'protected geographical indication': economic implications and their spatial dimension. Sustainability 12:1–21

Crescenzi R, De Filippis F, Giua M, Vaquero-Piñeiro C (2022) Geographical indications and local development: the strength of territorial embeddedness. Reg Stud 56:381–393. https://doi.org/10.1080/00343404.2021.1946499

Curzi D, Huysmans M (2022) The impact of protecting EU geographical indications in trade agreements. Am J Agric Econ 104:364–384. https://doi.org/10.1111/ajae.12226

Curzi D, Olper A (2012) Export behavior of Italian food firms: does product quality matter? Food Policy 37:493–503

Desquilbet M, Monier-Dilhan S (2015) Are geographical indications a worthy quality label? A framework with endogenous quality choice. Eur Rev Agric Econ 42:129–150

Fino MA, Cecconi AC, Bezzecchi A (2021) Gastronazionalismo. People

Food and Agriculture Organization of the United Nations (2019) Geographical Indications for sustainable food systems: preserving and promoting agricultural and food heritage

Gangjee DS (2007) Say cheese! A sharper image of generic use through the lens of feta. Eur Intellect Prop Rev 29:172–179

Gangjee DS (2017) Proving provenance? Geographical indications certification and its ambiguities. World Dev 98:12–24

Grandi A (2018) Denominazione di origine inventata. Le bugie del marketing sui prodotti tipici italiani. Mondadori

Höhn GL, Huysmans M, Crombez C (2024) Does terroir size matter? Protected geographical areas and prices of European hams. Reg Stud 58:1804–1817. https://doi.org/10.1080/00343404.2023.2187365

Hough C (2016) The EU tries to grab all the cheese. Politico.com

Hughes J (2006) Champagne, feta, and bourbon: the spirited debate about geographical indications. Hastings Law J 58:299–386

Huysmans M (2022) Exporting protection: EU trade agreements, geographical indications, and gastronationalism. Rev Int Polit Econ 29:979–1005. https://doi.org/10.1080/09692290.2020.1844272

Huysmans M, Swinnen J (2019) No terroir in the cold? A note on the geography of geographical indications. J Agric Econ 70:550–559

Jahn G, Schramm M, Spiller A (2005) The reliability of certification: quality labels as a consumer policy tool. J Consum Policy 28:53–73

Josling T (2006) The war on terroir: geographical indications as a transatlantic trade conflict. J Agric Econ 57:337–363

Li C, Bai J, Gao Z, Fu J (2017) Willingness to pay for "taste of Europe": geographical origin labeling controversy in China. Br Food J 119:1897–1914

Loureiro ML, McCluskey JJ (2000) Assessing consumer response to protected geographical identification labeling. Agribusiness 16:309–320

Mancini MC, Arfini F, Veneziani M, Thévenod-Mottet E (2016) Geographical indications and transatlantic trade negotiations: different US and EU perspectives. EuroChoices 16:34–40

Marette S, Clemens R, Babcock BA (2008) Recent international and regulatory decisions about geographical indications. Agribusiness 24:453–472

Marie-Vivien D, Biénabe E (2017) The multifaceted role of the state in the protection of geographical indications: a worldwide review. World Dev 98:1–11

Moser R, Raffaelli R, Thilmany-McFadden D (2011) Consumer preferences for fruit and vegetables with credence-based attributes: a review. Int Food Agribus Manag Rev 14:121–142

Raimondi V, Falco C, Curzi D, Olper A (2020) Trade effects of geographical indication policy: the EU case. J Agric Econ 71:330–356

Rickard BJ, McCluskey JJ, Patterson RW (2015) Reputation tapping. Eur Rev Agric Econ 42:675–701. https://doi.org/10.1093/erae/jbv003

Rippon MJ (2014) What is the geography of geographical indications? Place, production methods and protected food names. Area 46:154–162. https://doi.org/10.1111/area.12085

Sorgho Z, Larue B (2014) Geographical indication regulation and intra-trade in the European Union. Agric Econ 45:1–12

Stevenson T (2008) Definitive study of champagne's expansion plans. wine-pages.com

Swinnen J, Deconinck K, Vandemoortele T, Vandeplas A (2015) Quality standards, value chains, and international development: economic and political theory. Cambridge University Press, Cambridge

Teuber R (2011) Consumers' and producers' expectations towards geographical indications: empirical evidence for a German case study. Br Food J 113(7):900–918

Tietzel M, Weber M (1991) Von Betrügern, Blendern und Opportunisten—Eine ökonomische analyse. Z Wirtsch 40:109–137

Van Der Lans IA, Van Ittersum K, De Cicco A, Loseby M (2001) The role of the region of origin and EU certificates of origin in consumer evaluation of food products. Eur Rev Agric Econ 28:451–477. https://doi.org/10.1093/erae/28.4.451

van Noord D (2019) Geographic indications as Potemkin attributes: a new perspective on European food quality schemes. Utrecht University

Vandecandelaere E, Teyssier C, Barjolle D (2020) Strengthening sustainable food systems through geographical indications: evidence from 9 worldwide case studies. J Sustain Res 2:1–37. https://doi.org/10.20900/jsr20200031

Verbeke W, Pieniak Z, Guerrero L, Hersleth M (2012) Consumers' awareness and attitudinal determinants of European Union quality label use on traditional foods. Bio-Based Appl Econ 1:213–229

Watson KW (2016) Reign of terroir: how to resist Europe's efforts to control common food names as geographical indications. CATO Inst Policy Analysis 787:1–15

Winfree JA, McCluskey JJ (2005) Collective reputation and quality. Am J Agric Econ 87:206–213

Zappalaglio A (2021) The transformation of EU geographical indications law: the past, present and future of the origin link. Routledge

Zappalaglio A, Carls S, Gocci A et al (2022) Study on the functioning of the EU GI system. Max Planck Institute for Innovation and Competition

Zhao X, Finlay D, Kneafsey M (2014) The effectiveness of contemporary geographical indications (GIs) schemes in enhancing the quality of Chinese agrifoods—experiences from the field. J Rural Stud 36:77–86. https://doi.org/10.1016/j.jrurstud.2014.06.012

Chapter 3
PDO and Organic: Consumers' Willingness to Pay for Combined Labels

Tifenn Corre, Sylvette Monier-Dilhan, and Julie Regolo

Acronyms

GI	geographical indication
mkt share	Market share
NB	National brand
INAO	National Institute of Origin and Quality
Organic	Organic farming
PLB	Private label brand
PDO	Protected designation of origin
SIQO	quality and origin identification label
WTP	Willingness to pay

3.1 Introduction

Consumers' expectations regarding the quality of food products are manifold. They concern the intrinsic quality of the products, but also, and increasingly, the preservation of the environment, respect for animal welfare, the protection of know-how, the maintenance of employment in the territories, the remuneration of producers (Cartron and Fichet 2020). In response to these expectations, private labels (national, regional brands...) are multiplying as well as the number of

---

A longer French version of this study is available in Rural Economy (Corre et al. 2022)

T. Corre · S. Monier-Dilhan · J. Regolo (✉)
France's National Research Institute for Agriculture, Food and Environment (INRAE), Paris, France
e-mail: julie.regolo@inrae.fr

E. Vandecandelaere et al. (eds.), *Worldwide Perspectives on Geographical Indications*, https://doi.org/10.1007/978-3-031-71641-6_3

products benefiting from *quality and origin identification labels* (SIQO[1]). Many companies exploit one or more of these quality indicators on the same product (Monier-Dilhan 2018).

The accumulation of labels on the same product can generate confusion in consumer choices (Tagbata and Sirieix 2010; Janssen and Hamm 2012; Dekhili and Achabou 2013). Some labels are likely to have a halo effect on consumers' judgement of other attributes. For example, Sörqvist et al. (2015) show that the implementation of an eco-label on bananas induces a better taste perception. The association of several labels must be perceived as relevant, complementary, otherwise it can destroy value (Sirieix et al. 2013).

In this context, we are interested in the valuation by consumers of two official labels—the protected designation of origin (PDO) and the organic farming (organic)- and their ability to combine to ensure complementary differentiation on the market. These two quality schemes are regulated in France by the National institute of origin and quality (INAO), thus providing a common basis of trust for consumers towards these labels. Purchase intentions or consumer behaviours depend on the perceived reputation of the label or trust in this label (Larceneux 2001; Larceneux et al. 2012; Janssen and Hamm 2012).

The organic label is increasingly appearing on French products (the average value of organic food purchases per inhabitant has tripled between 2010 and 2022 (Agence Bio 2022) and reflects health and environmental concerns of consumers (Hughner et al. 2007; Mondelaers et al. 2009; Kriwy and Mecking 2011; Durham 2007; Monier-Dilhan and Bergès 2016). In the current context where regulatory frameworks aim to integrate agro-environmental dimensions into the SIQO, the question arises of the consumer's valuation of these dimensions on a product already PDO. This is the question we address in this study by estimating the effect of combining PDO and organic labels on the price the consumer is willing to pay.

Faced with this issue, we work on three main segments of the cheese market where these two labels are very present:[2] cooked pressed cheeses (with the PDO *Comté*), blue cheeses (with the PDO cow's milk blues and the PDO *Roquefort*) and camemberts (with the PDO *Camembert de Normandie*). Our study is conducted from the declared purchase data of a panel of about 20,000 households representative of the French population for the year 2017 collected by Kantar Worldpanel, a marketing and opinion research group. We adopt a hedonic price model to evaluate the implicit price that consumers are willing to pay for the PDO, organic and double labelling (PDO and organic), compared to a reference cheese without a label, given the other characteristics of the cheeses and the market structure. Our analysis takes

---

[1] SIQO includes the protected designation of origin (PDO), the protected geographical indication (PGI), the Label Rouge (Red Label), the guaranteed traditional specialties (STG) and products from organic farming (organic).

[2] In 2020, PDOs represent nearly 20% of French household cheese purchases in value. An increasing proportion of cheeses are from organic farming (INAO 2020). The PDOs considered in this article represent nearly 50% of the marketed production of PDO cheeses in volume (INAO-CNAOL 2021).

into account the existence of national or distributor brands, the distribution channel, the season and the sales format.

Our article is part of the literature that studies consumer preference for products under quality schemes and the WTP of consumers for quality schemes in the presence of other quality attributes (trade brands, intrinsic quality of the product...) or other private or official labels. The originality of our work is to evaluate the WTP for the accumulation of quality schemes, organic and PDO in the cheese sector.

Section 3.2 presents the data and method used. The results are presented in Sects. 3.3 and 3.4 concludes.

3.2 Data and Method

We are interested in French cheeses (excluding grated cheeses, to brown, in slices or cubes). After removing the aberrant observations,[3] the study covers 92,760 purchases of Emmental or Comté, 88,895 purchases of Camembert and 79,739 purchases of blue cheeses.

Emmental (non-PDO cheese) and *Comté* (PDO cheese) represent 75% of the cooked pressed cheeses market segment, and are considered as substitute goods (Colinet et al. 2006). *Comté* is the main French PDO cheese with 1/3 of French PDO production (INAO-CNAOL 2021). On this market, organic is weak in terms of market shares (ms) (0.8% of ms in value), it is better represented for the camemberts (2.44% of ms in value) and for the blue cheeses (1.69% in value terms). The high number of observations in each segment allows for a robust estimation. Corre et al. (2022) present the descriptive statistics (market shares, prices) on cheese purchases, distinguishing them according to the presence of organic and PDO labels.

The hedonic pricing method, developed by Rosen (1974), is commonly used in economics when goods are not homogeneous. It consists of explaining the prices of a good at equilibrium on the market by its characteristics, and not those of the buyers or sellers (Le Saout and Vignolles 2017; Pakes 2003). It differs from discrete choice models which are interested in the market share resulting from price, preferences and consumer characteristics. To account for the non-normality of the price distribution, we adopt the Box-Cox transformation, widely used in the literature, to estimate price determinants. The transformation of the price follows a Gaussian law, whose parameter λ is inferred from the model. It includes the linear form if $\lambda = 1$ and the logarithmic form if $\lambda = 0$.

---

[3] Outlier observations are those whose price is outside the range [Q1 − 1.5 * (Q3 − Q1), Q3 + 1.5 * (Q3 − Q1)], Q1 and Q3 being the first and third quartiles of the price distribution for each of the 4 organic/non-organic*PDO/non-PDO intersections. This rule is applied to the study of other cheeses.

$$\frac{p^{\lambda} - 1}{\lambda} = \alpha + \beta X + e \tag{3.1}$$

With p the price of a variety of cheese, X the vector of observed characteristics of the variety: national brand (MN), distributor brand (MDD), type of distribution network (hypermarkets, supermarkets, private brand dominant chains, traditional, drive and specialised stores), heat treatment of the milk (for camemberts), sales format, sales season and presence of one or two official quality signs (PDO, organic). We construct the variables of interest: PDO, equal to 1 if the variety is PDO, 0 otherwise; PDO* organic equal to 1 if the product is PDO and organic, 0 otherwise; and non PDO*organic equal to 1 if the product is not PDO but has an organic label, 0 otherwise. In the blue cheese market, we distinguish PDOs made from sheep's milk (*Roquefort*) from PDOs made from cow's milk (blues). The error term (e) includes the unobservable characteristics of the product and is assumed to be independent of the observed characteristics.

3.3   Results

Since all coefficients of interest are significant, the selected variables explain the variability of the price (the adjusted R^2 is between 0.70 and 0.86) (Corre et al. 2022). We observe a consistently positive effect of the organic label. The effect of the PDO is positive for *Comté, camembert and Roquefort*. It is negative for the blues; we will return to this result later. Unsurprisingly, private brands are on average cheaper than national brands, hypermarkets and hard discount stores offer lower prices than supermarkets, unlike traditional commerce, and the price per kilogram decreases with the sales format.

To facilitate interpretation, the estimated prices are represented in Graphs 3.1, 3.2 and 3.3, respectively for a reference good and for a similar PDO and/or organic good. For camemberts, we distinguish products according to the heat treatment of the milk. The results are presented "all other things being equal", i.e. for constant levels of the other explanatory variables of the model. For example, the results are presented for purchases in supermarkets of cheeses with national brands made in autumn and whose sales format is the most common.

In Graph 3.1, the price per kg of the reference good (Emmental) is 8.02 euros. That of the non-organic *Comté*, all other things being equal, is higher by 5.48 euros or 68% more. This reflects the higher WTP of consumers and therefore a perceived quality (taste, production conditions or other attributes) superior for this PDO product compared to its non-PDO counterpart. The WTP for the organic label is 7.63 euros for the reference good and 4.25 euros for the *Comté* (49% versus 24% of the price). This result is close to that of Hassan and Monier-Dilhan (2002a, 2006): the WTP for a PDO is lower on a good that already has a distinctive quality (brand, certification or other PDO).

In Graph 3.2, the price of the reference good is 5.94 euros. On the camembert market, the mode of production (pasteurised, thermised or raw milk) is the main

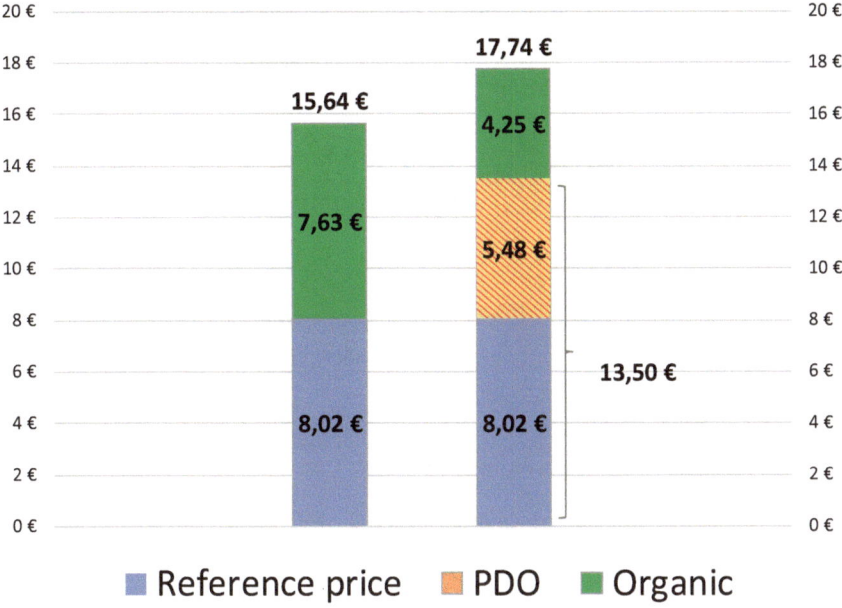

Graph 3.1 Results for Emmental and *Comté* with national brands in a supermarket in a 250 g portion in autumn

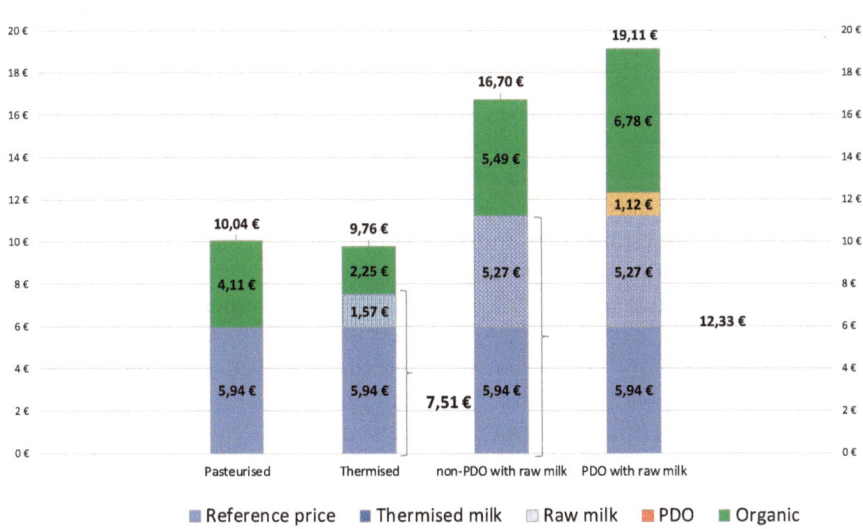

Graph 3.2 Results for non-light camembert with national brands in a supermarket in a 250 g portion in autumn

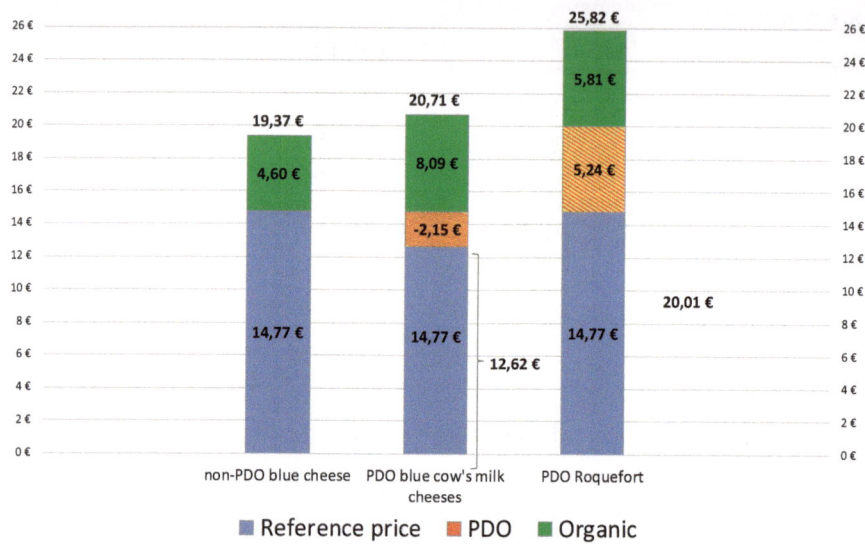

Graph 3.3 Results for blue cheeses with national brands in a supermarket in a portion of less than 150 g in autumn

differentiating factor. Indeed, all other things being equal, the price of camembert made with thermised milk is higher by 1.57 euros (+26%) compared to the reference good and that of camembert made with raw milk is higher by 5.27 euros (+89%). Thus, the mode of production "raw milk" is a quality indicator. The PDO, which imposes a production with raw milk, brings a small additional valuation of 1.12 euros, or about 10%. The mentions "made in Normandy" (for non-PDO camemberts) and "*Camembert de Normandie*" (for PDO camemberts) can be a source of confusion for the consumer as, like the *Camembert de Normandie*, some camemberts made in Normandy are made with raw milk.

In Graph 3.3, the estimated price per kg of the reference good is 14.77 euros. Our estimates show that the selling price of non-organic PDO blues is 2.15 euros lower (around 15%) than that of non-PDO non-organic blues, all other things being equal. They are consistent with the results of Callois et al. (2019) who find, from field surveys, that the PDO *cheeses of Auvergne* (main French PDO blues) have a reputation for low-quality products. This result can also be explained by the strong reputation of certain blues with national brands predominantly non-PDO as shown by Hassan and Monier-Dilhan (2002b). The impact of the organic label appears in these cases much higher for PDO blues (+8.09 euros or 64%) compared to the reference good (+4.60 euros or 31%) and allows these cheeses to "catch up" with the price of non-PDO blues also having the organic label. This effect can be explained by a lower initial price as well as by the profile of PDO consumers potentially more sensitive to the organic label.

The PDO *Roquefort* displays a price significantly higher than the non-PDO reference good price, by +5.24 euros (+35%). The reference good here is a non-PDO cow's milk blue, as there is no significant purchase of non-PDO sheep's milk blue on the French market in 2017. This price difference can therefore partly reflect the differences in production costs between sheep's milk and cow's milk.

According to our results, the effect of the organic label on this cheese is 5.81 euros (+29%) and is therefore weaker, in percentage, than on the other cheeses in this segment.

Table 3.1 summarises the estimates results of the effects on prices of PDO and organic labels taken individually or simultaneously in comparison with the reference good, on each market segment. For the camembert, the most relevant reference good is the raw milk camembert as the PDO camembert is raw milk. The prices are estimated from Eq. (3.1) for a product with a national brand sold in autumn in a supermarket in the best-selling format.

Our results show that the effect of the PDO (column (1)) on the price of the reference good is heterogeneous according to the market segments studied. The effect of the PDO is very significant in the case of Comté compared to Emmental (+68%), moderate for *Roquefort* and *Camembert de Normandie* (respectively +35% and +10%) and negative for PDO blue cow's cheese (−15%).

The effect of the organic label (column (2)) on the reference good price is always positive and often of greater magnitude than the effect of the PDO (between 30% for thermised camembert and 95% for Emmental).

Finally, the effect of the organic label on an PDO cheese is also always positive and of great magnitude (between 29% and 64%) (column (3)). By comparing columns (2) and (3) we observe that the effect of the organic label differs depending on the good (reference or the PDO good). This difference seems to be linked to the valuation of the PDO good. The organic label has a significantly weaker effect on *Comté* (+31% versus +95%), equal for Roquefort (+29% versus +31%), slightly higher for *Camembert de Normandie* (+55% versus +49%) and much higher for PDO blues (+64% versus +31%) compared to reference goods. In these latter cases, the organic label has a complementary effect on the PDO by reinforcing its power of differentiation: the sum of the individual effects of the PDO (column 1) and the organic (column 2) is less than the effect of the combination of these labels (column 5). These differences in the effect of the organic label on PDO cheeses between segments can be explained by the fact that some denominations are already perceived as environmentally friendly. (e.g. pastures) and associated with more "natural" cheeses.

3.4   Conclusion

The results of this work suggest that the official quality signs of PDO and organic generally reflect a higher quality, particularly for the organic label (between 30% and 95%). The PDO effect is very variable depending on the products, it is very important in the case of *Comté* (+68%), moderate for *Roquefort* and *Camembert de Normandie* (respectively +35% and +10%) and negative for the PDO cow blues (−15%). These differences may reflect the reputations of these cheeses.

Table 3.1 Estimated effects of organic and PDO labels, and their cumulative effect on the three market segments in 2017

Market segment	Single quality sign/Reference good		Two quality signs/Single quality sign		Two quality signs/Reference good
	PDO price/Reference good price (1)	Organic price/Reference good price (2)	PDO + Organic price/PDO price (3)	PDO + Organic price/Organic price (4)	PDO + Organic price/Reference good price (5)
Comté (reference good: Emmental)	+68%	+95%	+31%	+13%	121%
"Camembert de Normandie" (reference good: non-PDO raw milk camemberts)	+10%	+49%	+55%	+14%	71%
Blue cheese: PDO cow (reference good: non-PDO cow's milk)	−15%	+31%	+64%	+7%	40%
Blue cheese: PDO *Roquefort* (reference good: non-PDO cow's milk)	+35%	+31%	+29%	+33%	75%

Our results show that PDO and organic are two weakly substitutable labels and present effects of complementarity particularly on markets where PDO alone is weakly differentiating. The more the perceived quality of the PDO good is important, the more the effect of the organic on the price of this product is weak, it remains however always at high levels (more than 29%). These results are in line with those of Dufeu et al. (2014) who show that the WTP for an additional label depends negatively on the valuation of the initial label. Moreover, some PDOs may be more respectful of the environment or perceived as such and associated with more "natural" cheeses compared to similar conventional products. This could explain stronger substitution effects between PDO and organic (lower additional), for example for *Comté* or *Roquefort*.

The results confirm the relevance of combining these labels which reflect overall coherent and weakly substitutable values; and thus do not reveal any negative effect related to the accumulation of information (Janssen and Hamm 2012; Dekhili and Achabou 2013; Sirieix et al. 2013). These conclusions are consistent with a context of strengthening the principles of agroecology in geographical indications initiated by the Ministry of Agriculture and Food.

Our results are also to be put into perspective in relation to the characteristics of consumers of organic and PDO products who have a higher level of education and a higher socio-professional category compared to consumers of conventional products (Goudis and Skuras (2021), Magnusson et al. (2003), Monier-Dilhan et al. (2009) and Wier et al. (2008)). Thus, the complementarity of these labels may reflect a higher WTP for the organic label among consumers already sensitive to the PDO label.

Finally, the estimated implicit prices of PDO and organic can also be influenced by the costs related to the constraints of the specifications and the industrial organisation of the segment. On this subject, Monier-Dilhan et al. (2019) and Bonnet and Bouamra-Mechemache (2016) have shown that PDO and organic labels provide higher margins on all the links in the production chain, particularly for organic products. It would be interesting to study whether the costs generated by the accumulation of labels are decreasing: the increase in costs following the adoption of a new specification could be stronger for a company that starts to produce origin and quality signs than for a company which already meets a specification, in terms of production, control and organization.

References

Agence Bio (2022) Le marché alimentaire de la bio en 2022. https://www.agencebio.org/wp-content/uploads/2023/11/RAPPORT-2023-donne%CC%81es-2022-FINAL-nov23.pdf

Bonnet C, Bouamra-Mechemache Z (2016) Organic label, bargaining power, and profit-sharing in the French fluid milk market. Am J Agric Econ 98(1). https://doi.org/10.1093/ajae/aav047

Callois J-M, Farsti I, Ngoulma J, Jeanneaux P (2019) Perception de la qualité par la distribution et dynamique des ventes. Le cas des AOP fromagères d'Auvergne. Économie rurale, Octobre–Décembre 370:7–28

Cartron F, Fichet J (2020) Rapport d'information fait au nom de la délégation sénatoriale à la prospective sur « Vers une alimentation durable : Un enjeu sanitaire, social, territorial et environnemental majeur pour la France »

Colinet P, Desquilbet M, Hassan D, Monier-Dilhan S, Orozco V, Réquillart V (2006) Economic analysis of food quality assurance scheme: case study Comté. European Technico-Economic Policy Support Network 91. Institute for Prospective Technological Studies (IPTS) of the European Commission's Joint Research Centre (JRC)

Corre T, Monier-Dilhan S, Regolo J (2022) AOP et AB : quelle disposition à payer des consommateurs pour la double labellisation ? Économie rurale 381:39–60. https://doi.org/10.4000/economierurale.10275

Dekhili S, Achabou M-A (2013) Pertinence d'une double labellisation bio/ écologique auprès des consommateurs. Une application au cas des œufs. Économie Rurale 336:41–59

Dufeu I, Ferrandi J-M, Gabriel P, Gall-Ely M (2014) Multi-labellisation socio-environnementale et consentement à payer du consommateur. Rech Appl Mark 29:34–55. https://doi.org/10.1177/0767370114527667

Durham CA (2007) The impact of environmental and health motivations on the organic share of purchases. Agric Resour Econ Rev 36(2):304–320

Goudis A, Skuras D (2021) Consumers' awareness of the EU's protected designations of origin logo. Br Food J 123(13):1–18. https://doi.org/10.1108/BFJ-02-2020-0156

Hassan D, Monier-Dilhan S (2002a) Signes de qualité et qualité des signes. Cahiers d'économie et sociologie rurales, n° 65

Hassan D, Monier-Dilhan S (2002b) Valorisation des signes de qualité dans l'agro-alimentaire : exemple des fromages à pâte persillée. Recherches Pour et Sur le Développement Régional. Séminaire DADP des 17 et 18 décembre 2002 à Montpellier

Hassan D, Monier-Dilhan S (2006) National brands and store brands: competition through public quality labels. Agribusiness 22(1):21–30

Hughner RS, McDonagh P, Prothero A, Clifford J, Shultz CJ, Stanton J (2007) Who are organic food consumers? A compilation and review of why people purchase organic food. J Consum Behav 6(2–3):94–110

INAO (2020) Les produits sous signe d'identification de la qualité et de l'origine : Chiffres clés 2019. https://www.inao.gouv.fr/Publications/Donnees-et-cartes/Informations-economiques

INAO et CNAOL (2021) Chiffres clés 2020 des produits laitiers AOP et IGP. https://www.inao.gouv.fr/Publications/Donnees-et-cartes/Informations-economiques

Janssen M, Hamm U (2012) Product labelling in the market for organic food: consumer preferences and willingness-to-pay for different organic certification logos. Food Qual Prefer 25:9–22

Kriwy P, Mecking RA (2011) Health and environmental consciousness, costs of behaviour and the purchase of organic food. Int J Consum Stud 36:30–37

Larceneux F (2001) Proposition d'une échelle de mesure de la crédibilité d'un signe de qualité. Centre de recherche DMSP Dauphine Marketing Strategie Prospective, Cahier N°289, Avril

Larceneux F, Benoit-Moreau F, Renaudin V (2012) Why might organic labels fail to influence consumer choices? Marginal labelling and brand equity effects. J Consum Policy 35(1):85–104

Le Saout R, Vignolles B (2017) Les indices de prix hédoniques : Principes et Illustration à partir du Prix des Terrains à Bâtir. INSEE SMS, 24 Mars 2017.

Magnusson MK, Arvola A, Koivisto Hursti UK, Aber L, Sjoden PO (2003) Choice of organic foods is related to perceived consequences for human health and to environmentally friendly behaviour. Appetite 40:109–117

Mondelaers K, Verbeke W, Van Huylenbroeck G (2009) Importance of health and environment as quality traits in the buying decision of organic products. Br Food J 111(10):1120–1139

Monier-Dilhan S (2018) Food labels: consumer's information or consumer's confusion. OCL. https://doi.org/10.1051/ocl/2018009

Monier-Dilhan S, Bergès F (2016) Consumers' motivation driving organic demand: between self-interest and sustainability. Agric Resour Econ Rev 45(3):522–538

Monier-Dilhan S, Hassan D, Nichèle V, Simioni M (2009) Organic food consumption patterns. J Agric Food Ind Organ 7(2):1–23. https://doi.org/10.2202/1542-0485.1269. ISSN (Online) 1542-0485

Monier-Dilhan S, Poméon T, Böhm M, Brečić R, Csillag P et al (2019) Do food quality schemes and net price premiums go together? J Agric Food Ind Organ 20190044. https://doi.org/10.1515/jafio-2019-0044. eISSN 1542-0485, ISSN 2194-5896

Pakes A (2003) A reconsideration of Hedonic Price Indexes with an application to PC's. Am Econ Rev 93(5):1578–1596

Rosen S (1974) Hedonic prices and implicit markets: product differentiation in pure competition. J Polit Econ 82(1):34–55

Sirieix L, Delanchy M, Remaud H, Zepeda L, Gurviez P (2013) Consumers' perceptions of individual and combined sustainable food labels: a UK pilot investigation. Int J Consum Stud 37:143–151

Sörqvist P, Haga A, Linda L, Mattias H, Maria W, Nöstl A, al. (2015) The green halo: mechanisms and limits of the eco-label effect. Food Qual Prefer 43:1–9

Tagbata D, Sirieix L (2010) L'équitable, le bio et le goût. Quels sont les effets de la double labellisation bio- équitable sur le consentement à payer des consommateurs ? Cahiers d'Agriculture 19:1

Wier M, Jensen KD, Andersen LM, Millock K (2008) The character of demand in mature organic food markets: Great Britain and Denmark compared. Food Policy 33(5):406–421

Chapter 4
Innovating the Link to Origin: Is There a Difference Between PDOs and PGIs?

Maurizio Crupi

Abbreviations

EU   European Union
GI   Geographical Indication
PDO   Protected Denomination of Origin
PGI   Protected Geographical Indication
TSG   Traditional specialty guaranteed

4.1 Introduction

The European Union (EU) policy rationale for the protection of Geographical Indications (hereinafter GIs) is to preserve local traditions and cultural diversity, establishing permanent communal rights (Recitals 1 and 2, Regulation (EU) No 1151/2012 of the European Parliament and of the Council of 21 November 2012 on quality schemes for agricultural products and foodstuffs, hereinafter Regulation No 1151/2012). This justification is used to include GIs as part of the broader category of intellectual property rights, this last being aimed at fostering innovation and creativity by granting a temporary monopoly (Gervais 2012). That being said, the purpose of the GI system is not to reward innovation, but rather to reward members of a group of producers complying with practices and methods belonging to their local traditions.

Continuous striving for innovation and mechanisation of the process of production has turned many traditional methods into outdated and uncompetitive

M. Crupi (✉)
University of Alicante, Alicante, Spain

Boards of Appeal, European Union Intellectual Property Office, Alicante, Spain
e-mail: maurizio.crupi@ua.es

© The Author(s) 2025
E. Vandecandelaere et al. (eds.), *Worldwide Perspectives on Geographical Indications*, https://doi.org/10.1007/978-3-031-71641-6_4

knowledge. In recent times, small enterprises and family-owned businesses recovered and valorised traditional know-how as a strategic element of rural development. The aim is to produce value for local communities protecting local traditions and creating new job opportunities (Recital 5, Regulation No 1151/2012).

Research has already been conducted on the amendments to EU GIs, that is on the changes to the product specifications taking place after registration, when the product no longer complies with the original specifications (Recital 61 and Article 53, Regulation No 1151/2012). Having regard to fruits and vegetables, for example, it has been observed that Italian and Spanish amendments tend to include more flexibility and innovation, while French amendments tend to adopt stricter rules for strengthening the product's identity (Marescotti and others 2020). More precisely, this research analysed and qualified the changes in the geographical area, farming and processing, finding out whether the amendments resulted in more flexible or more restrictive rules, for example by allowing for a higher number of varieties, or an intensification of the production system by increasing the maximum number of plants per hectare. Looking at the respective justifications given by producers, market changes, new technologies and quality of the product appear to be the most common reasons for the adoption of the amendments, while environmental concerns seem less relevant.

Another study on the amendments of EU GIs for processed meat products noticed that most amendments are justified by the need to implement new legal provisions, mainly impacting the contents of the method of production; the nature and sourcing of the raw materials and the rules on packaging and labelling (Zappalaglio and others 2022).

The present research aims at providing a better understanding of the concepts of 'tradition' and 'innovation', which have not been defined by the EU legislator, together with an empirical analysis of the amendments for processed meat products. Differently from previous studies, the present research is aimed not only at examining the different types of amendments, together with their justifications, but also at comparing the results obtained for PDOs and PGIs.

To this extent, it must be recalled that both quality schemes protect the name of a product that originates from a specific place. However, PDOs have a stronger link to origin, requiring qualities and characteristics essentially or exclusively due to that geographical environment, including natural and human factors. On top of that, every part of the production, processing, and preparation process, including sourcing of raw materials, must be carried out in the specific place (Article 5(1) Regulation No 1151/2012). As for PGIs, they have a looser link to origin, since at least one of the stages of production, processing or preparation takes place in the region but the product must still possess a quality, reputation, or characteristic essentially attributable to its geographical origin (Article 5(2) Regulation No 1151/2012).

Given this twofold link to origin, the present research aims to understand if there is a correlation between the two quality schemes and the nature of the amendments: whether PDOs and/or PGIs tend to adopt more flexible amendments, providing a wider range of options to producers, or whether PDOs and/or PGIs adopt stricter amendments, reducing the range of options available to producers.

4.2   GIs and Tradition

Before analysing the various amendments of GIs for processed meat products, in order to better understand the role and nature of the amendments, it is necessary to: firstly, understand the meaning of tradition, which is an important part of the EU quality policy (Whereas 1 and 2 Regulation No 1151/2012); and, secondly, clarify its (apparently incompatible) relationship with innovation. To this aim, it is required to zoom out from the small group of GIs for processed meat products, which are the object of this study, and provide an overview of the meaning of tradition for all GIs in the various classes of e-Ambrosia, the publicly available database listing the names of agricultural products and foodstuffs, wine, and spirit drinks that are registered and protected across the EU.

Even though 'tradition' does not serve as link to origin under Article 5 Regulation No 1151/2012, it appears in almost all single documents for EU GIs. 'Traditions' and 'knowledge' are only briefly mentioned in Art. 3(3) of that Regulation, which defines 'traditional' as intergenerational usage within the domestic market for at least 30 years, in relation to Traditional Specialities Guaranteed (TSG).[1]

For the purpose of this chapter, tradition is interpreted as an expression identifying knowledge, culture, and creations of a given local community handed down from one generation to the other (WIPO 2001; Gervais 2003). In particular, tradition does not refer only to the antiquity of the knowledge itself but means that a given knowledge represents the cultural values of a community and is held collectively as part of its cultural traditions.

The qualitative analysis of over 1300 single documents (621 PDOs and 711 PGIs) aimed at understanding how all EU registered GIs for agricultural products and foodstuffs are linked to their origin, allows a better understanding of the relationship between GIs and the culture of their production (Crupi 2022). The quotations contained in the single documents under the section named 'link with the geographical area', describe the connection of a product with the know-how of a community of producers, referring to the human factors and the traditional character of a product with the use of keywords like 'traditional', 'longstanding production techniques', and 'handed down from one generation to the other'. These quotations emphasise the traditional character of the product and its cross-generational production.

The Chart 4.1 shows that there is a high occurrence of quotations on tradition. Starting from 2006 almost every PDO and PGI has a reference to traditional know-how. Tradition seems to have an important role in linking the product to its origin, although slightly more frequent for PGIs rather than for PDOs. A possible explanation involves the nature of the link to origin for PGIs. Because the link to origin through natural factors and local physical resources is weaker for PGIs, more

---

[1] Differently from PDOs and PGIs, Traditional specialty guaranteed (TSG)s protect names describing a specific product or foodstuff that results from a mode of production, processing or composition corresponding to traditional practice for that product or foodstuff; or is produced from raw materials or ingredients that are those traditionally used (Article 18 Regulation No 1151/2012).

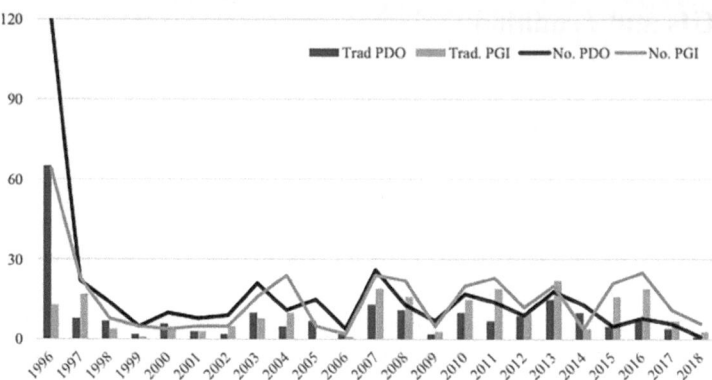

Chart 4.1 Number of quotations on tradition for PDOs and PGIs. (Source: author)

emphasis is placed on processing methods and tradition to justify the protection conferred to the PGI.

The point is made more complex by the fact that tradition can be approached not only as a 'static' notion, as in the EU legislation, but also as a 'dynamic' notion. In particular, 'traditional' does not mean that the know-how is old but deals with the process of sharing and learning. The fact that traditional know-how is passed down from one generation to another does not exclude that it undergoes a process of incremental development (alias innovation), where each generation adds new layers of knowledge to the inherited traditions (Dutfield 2003; Montanari 2006).

Changes that occurred to the 'traditional' recipes over the centuries must be carefully considered. It would be incorrect to systematically consider the impact of changes on product characteristics in a negative light. In particular, the product may undergo positive and even desirable changes, a natural consequence of a changing society and environment (Bromberger 2006). In particular, innovation is welcome as afar as it preserves the link to origin and the final specific quality of the product which has made its reputation.

4.3   GIs and Innovation

For the purpose of this chapter, the term innovation is interpreted according to the definition of 'business innovation' provided by the OSLO Manual (OECD 2018):

> *A business innovation is a new or improved product, or business process (or combination thereof) that differs significantly from the firm's previous products or business processes and that has been introduced on the market or brought into use by the firm.*[2]

---

[2] OECD/Eurostat (2018), 68.

This broad definition encompasses a range of sub-categories, dealing with the implementation of one or more types of innovations, such as product and process innovations. The minimum requirement for a product or process to be considered innovative is that it must be new to the company/organisation that adopted it.

In other words, new products and processes are those that differ significantly in their characteristics from the ones previously produced/adopted by the company. This means a change in materials, components, and other characteristics of the product. Therefore, innovation is not limited to an enhancement of the quality of the product but includes all various amendments, including those consisting of a decrease in production costs and/or maintaining a certain quality level despite the change of external conditions (e.g. climate change), through a modification in the technical methods of production and the equipment used.

The entire GI system is based on the recognition of a name linked to specific production practice (INAO 2017). Therefore, product innovation for GIs does not refer to the creation of 'new' products but mainly deals with the 'improvement' of existing ones.[3]

The analysis of the single documents for all EU registered GIs for agricultural products and foodstuffs makes it possible to distinguish three different types of innovation for GIs, depending on when the innovation took place. Firstly, it is possible to identify the innovation that occurred in the past before drafting the product specifications (even before the GI application). This is the 'historical innovation' that happened to the 'traditional' recipes over the centuries as a natural consequence of a changing society and environment. It is interesting to note that around 20% of the single documents for EU foodstuffs contain a reference to 'innovation', meaning the modernisation of the process of production that took place over the centuries. In particular, many innovations became the new standard of production, being indissolubly linked to the product as we know it. This code is often observed in the single document as part of process innovation, showing the ability of the local community of producers to innovate within the traditional process of production, advocating in favour of a dynamic concept of tradition.[4]

Secondly, the 'collective innovation' that occurred at the time when the product specifications were drafted. The definition of the production standards is a complex procedure that requires drafters and decision-makers to represent the interests of all legitimate GI beneficiaries (Marie-Vivien and others 2019). In this phase, the community of producers does not only limit confirmation of the existing practice but needs to reconcile different perspectives and come to commonly agreed product specifications, depending on the producers' objectives and priorities. An example

---

[3] The system allows the recognition of 'invented' products [*sic* in the single document] and products with a recent history of production. See Tekovský Salámový Syr PGI [2010], whose "production process was invented in 1921". And the kiwi Aktinidio Pierias PGI [2002], whose first plant was grown in Greece only in 1973.

[4] As an example of historical innovation see Miel de Galicia PGI [2005] OJ C30/16 [4.6]. Reference is made to the creation of movable hives in 1880 and the construction of the first 'nursery hive' designed for breeding by division and for the breeding of queens.

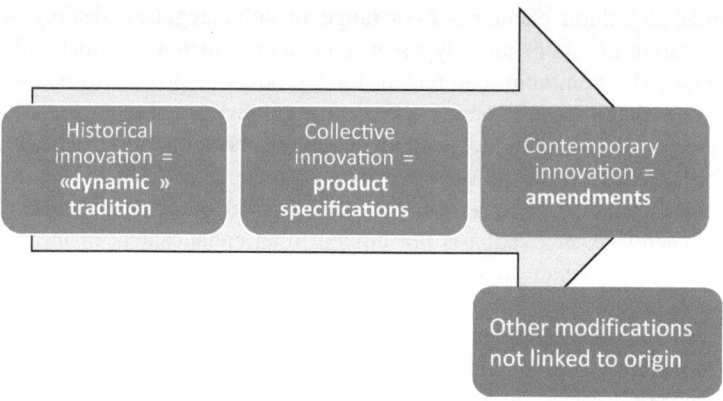

Fig. 4.1 Different types of innovation for GIs. (Source: author)

concerns the adoption of new production methods versus pre-existing ones, often putting farmers against industrial producers.

Thirdly, 'contemporary innovation' occurs with the amendments of the product specifications, after the registration of the product. These may be classified into two sub-categories: compulsory amendments requiring compliance with legal provisions, such as the entry into force of new safety regulations that imposes changes in the existing practices,[5] and voluntary ones adopted by producers for strategic reasons, such as a change in consumers' needs,[6] or technological developments. The latter can be heavily influenced by other external factors like climate change and technical restrictions to the production according to the traditional recipe, and outside producers' control. All these issues could force producers to amend the traditional process of production, deviating from their traditional know-how.

For the scope of this chapter, product innovation for GIs is further limited to those modifications dealing with a specificity of the product linked to its origin, for which the modification can take place while drafting the product specifications or later on with the amendments. Other modifications not linked to the origin in production/ process, referring to generic methods not described in the specifications, are not taken into account (Fig. 4.1).

---

[5]Schwarzwalder Schinken PGI [2012], OJ C 274/2, no. 3.e. The product is no longer cured in wooden vats but in stainless steel containers for reasons of hygiene.

[6]*Pecorino Toscano* PDO [2015] OJ C18/12, n.3. The amendment of the method of production of *Pecorino Toscano* allows the use of vegetable rennet in the production of cheese in compliance with the Kosher certification. The possibility is added of using vegetable rennet, a long-standing practice in Tuscany for the production of Pecorino (already mentioned in the national registration application submitted in 1985). This practice has been taken up again in the last few years both as practice that is typical of the territory and for the production of kosher cheeses.'

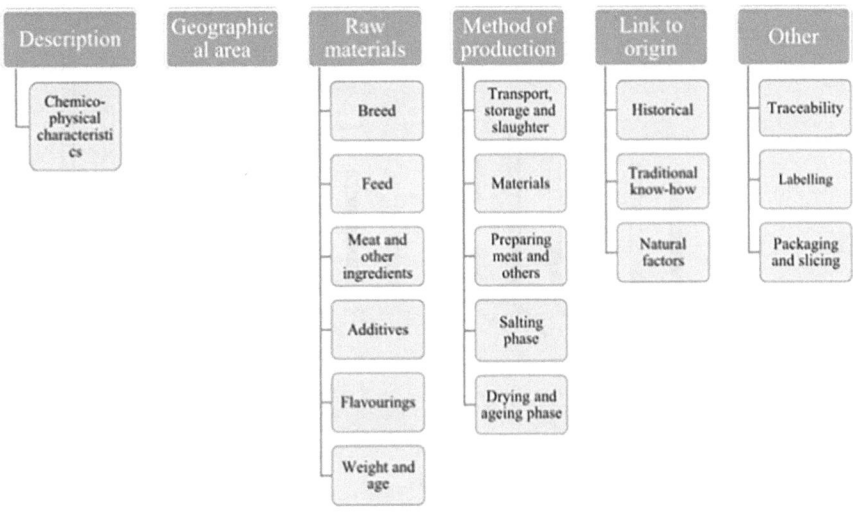

Fig. 4.2 Code structure. (Source: author)

4.4   Methodology

The methodology consists of a directed content analysis of 25 amendments for PDOs and 29 for PGIs for processed meat products (out of a total of 36 PDOs and 140 PGIs registered for processed meat products, class 1.2 pursuant to Article 2(1) EU Regulation No 1151/2012). The unit of meaning for this analysis are both minor and not minor amendments (now Standard and Union amendments) published in the EU database and approved before 1 November 2019 (last amendment published on 24 June 2019[7]). Categories and codes have been developed after multiple readings of the single documents, starting from the Regulation No 1151/2012 and the EU guide to applicants on how to compile the single documents, as detailed in Fig. 4.2.

Despite the relatively small number of amendments, processed meat products are indeed relevant in the analysis of the link of PDOs and PGIs with their territory. Class 1.2 has been chosen because it deals with processed products, allowing a more complete analysis of both the origin of raw materials and the process of production of the final transformed products. This differs from other classes, for example class 1.6 on fruit and vegetables, where the focus is placed on the production of raw materials and not on their transformation.

In addition, some processed meat products could benefit from the exception of Article 5(3) Regulation No 1151/2012, which allows for a name to be registered as a designation of origin even though the raw materials (limited to live animals, meat and milk) come from a geographical area that is larger than the defined geographical area. This exception can be applied to designations of origin recognised in the

---

[7] *Tiroler Speck* PGI [2019], OJ L 167/21.

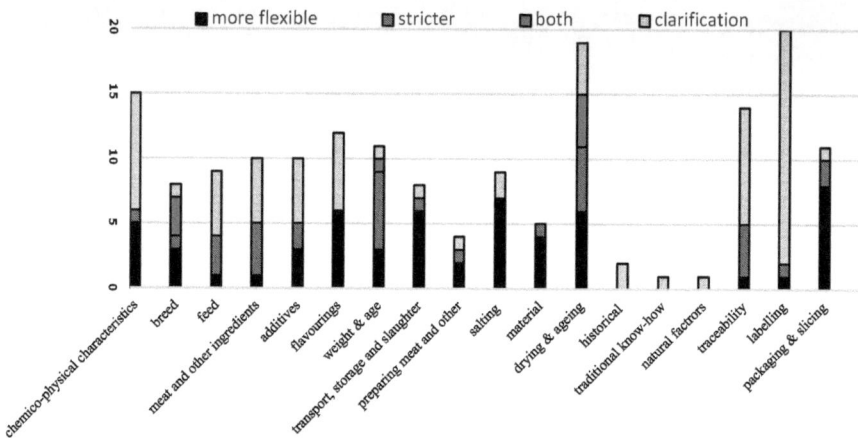

Fig. 4.3 Amendments for PDOs. (Source: author)

country of origin before 1 May 2004, as soon as the production area of the raw materials is defined and there are special conditions for their production. Therefore, the analysis of products in class 1.2 allows for a deeper understanding of the evolution of the link to origin, highlighting the cases and the reasons for using local raw materials for those PDOs that could benefit from a broader source of raw materials.

Similarly to previous research conducted on the amendments for fruit and vegetables (in class 1.6) (Marescotti et al. 2020), the amendments have been classified using the labels 'more flexible' (when they provide a wider range of options), 'stricter' (when they reduce the options available to producers), 'both' (when the amendment provides both options), depending on if they increase or reduce the options available in the previous product specifications.[8] The code 'clarification' is used to classify those amendments that simply correct mistakes or provide additional information on the common practice of the sector (Figs. 4.3 and 4.4).

---

[8] An example of a 'more flexible' amendment, *Coppa piacentina* PDO [2010] OJ C311/24, n.8. The temperature in the drying stage range from 15 to 25 °C (originally ranged from 17 to 20 °C).

An example of a 'stricter' amendment, *Prosciutto di Carpegna* PDO [2009] OJ C189/03, n.3.2. In Article 5 of the product specifications, the description of the maturing period as 'on average 14 months and never less than 12' has been replaced by 'is not less than 13 months', therefore raising the minimum requirements.

An example of 'both', *Pancetta Piacentina* PDO [2010] OJ C64/32, n. 10,11. Here the maturation phase has been increased from at least 2 months to at least 3 months from the date of salting, therefore imposing stricter requirements for the drying and ageing phase. At the same time, the range of relative humidity has been increased from 70–80% to 70–90%, therefore giving more flexibility to producers with regard to the maximum percentage of relative humidity.

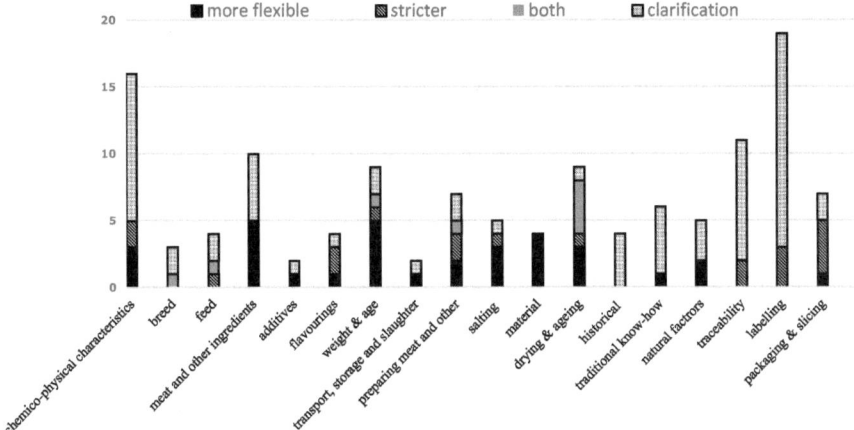

Fig. 4.4 Amendments for PGIs. (Source: author)

4.5   Research Findings

4.5.1   *Amendments to the Area of Production*

The geographical area for PDOs has been amended in 12% of the documents considered. The list of amendments contains two clarifications: one is aimed at better defining the area of production of the raw materials and the area of production and maturing of the final product;[9] the other one is aimed at correcting mistakes contained in the previous product specifications.[10] In addition to that, this section contains an amendment that brings together the criteria for identifying the final stage of production and the curing facility. This method, more accurate than the previous one, refers to the absolute altitude of the facilities entered in the PDO registers and not to the altitude of the municipalities. A margin of tolerance has been adopted to keep the facilities registered under the previous product specifications.[11]

As for PDOs, the geographical area for PGIs has rarely been amended. 13,8% of the documents considered contain a clarification rectifying the list of municipalities, including some municipalities at first excluded by an oversight, correcting some spelling mistakes and describing the boundaries of the geographical area with a higher degree of accuracy.[12]

---

[9] *Dehesa de Extremadura* PDO [2016] OJ C 207/21, no. 3.

[10] *Guijuelo* PDO [2015] OJ C 329/7, no. 3.

[11] Jamón de Teruel PDO [2013] OJ C 242/18 no. 3.

[12] *Saucisson de l'Ardèche* PGI [2015] OJ C437/9, no. 5.2. *Jambon de l'Ardèche* PGI [2015] OJ C330/3, no. 5.1. *Gailtaler Speck* PGI [2018] OJ C195/47, no. 5. *Schwarzwälder Schinken* [2012] OJ C274/2, no. 3.c.

4.5.2 Amendments to the Raw Materials

As a general rule, PDOs require raw materials to come from within the area of production, while for PGIs there is no such requirement. The research shows that PDOs did not change the provisions concerning the origin of raw materials, none of the single documents abandoned the exception under Article 5(3) Regulation No 1151/2012, which allows raw materials for PDOs to come from a larger area, in favour of a stricter requirement. In other words, none of the PDOs benefiting from the exception under Article 5(3) opted for a stricter link to origin as regards the origin of the raw materials.

The category of raw materials for PDOs presents stricter amendments when it comes to 'feed', 'meat and other ingredients', and 'weight and age'. It is possible to observe how feed and the fattening phase have been strictly regulated in three amendments, introducing a minimum fattening time, a stocking density, and replacing the maximum permitted average weight with a maximum weight for the individual carcasses. Four amendments present a stricter definition of the characteristics of the meat, such as the EU classification scale used to distinguish the carcasses of the animals, together with the definition of minimum and maximum limits of fat and the use of specific meat cuts. Other sub-categories such as 'breed' and 'flavourings' present more flexible amendments aimed at allowing producers to personalize the recipe, adapting the product specifications to the growing trends of reducing the amount of salt in food.

As regards PGIs, it is possible to identify a trend, which favours flexibility. In particular, 4 out of 28 PGIs allowed raw materials, originally coming from within the area of production, to come from outside the area of production.[13] The most common reason for these amendments is that product's characteristics and appearance are not affected by the extent to which the ingredients originate in the region.

Contrary to PDOs, PGIs are characterised by a higher degree of flexibility when it comes to 'feed', 'meat and other ingredients', and 'weight and age' of the animals. In particular, various amendments allow a change in the muscle ratio, considering the advance in breeding techniques, and the use of nitrates.

As to the raw materials, the analysis of the above-mentioned categories shows that PDOs seem to 'balance' the broad exception of the origin of raw materials provided by Art. 5(3) Regulation No 1151/2012 with the adoption of stricter requirements for the characteristics and use of raw materials, in particular 'feed', 'characteristics of the meat', and 'weight and age' of the animal. Conversely, PGIs tend to amend the same categories by allowing more flexibility.

---

[13]Thuringer Rostbratwurst; Gailtaler Speck; Jambon Sec des Ardennes; Eichsfelder Feldgieker.

4.5.3 Amendments to the Method of Production

The majority of the amendments regarding the method of production for PDOs provide some flexibility to producers. For example, the sub-category 'transport, storage and slaughter' gives flexibility regarding the minimum time that meat has to stay at the slaughterhouse before slaughter. The amendments contained in the sub-category 'material' allow the introduction of non-traditional material in the process of production,[14] to better reflect current market conditions, such as the possibility to use elastic twine and not only natural one in the tying process, with the result of facilitating tying and improving the processing of the product. The amendments of the sub-category 'salting' are characterised by a difference in the salting period, usually a decrease of the minimum period;[15] this complies with modern salting techniques and avoids the meat absorbing too much salt, in line with consumers' current food requirements.

Similarly, PGIs provide a higher degree of flexibility to producers. The subcategory 'transport, storage and slaughter' broadens the temperature range used in the delivery of the meat.[16] Some amendments allow the use of machines in the salting process, together with the traditional hand-made process.[17] Tradition is partly overcome also in the sub-category 'material', which allows the use of non-traditional wood in the smoking process and expands the type of casings permitted in the production.[18] The sub-category 'drying and ageing' broadens the length of the process and introduces more flexible temperature and ageing conditions.[19]

As to the method of production, both PDO and PGI producers are interested in achieving more flexibility to adapt their products to a new market and consumers' needs, modern practices of production, and new food safety standards. In other words, the amendments to the method of production do not seem to be influenced by the specific quality scheme chosen.

---

[14] *Capocollo di Calabria* PDO [2015] OJ C82/12, no. 3. *Pancetta Piacentina* PDO [2014] OJ C 86/9, no. 3. *Coppa Piacentina* PDO [2014] OJ C 88/21, no. 3. *Salame Piacentino* PDO [2014] OJ C 88/26, no. 3.

[15] *Crudo di Cuneo* PDO [2016] OJ C188/54, no. 5. *Jabugo* PDO [2016] OJ 415/9 no.5. *Prosciutto di Carpegna* PDO [2009] OJ C189/03, no. 3.2. *Coppa Piacentina* PDO [2014] OJ C 88/21, no. 3. *Guijuelo* PDO [2015] OJ 329/3 no. 5.

[16] *Speck Alto Adige* PGI [2016] OJ C334/9, no. 5.

[17] *Jambon Sec des Ardennes* PGI [2014] OJ C444/26, no. 3.4. *Schwarzwälder Schinken* [2012] OJ C274/2, no. 3.e.

[18] *Holsteiner Katenschinken* PGI [2017] OJ C247/8, no. 5. e. *Gailtaler Speck* PGI [2018] OJ C195/ 47, no. 5. *Salchichón de Vic* PGI [2017] OJ C368/11, no. 5. *Halberstädter Würstchen* PGI [2014] OJ C270/5, no. 3. e.

[19] *Kranjska Klobasa* PGI [2015] OJ C441/5, no. 5.1. *Speck Alto Adige* PGI [2016] OJ C334/9, no. 5. *Breasaola della Valtellina* PGI [2010] OJ C321/25, no. 3.2.

4.5.4   Amendments Per Member State

Apart from the analysis of the amendments for products of class 1.2 grouped into PDOs and PGIs, further research has been conducted on the products originating from those Member States with the highest number of amendments: namely, Italy (19 products amended), Spain (6 products), Germany (6 products) and France (5 products).

This analysis has been limited only to the categories that present a higher number of amendments ('raw materials' and 'method of production'). Due to the limited number of products, the results have not been classified into PDOs and PGIs but have been considered altogether in order to identify possible national trends towards more flexible or stricter amendments (Fig. 4.5).

While Italian, Spanish and German amendments resulted in more flexible product specifications, it is worth noting that French amendments appear more 'balanced', with an equal number of more flexible and stricter amendments (6 each).

However, these results may be biased by the small samples considered and cannot be used to claim a clear direction of the amendments for each Member State. Nonetheless, it is important to note that these results seem in line with previous research in the field, showing a similar directionality for Italian, Spanish and French amendments for fruit and vegetables in class 1.6 (Marescotti et al. 2020).

4.6   Interpretation of Results

The fact that sometimes tradition and innovation coexist within the same product, as reasons for different amendments, can be used to prove that they are deeply intertwined in GIs, arguing in favour of the fact that tradition and innovation are 'two sides of the same coin'. An example can be found in the amendments of Capocollo di Calabria where, on the one hand, the extension of the maximum curing time from 8 to 14 days takes traditional processing methods into account and, on the other hand, the removal of the requirement to use only natural twine in the tying process (enabling manufacturers to use elastic twine) facilitates tying and improves the processing of the product.[20]

The reasons for the amendments, published in the EU database together with the amendment itself, provide a better understanding of the results of the qualitative analysis.

For PDOs, the majority of the amendments concerning raw materials are aimed at providing a more accurate description of the specifications, correcting mistakes and using more precise wording, avoiding misinterpretation. Other frequent reasons concern producers' willingness to modify the section on raw materials to make them more in line with traditional practices, to improve the quality of the final

---

[20] *Capocollo di Calabria* PDO [2015] OJ C82/12, no. 3.

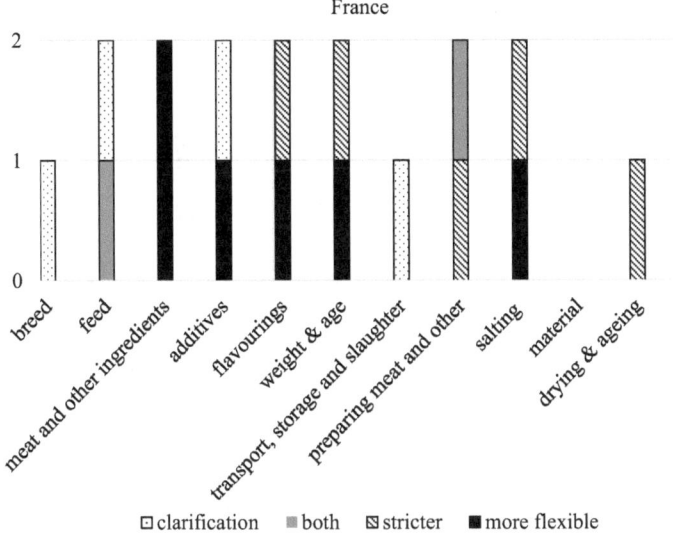

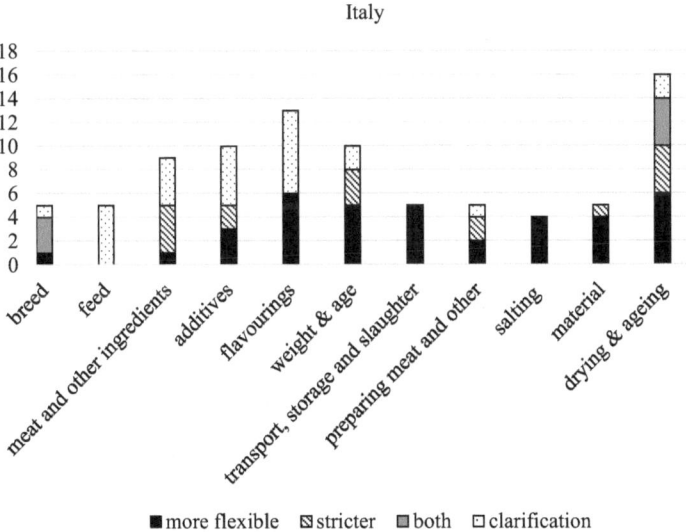

Fig. 4.5 Amendments to French, Italian, Spanish and German products. (Source: author)

product, and to comply with both national and EU legal provisions. When it comes to PGIs, producers' interest, apart from a more accurate description and a focus on quality, is to adapt product specifications to the changes in raw materials, to adhere

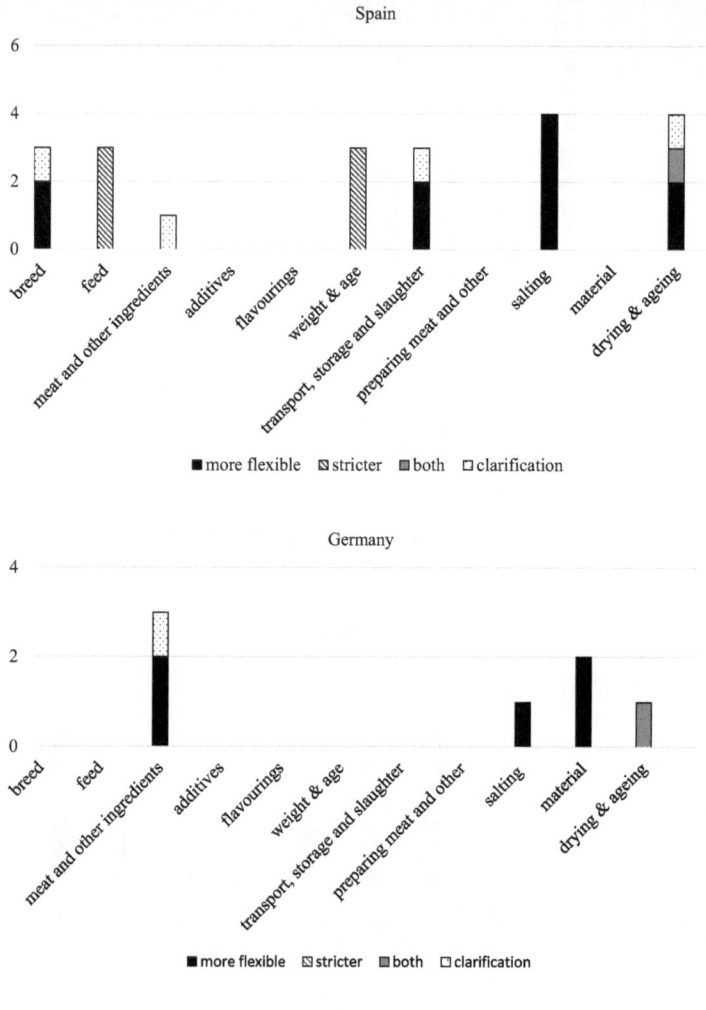

Source: author

Fig. 4.5 (continued)

to common practices without specifying whether these are traditional practices that have not been codified in the previous version of the specifications or whether these are innovations commonly adopted by the majority of producers, and market needs.

As to the method of production, both PDOs and PGIs aim at meeting new market and consumers' needs, modern practices of production, and new food safety standards. For PDOs, a relevant percentage of amendments have been adopted to adapt to climate change (4 amendments[21]) and to increase compliance with traditional practices (12 amendments).

The results of this research show that the category PDO or PGI does not have a significant impact on the amendment of the link to origin for processed meat products, except for the origin/production of raw materials. Here a correlation has been found, showing that PDOs are more conservative than PGIs, reducing the range of options available to producers.

Further research is required to prove whether these results can be verified also for other categories of products, in particular those that do not benefit from the exception of Art. 5(3) Regulation No 1151/2012.

References

Bromberger BE (2006) Aged, but not old: local identities, market forces, and the invention of "traditional" European cheeses. In: Hosking R (ed) Authenticity in the kitchen: proceedings of the Oxford symposium on food and cookery. Prospect Books, pp 89–102

Crupi M (2022) A pragmatic approach to the link to origin: EU PDOs and PGIs for registration, innovation and trade in origin products. Doctoral thesis, Maastricht University, Universidad de Alicante. EIPIN-Innovation Society. https://doi.org/10.26481/dis.20220926mc

Dutfield G (2003) Protecting traditional knowledge and folklore. A review of progress in diplomacy and policy formulation. International trade & sustainable development series. Intellectual Property Rights No. 4. ICTSD—UNCTAD Project on IPRs & Sustainable Development

Gervais DJ (2003) Spiritual but not intellectual? The protection of sacred intangible traditional knowledge. Cardozo J Int Comp Law 11(2):467–496

Gervais DJ (2012) Traditional innovation and the ongoing debate on the protection of geographical indications. In: Drahos P, Frankel S (eds) Indigenous peoples' innovation: intellectual property pathways to development. ANU Press, New York, pp 121–146

INAO (2017) Guide du demandeur d'une appellation d'origine protégée (AOP) ou d'une indication géographique protégée (IGP) à l'exception des vins, boissons alcoolisées et boissons spiritueuses. https://www.inao.gouv.fr/Espace-professionnel-et-outils/Produire-sous-signes-de-qualite-comment-faire/Guides-pratiques. Accessed 15 Dec 2023

Marescotti A et al (2020) Are protected geographical indications evolving due to environmentally related justifications? An analysis of amendments in the fruit and vegetable sector in the European Union. Sustainability 12(9):3571. https://doi.org/10.3390/su12093571

Marie-Vivien D et al (2019) Controversies around geographical indications—Are democracy and representativeness the solution? Br Food J 121(12):2995–3010. https://doi.org/10.1108/BFJ-04-2019-0242

Montanari M (2006) Food is culture. Columbia University Press, New York

OECD/Eurostat (2018) Oslo manual 2018: guidelines for collecting, reporting and using data on innovation, 4th edition. The measurement of scientific, technological and innovation activities. OECD Publishing/Eurostat, Paris/Luxembourg. https://doi.org/10.1787/9789264304604-en

---

[21] *Coppa piacentina* PDO [2010] OJ C311/24, n. 3.9. The amendment deals with a change in the temperature and relatvie humidity of the premises where ageing takes place. The new value range takes into account the climatic variations in the province of Piacenza which influence the ambient conditions prevailing in the natural cellars where coppa is matured.

WIPO (2001) Intellectual property needs and expectations of traditional knowledge holders WIPO report on fact-finding missions on intellectual property and traditional knowledge (1998–1999) Available at https://www.wipo.int/edocs/pubdocs/en/tk/768/wipo_pub_768.pdf. Accessed 15 Dec 2023

Zappalaglio A et al (2022) Study on the functioning of the GI system. Max Planck Institute for Innovation and Competition, Research Papers. Available at SSRN: https://ssrn.com/abstract=4061160. Accessed 15 Dec 2023

Chapter 5
The Traditional Specialty Guaranteed or the Protected Geographical Indication as Quality Schemes for the Protection of *Jamón Serrano* (*Serrano Ham*)

Vicente Gimeno Beviá

Acronyms

TSG	Traditional Specialty Guaranteed
PGI	Protected Geographical Indication
PDO	Protected Designation of Origin
EU	European Union
ANICE	National Association of Meat Industries of Spain
AENOR	Spanish Association for Standardization and Certification

5.1 Introduction

Jamón Serrano, in addition to being one of the most typical dishes of Spanish cuisine, is undoubtedly one of the most marketed meat products both nationally and internationally.[1] This has been greatly contributed to by the Spanish ham sector, which ensures the maintenance of quality standards and jointly carries out an extremely important task of projecting and promoting *Jamón Serrano*.

Precisely, the protection that *Jamón Serrano* received from the European Commission at the end of the last century was fundamental to achieving these goals.

---

[1] According to data from the Spanish *Jamón Serrano* Consortium, a total of 59,850.29 tons were exported in 2023 for a value of 657,181,993.84 euros, which represents an increase of 10.80% compared to the previous year. More information at https://consorcioserrano.es/el-sector-del-jamon-curado-exporto-59-850-toneladas-en-2023/

V. Gimeno Beviá (✉)
University of Alicante, Alicante, Spain
e-mail: vicente.gimeno@ua.es

© The Author(s) 2025
E. Vandecandelaere et al. (eds.), *Worldwide Perspectives on Geographical Indications*, https://doi.org/10.1007/978-3-031-71641-6_5

Specifically, on November 13, 1999, the request of several Spanish meat associations to register this product as TSG was granted.[2] This, as is known in Europe, means that *Jamón Serrano* is recognized by one of the EU's quality regimes that is not considered an intellectual property right and, although it is not linked to a specific geographical area, it highlights the traditional aspects of the product and protects producers against improper use of the term.

However, this protection is not entirely satisfactory for the majority of *Jamón Serrano* producers who see no legal obstacles to the aforementioned product being registered as PGI.

5.2 The Traditional Specialty Guaranteed and the Protected Geographical Indication as Differentiated Quality Schemes

The conflict over the appropriate protection of *Jamón Serrano* takes place between two distinctive quality signs belonging to the category of differentiated quality schemes, regulated by European Union legislation: the traditional specialty guaranteed -TSG- and the protected geographical indication -PGI-.

The first of these, provided for in Title III of Regulation (EU) No 1151/2012 of the European Parliament and of the Council, of November 21, 2012, on quality schemes for agricultural products and foodstuffs, aims to protect the methods of production and traditional recipes so that producers market their products and inform consumers of such characteristics that give them added value. The TSG, unlike the geographical indication, is also not geographically limited to any territory. In Spain, currently, there are four guaranteed traditional specialties: the *"Tortas de aceite de Castilleja de la Cuesta"* (oil tart), the "panellets" (a bakery product), la "Leche certificada de granja" (certified farm milk) and the *"Jamón Serrano"* (*Jamón Serrano*) as the first of all of them in chronological order. In other European countries, the most well-known examples are the Neapolitan pizza, the mozzarella or the Portuguese cod or recognized Belgian beers.[3]

For the registration of a product or food as a TSG the determining criterion is tradition, required both in the manufacturing process and in the raw materials and ingredients used as well as in its identification, that is, that the name is the one that, for a long time ago, has been used to refer to it (Seville 2009). In addition, it is necessary that they comply with a specification that establishes the necessary requirements for the producer to display the registered name as TSG and its

---

[2]In particular, the applicant groups were the Association of Meat Industries of Spain (AICE), the Catalan Federation of Meat Industries (FECIC) and the Professional Association of Slaughterhouses and Meat Companies (APROSA-ANEC).

[3]As an example, the Geuze ETG as a traditional beer generally produced in Belgium that stands out for its spontaneous fermentation. Geuze ETG (1997) OJ L 318/21.

characteristic symbol. Precisely, the TSG "*Jamón Serrano*" is reserved for osteomuscular pieces corresponding to the hind limbs of the pig with a minimum weight determined, transported at a temperature not exceeding three degrees, with a curing time not less than seven months, among many other requirements. Those hams that, even when they are similar, do not meet such conditions of the specification will be marketed, mostly, as "cured ham".

The PGI, also regulated in the aforementioned Regulation (EU) no 1151/2012 serves to protect those products originating from a specific place, a region or a country that possess a specific quality, a reputation or another characteristic that can essentially be attributed to its geographical origin and of whose production phases, at least one takes place in the defined territory (Montero 2021).[4] The difference with the protected designation of origin -PDO- lies in that it has a lesser link with the territory, since the requirements relating to the production phases are not cumulative but alternative, that is, it is enough that at least one of them takes place in the defined geographical area (Gallego and Fernández 2019).

For registration as a PGI, it is necessary for the product to comply with the provisions in a specification that contains the name to be protected, a specific description of the raw materials used and their main characteristics, detail of the production method and, where appropriate, of the packaging, the delimitation and specification of the geographical area of the indication, as well as the justification of the link existing between said territory and the product. Precisely, this last requirement about the relationship with geography is the fundamental difference existing between this distinctive quality sign and the guaranteed traditional specialty. Thus, for example, hams like the of *Trévelez or Serón* as PGI or others like that of *Teruel, Los Pedroches, Jabugo, Guijuelo or Dehesa de Extremadura* as PDO, are linked to such homonymous regions.

5.3 Legal Protection of *Jamón Serrano*

5.3.1 Current State of Affairs

Jamón Serrano, as has been previously demonstrated, has been protected as a TSG since 1999. Only hams that meet the requirements provided in the specification can benefit from the distinctive of the aforementioned differentiated quality scheme and will be authorized to use the name "Jamón Serrano" alongside their individual brand.

Other hams that, like those covered under the TSG *Jamón Serrano*, come from white pig but do not meet the requirements provided in the specification (for example, because they do not meet the minimum period of seven months of curing)

---

[4]On the exclusive coverage of agri-food products and the existing debate around the legal regime of geographical indications and their possible extension to non-agricultural products, see MONTERO GARCÍA-NOBLEJAS (2021), p. 427.

as, obviously, they cannot be identified with the term "serrano", use the reference "cured ham" in alluding to the manufacturing process.

However, it seems that this is not enough for a large part of the ham sector in Spain, such as the Spanish *Jamón Serrano* Foundation, which brings together more than eighty percent of the production of *Jamón Serrano*, or the National Association of Meat Industries of Spain, which has over six hundred business owners from the meat sector as members. Both promote the creation of a PGI that replaces the current TSG as a sign that identifies *Jamón Serrano*.[5]

And on this path they have broad parliamentary support. In this regard, in 2014 a parliamentary initiative on Spanish *Jamón Serrano* was unanimously approved, presented by the Socialist Parliamentary Group with the aim of recognition of *Jamón Serrano* as a PGI.[6] This crystallized, later, in the express support of the Ministry of Agriculture, Fisheries and Food which collaborated in such an initiative that led in September 2016 to the application for registration of the Protected Geographical Indication (PGI) "*Jamón Serrano*".[7] Precisely, that same year, the National Association of the Meat Industry submitted to the European Commission a request for annulment of the Traditional Specialty Guaranteed (TSG) "Jamón Serrano" as long as it is a necessary requirement for registration as a PGI, in accordance with the provisions of Article 6.3 of Regulation (EU) No 1151/2012, that there is not a homonymous name of another that is already registered in the register. But, in case it finally does not prosper, they condition the request for annulment to the simultaneous registration of *Jamón Serrano* as a protected geo-graphical indication.[8]

After the exchange of communications between the services of the European Commission and the General Directorate of the Food Industry as an agency attached to the Ministry of Agriculture, Fisheries and Food of Spain, in November 2020 a resolution was issued by which they publicized the application for registration of the Protected Geographical Indication (PGI) "*Jamón Serrano*".[9] The text indicated that, once the technical review process of the file by the European Commission was

---

[5] Although the *Jamón Serrano* Foundation went into bankruptcy, its associates have created the "Serrano Consortium" association which maintains the same position regarding the creation of a PGI for *Jamón Serrano*.

[6] More information about the PNL and its processing, available at https://www.congreso.es/web/ guest/busqueda-de-iniciativas?p_p_id=iniciativas&p_p_lifecycle=0&p_p_state=normal&p_p_ mode=view&_iniciativas_mode=mostrarDetalle&_iniciativas_legislatura=X&_iniciativas_id=1 61/002467

[7] File on the application of *Jamón Serrano* as PGI, available here https://ec.europa.eu/info/food-farming-fisheries/food-safety-and-quality/certification/quality-labels/geographical-indications-reg ister/details/EUGI00000017021

[8] Cancellation request in accordance with Article 54, paragraph 1, of Regulation (EU) No 1151/ 2012, available at https://www.mapa.gob.es/es/alimentacion/temas/calidad-diferenciada/ solicituddeanulacionetg_tcm30-526237.pdf

[9] Resolution of November 12, 2020, of the General Directorate of the Food Industry, which publicizes the registration request of the Protected Geographical Indication (PGI) "*Jamón Serrano*", available at https://www.boe.es/diario_boe/txt.php?id=BOE-B-2020-42142

concluded and in order to ensure a transparent processing procedure that does not generate defenselessness in third parties, published the terms and conditions resulting from said file, a period of two months was granted for any natural or legal person established or residing in Spain and who has a legitimate interest can oppose the application for registration. After this period, about twenty interested parties expressed their opposition, all from Spain, among which are several regulatory councils of PGIs and PDOs, public administrations such as the Provincial Council of Granada or the Spanish Association of Origin Denominations "Origin Spain", among others. After various modifications, the Ministry of Agriculture, Fisheries and Food sent on July 20, 2022 the complete file requesting the European Commission to register the Protected Geographical Indication (PGI) *Jamón Serrano* and the cancellation of the TSG *Jamón Serrano*.

However, in September the producers of the Protected Geographical Indication (PGI) "*Jamón de Trevélez*" (Ham of Trevelez), as they consider that such recognition would devalue their distinctive quality sign, filed an appeal before the Sala de lo Contencioso-Administrativo del Tribunal Superior de Justicia de Madrid (Administrative Litigation Chamber of the Superior Court of Justice of Madrid) against the registration of a PGI for *Jamón Serrano* and the Ministry of Agriculture, Fisheries has halted the processing procedure until the sentence takes place. To this day the file remains pending resolution. Therefore, seven years after the application for the PGI, *Jamón Serrano* remains registered as a TSG.

5.3.2 Thesis in Favor of Registering Jamón Serrano as a Protected Geographical Indication

Supporters of the change of protection of *Jamón Serrano* as a Protected Geographical Indication represent the majority of the Spanish ham sector[10] justify their position based on the following arguments.

One of the reasons for the change from TSG to PGI is the territorial limitation of protection because, with the Protected Geographical Indication, the name "*Jamón Serrano*" would be reserved, exclusively, for those hams from Spain that meet the characteristics indicated in the specification (Huerta 2021) although, PGIs whose extension coincides with a territory are not common. But the most significant change from TSG to PGI as a globally recognized intellectual property right is, precisely, the possible protection of the product in third countries, through bilateral agreements. With this, on the one hand, they would restrict the use of the term "Jamón Serrano" to Spanish producers and, on the other, they would avoid confusion among consumers about the characteristics of the product. In this sense, in line with what was

---

[10] It includes both large and small and medium entrepreneurs who, in addition, are members of the main associations in relation to ham such as the National Association of the Meat Industry (ANICE) or the Serrano Consortium.

provided in the previous paragraph, while the traditional specialty guaranteed is linked with the traditional method of preparation, the determining note of the Protected Geographical Indication is the relationship with the territory, so that the distinctive quality sign can be circumscribed to a specific geographical area, not larger in extension than that of a country, while the TSG is open to producers from other States so that no unfair competition conditions are created.[11] In fact, currently, there are producers in third countries that use this TSG, which does not pose any problem as long as they comply with the specification (Guillem 2021).

However, the main problem with the TSG is that it does not exist in other legal systems, unlike what happens with PDOs and PGIs as intellectual property rights and this implies their impossible recognition in bilateral agreements that protect products covered by differentiated quality schemes in third countries. Thus, agreements like those signed with Switzerland, the United Kingdom or Ukraine recognize the mutual protection of the geographical indications of the contracting parties. And the same is true with other countries outside the European continent, like Mexico or, recently, China, which establish mutual protection of their PDOs and PGIs. As an example, the treaty signed with the Asian country included mutual recognition of a hundred geographical indications and appellations of origin—a dozen of them Spanish—which aims to be expanded in the coming years and will protect hams like *Jabugo* or *Guijuelo*, but not *Jamón Serrano* that as a TSG is left out of the European Union's trade agreements with third parties.

In addition to these reasons, there are also others related to the management of the differentiated quality scheme, more professional in the case of the PGI. As proof of this, while the constitution of a management body is not provided for the TSG, in the PGI this task is entrusted to the Regulatory Council (Montero 2016). This entity, constituted as a public or private law corporation as the case may be, previously authorized by the Ministry of Agriculture, Fisheries and Food or by the respective autonomous councils depending on their territorial scope, has its own legal personality, a governing body in which the economic and sectoral interests linked to the obtaining of the product are represented and they have the necessary means for the development of their functions.[12]

The PGI maintains greater control over entrepreneurs as the Regulatory Council keeps a register in which the covered operators who are authorized to use the scheme are registered. quality as long as they remain in it. In the case of hams, there will be a record of all the transforming and processing industries that include salting houses, drying houses, slaughterhouses, cutting rooms, etc. As for its control, in addition to the usual delegation to specific certification entities -for example, The Spanish Association for Standardization and Certification (AENOR)- they certify

---

[11] This is specified in recital 39 of Regulation (EU) No 1151/2012 of the European Parliament and of the Council, of 21st November 2012, on quality schemes for agricultural products and foodstuffs.

[12] Thus, see Art. 15 of Law 6/2015, of May 12, on Denominations of Origin and Protected Geographical Indications of supra-autonomous territorial scope. Or, as an example of the regional legislation, in the same sense, Art. 12 of Law 2/2011, of March 25, on Agri-food and Fisheries Quality of Andalusia.

compliance with the requirements of the TSG *Jamón Serrano*[13] -, the existence of a registry, the requirement for periodic information from the registered about compliance with the conditions of the specification and the role of the Regulatory Council itself in its work *in vigilando* and in collaboration with certification entities and competent authorities in agri-food matters are circumstances that determine the existence, generally, of more intense control in protected geographical indications.

And the same is also predictable with respect to the organization of the differentiated quality scheme itself. The greater formalities or requirements of the PGI in relation to the TSG contribute to better functioning that positively affects the promotion and protection of the product. For this, the aforementioned Regulatory Council carries out important work in carrying out dissemination and advertising campaigns, as well as surveillance against possible infringements of current regulations that affect its interests, especially in matters of intellectual property or unfair competition. The greater formality of the PGI is also evident in the self-financing system by which management entities provide themselves with the necessary resources from the contributions of registered operators, without prejudice, also, to the possible aids to which they opt within the framework of the Common Agricultural Policy. In the case of TSGs, on the contrary, it is more difficult to develop a joint strategy that will depend, in each particular case, on the commitment of the groups that request it and their associates.[14]

5.3.3 Thesis Contrary to the Registration of Jamón Serrano as a Protected Geographical Indication

Against the previous claim, Spanish producers rise who believe that such modification harms other national PGIs related to ham. In particular, among the protected geographical indications, the PGI "*Jamón de Trévelez*" (ham of Trevelez) and the PGI *Jamón de Serón* ("ham of Serón") have publicly expressed their opposition through their Regulatory Councils as well as the Plenary of the Provincial

---

[13] More information about the minimum control requirements of the specific characteristics of *Jamón Serrano*, available at http://www.fundacionserrano.org/jamon-serrano-informacion-al-consumidor-vida-sana/la-etg/la-etg_1_45_2_0_1_in.html

[14] In the particular case of the TSG *Jamón Serrano*, it was requested by the Association of Spanish Meat Industries. However, it is the *Jamón Serrano* Foundation who stands as the Representation Body of the TSG *Jamón Serrano* and requires an annual donation of one thousand one hundred and ten euros to be a member of the foundation. Unlike the relationship between the operators registered in the PGI, membership in the said foundation does not grant the right to use the TSG. In the case that it complies with the requirements of the foundation's counter-labels, that are different from the differentiated quality scheme, yes, they will be able to make use of them.

Deputation of Granada that approved an institutional declaration in which it opposed the granting of the PGI *Jamón Serrano*.[15]

Such protected geographical indications consider that the possible concession to *Jamón Serrano* of the requested sign will imply a devaluation of their PGIs. To justify such an argument, they maintain that the specification of the PGI *Jamón Serrano* has much less demanding requirements compared to those of the Andalusian geographical indications. As an example, the curing time of Trévelez ham is more than double that of Serrano and takes place in natural drying houses. In relation to this argument, they also point out that the proposed specification does not propose any change with respect to the one that currently governs the guaranteed traditional specialty.

Another of the reasons that underpin their opposition, from their point of view, lies in the lack of linkage of the product with a geographical origin. To the extent that it is protected as a TSG, they consider that *Jamón Serrano*, strictly speaking, identifies a traditional production method but does not convey to the consumer any specific origin. They also point out that the fact that the application for the PGI *Jamón Serrano* identifies the delimited geographical area with the entire administrative territory of the Spanish State.[16] It is misleading to consumers due to its lack of homogeneity. They also delve into the possible confusion with reference to the semantics of the PGI, as the term "serrano" evokes the mountain range as a space for the production and drying of hams, it should not be allowed, according to them, the use of artificial environmental temperature control devices.

Finally, they also base their refusal on the impossibility of registering it by virtue of the provisions of article 6.1 which prevents the registration as a protected designation of origin and as a protected geographical indication of those generic terms. In this sense, they maintain that the word "serrano" as relative to the mountain range is a generic name that, according to the Dictionary of the Royal Spanish Academy, preceded by the term "ham", means "cured ham" and, therefore, is considered generic.

---

[15] Institutional statement in which they present their arguments against the PGI *Jamón Serrano*, available here https://www.politico.eu/wp-content/uploads/2021/01/26/D.-I.-Jamo%CC%81n-Treve%CC%81lez.pdf

[16] In the institutional declaration of the plenary of the Granada Provincial Council, they forget that the first specification of the PGI excludes the autonomous cities of Ceuta and Melilla. Document available here https://www.politico.eu/wp-content/uploads/2021/01/26/D.-I.-Jamo%CC%81n-Treve%CC%81lez.pdf The previous specification contained an express exclusion of municipalities that do not have the necessary conditions for the production of *Jamón Serrano*. Available at https://carnica.cdecomunicacion.es/images/descargas/carnica/Pliego_de_condiciones_IGP_Jam%C3%B3n_Serrano__Documento_%C3%9Anico.pdf

5.3.4 Comment on the Feasibility of Registering **Jamón Serrano** *as a Protected Geographical Indication*

Despite the existing resistance from both national producers and associations and even provincial public administrations to the possible recognition of the protected geographical indication, the Ministry of Agriculture, Fisheries and Food has dismissed the objections raised and has made public the favorable decision to continue with the registration procedure of the aforementioned PGI. Undoubtedly, this is a complex case that requires a specific analysis of the necessary requirements for a product to be registered as a PGI.

The criterion relating to the lesser requirement of the specification provided in the application for the PGI *Jamón Serrano* compared to other PGIs does not have a legal basis to support it. It is easily verifiable that the curing time of *Jamón Serrano* is less than that of other PGIs or PDOs and also that the use of suitable devices for maintaining relative humidity is contemplated, unlike what happens with other hams that have stricter requirements and more costly procedures in economic terms. However, there is no rule that indicates that common standards should be noticed among the protected geographical indications of a certain product, even though there are Regulatory Councils that consider that this affects their reputation or entails a dilution of their distinctive quality sign.[17] Moreover, there are no obstacles for the conditions established in the specification of the traditional specialty guaranteed "*Jamón Serrano*" to be maintained in the PGI, it is not necessary for them to make any modification if they comply with the provisions of the current regulations. For the application to succeed, it is sufficient for the specification to comply with the minimum content provided for in Article 7 of Regulation (EU) no 1151/2012 of the European Parliament and of the Council, of 21 November 2012, on quality schemes for agricultural products and foodstuffs.

Nor does the argument that points to the non-existence of the link between a certain quality, reputation or other characteristic of the product and the geographical origin, established as an essential element of this quality scheme according to art. 5.2 of the aforementioned Community Regulation, seem conclusive. In this case, the application focuses on reputation as the main link with the geographical environment, a criterion that is flexible and relatively easy to prove (Zappalaglio 2022). The relationship of the *Jamón Serrano* with Spain is justified both by its external recognition and consideration abroad as a typical product of our gastronomy and by its presence, from old, in Spanish culture. For this, they provide as evidence references from international press and magazines, recipe and gastronomy websites

---

[17] The period of production determines the optional mention as "cellar", "reserve" or "grand reserve" for non-Iberian hams or shoulders based on the provisions of Royal Decree 474/2014, of June 13, which approves the quality standard for meat derivatives, but such issue is not relevant for the purpose of this intellectual property right.

or multiple works of our literature[18] where, in many cases, the expression "*Jamón Serrano*" is followed by the qualifying adjective "Spanish", so that the public circumscribes the product to a certain territory. This circumstance should not be interpreted in the sense of a possible genericity of the term "serrano", since outside Spain it is not a usual name in the designation of a type of ham -even in Spanish-speaking countries-, but it reinforces the product's link with its origin.

Precisely, in relation to the above, there are those who maintain the thesis that *Jamón Serrano* has a weak geographical link (Gonzálvez 2022) and that it only alludes to a traditional preparation method, since *Jamón Serrano* "does not have implicit geographical connotations".[19] It seems that the fact that they opt for a much larger extension than that of any other PDO or PGI and that *Jamón Serrano* cannot be identified with a more limited area, with a climate or physical characteristics of the geography that is homogeneous, is enough, according to its detractors, to deny it the link with the territory.

However, if in various regions of Spain, simultaneously, it is possible to meet the conditions of the specification in relation to the characteristics of the product, the minimum requirements of the raw material, the preparation process or the presentation of *Jamón Serrano*, there will be no problem in its registration regardless of the obvious climatic or orographic variations that, of course, exist, but are also observed in other regional PDOs or PGIs of considerable extension -for example, the PGI "Queso Castellano" which includes all the municipalities of Castilla y León, or some meats that also cover the entire autonomous territory.[20]

In addition, unlike the time when the TSG *Jamón Serrano* was registered, there are now several protected geographical indications that cover the entire extent of a country. This is the case, for example, of the Netherlands and the protected geographical indications "*Gouda Holland*" and "*Edam Holland*". Recently, even, a protected designation of origin, "*Halloumi*", for cheeses, that covers the entire extent of Cyprus was admitted.

It could be criticized, in relation to the previous argument, that such PGIs or the PDO cover a territory notably smaller than that of the specification of the PGI *Jamón Serrano* -which includes, the peninsular and insular territory of Spain-. While it is true that in Europe geographical indications covering such a wide territory are not common, it is no less true that there are registered in the European Union other protected geographical indications from third countries that have an extension greater than the Spanish one, such as the PGI *Café de Colombia* which, even though it does not include the entirety of the territory of that Republic, uses its name in the

---

[18] Among the international press, they cite widely circulated newspapers such as the German daily *Frankfurter Allgemeine*, or the British tabloid *The Telegraph*. They also echo the reference to *Jamón Serrano* in works by such relevant authors as Miguel Delibes or Nobel laureate Camilo José Cela.

[19] In these terms, the aforementioned institutional declaration of the plenary of the Granada Provincial Council in defense of the ham of Trévelez and Serón is expressed.

[20] As an example, the PGI "*Extremadura cow*" or the PGI "*Lechazo de Castilla y León*".

denomination.[21] In addition, other published applications, such as those relating to Basmati that pits India and Pakistan against each other, also cover vast geographical areas. And, although they are subject to another regulation[22] the same happens also in the case of some spirits.[23]

Another reason given for the application not to succeed is the supposed generic nature of the expression "*Jamón Serrano*". If the lack of differentiating capacity of the requested term is appreciated, it must be rejected based on the prohibition established in Article 6.1 of Regulation (EU) No 1151/2012 of the European Parliament and of the Council, of 21 November 2012, on the quality schemes for agricultural products and foodstuffs (Ribeiro de Almeida and Carls 2021). Ultimately, in accordance with the definition of "generic" in the said Regulation, it will be a matter of proof whether the proposed term, despite referring to the place, region or country where a product was originally produced or marketed, whether or not it has become the common name of that product in the European Union. In any case, the casuistry in the registration of quality schemes reveals a favorable position towards the admissibility of the proposed terms (Montero 2016), as was evident in the well-known judgment of the Court of Justice of the European Union of October 25, 2005 on the supposed generic nature of the name "*Feta*"[24] and in other subsequent cases.

---

[21] Although in the registration application they specify that the geographical area does not include the entire territory of the Republic of Colombia, but they limit it to coordinates relative to the coffee axis, they dispense with regional names and link it to the name of the country based on historical, traditional, cultural and social factors. Registration application available at https://eur-lex.europa. eu/LexUriServ/LexUriServ.do?uri=OJ:C:2006:320:0017:0020:EN:PDF In resolution No. 4819 of 2005, which registers at the national level the denomination of origin for Colombian Coffee, they distinguish between political delimitation, which they identify as Colombia, and the geographical delimitation, which they circumscribe to a specific area. Resolution available here https://www.sic. gov.co/sites/default/files/files/Denominacion%20de%20Origen/Agro%20-%20Alimenticios/Caf% C3%A9%20de%20Colombia/cafe_de_colombia.pdf

[22] Specifically, Regulation (EU) 2019/787 of the European Parliament and of the Council, of April 17, 2019, on the definition, designation, presentation, and labeling of spirits, the use of the names of spirits in the presentation and labeling of other food products, the protection of geographical indications of spirits and the use of ethyl alcohol and distillates of agricultural origin in alcoholic beverages, and repealing Regulation (EC) No. 110/2008. Regarding the risk that geographical indications with the country's designation may pose an obstacle under trademark law, see Corthésy (2021), p. 358.

[23] As the most significant case, two years ago the registration of the geographical indication *Tequila* took place, which covers the entire State of Mexico. Pisco has also been protected since 2012 and Scotch Whisky has been protected as a geographical indication since 1989.

[24] The present case began after the registration of "*Feta*" as a denomination of origin in the European Union in 1996 in the annex of Regulation No. 1107/96. Three years later, following the appeal filed by several European countries, the Court of Justice annulled Regulation No. 1107/96, insofar as it registered the denomination "feta" as a protected denomination of origin. In compliance with such judgment, the European Commission adopted, on May 25, 1999, Regulation (EC) No. 1070/99, by which it removed the denomination "feta". Subsequently, after two years of conducting surveys among the Member States about the manufacture, consumption, and notoriety of "*Feta*" cheese in the EU, they concluded that this term was not generic and therefore they registered this denomination in the register of protected designations of origin and protected

To warn of this circumstance, indications such as relevant legislation, the perception of the average consumer of the supposedly generic denomination, whether the product has been legally marketed with that denomination in the common market, whether the production in the country of origin has respected the traditional method, the market share of products that use such denomination but have been produced outside of such method compared to those that do comply with the traditional criteria or the number of Member States that invoke the supposed generic nature of the proposed term.[25] And in the specific case of *Jamón Serrano*, in light of such indicators, the answer to the possible genericity of the term is negative. The reason against argued by the detractors of the PGI, consisting in the fact that the Dictionary of the Royal Spanish Academy identifies *Jamón Serrano* with cured ham, in addition to having no legal significance, is based on an incorrect statement. *Jamón Serrano* is not generic nor, much less, is it synonymous with cured ham, which refers, only, to a process of product preparation that is common to all hams covered under a differentiated quality scheme -both Iberian hams and white pork hams protected by PGI or PDO-. This is easily deduced from specific Spanish legislation, such as Royal Decree 474/2014, of June 13, which approves the quality standard for meat derivatives, which has an express definition of the term "cured" linked with one of the specific techniques for the preparation of meat derivatives.[26] But the distinction between "Serrano" and "cured" can also be checked in any supermarket or point of sale, which demonstrates compliance with the criterion related to the correct way of marketing the product. In this sense, in most cases, the consumer clearly differentiates hams protected by a specific PGI or PDO, others that use the TSG *Jamón Serrano* and, finally, hams and shoulders of other brands, visually similar, but that identify their product with the expression "cured ham".

Finally, it is worth considering what would happen in the case of prior rights granted to producers from third countries. So far there has been no opposition, beyond the one mentioned within Spain, to the change of differentiated quality scheme, from the TSG to the PGI. From a quantitative point of view, unlike other TSGs such as "Neapolitan pizza" used outside the original origin of this quality scheme -Naples-, there are no producers who have publicly shown their rejection, so the impact of the present measure cannot be assessed. Neighboring countries that, due to geographical and cultural connection, could have producers of the TSG

---

geographical indications provided for in paragraph 3 of Article 6 of Regulation (EEC) No. 2081/92 as a protected denomination of origin (PDO). Denmark and Germany appealed, again, the new Regulation but, in this case, the CJEU dismissed the appeal as it considered that the term "Feta" was not generic.

[25] See, in this regard, the CJEU judgment of September 12, 2007 in the "*grana padano*" case or the CJEU judgment of March 16, 1999 on the "*Feta*" cheese.

[26] It defines this term as "Treatment with salt, which may be accompanied by the use of nitrites, nitrates, and other components or a combination of them, which must respond to a technological need, resulting in compounds derived from the combination of these preservatives with the proteins of the meat. The treatment can be carried out by dry application, to the surface of the meat, of the curing mixture, by immersing it in the curing solution or by injecting the curing solution into the meat piece".

"*Jamón Serrano*" have opted for the production of other types of products. This is the case, for example, of Portugal, where the production of Iberian pork ham predominates, such as the PDO "*presunto do Alentejo*" or the PDO "*presunto Barrancos*". The same is true in the south of France, where in the region of Nouvelle Aquitaine they produce the PGI "*jambon de Bayonne*". In any case, in the event that there were foreign producers adhered to the TSG, this is not an obstacle for the cancellation request that has been made by the same legal entity that requested its registration—the National Association of Meat Industries of Spain (ANICE) in Pursuant to Article 54, paragraph 1, of Regulation (EU) No 1151/2012. Occasionally, modifications to specifications exclude producers who do not meet the new requirements, so it does not seem that the position that, if applicable, would be minority, poses an obstacle to the annulment of the aforementioned differentiated quality scheme as a necessary requirement according to Article 6 paragraph 3, of Regulation (EU) No 1151/2012 for the protection of *Jamón Serrano* as a PGI. Another possibility, more remote, in the event that there were producers from border areas of neighboring countries would be the extension of the territorial scope of the PGI through a cross-border joint registration request, as is already the case with two of these intellectual property rights that share areas of Spain and France, although this possibility has not been mentioned at any time.[27]

5.4 Conclusions

From an economic point of view, the TSG, although it was useful at the time for the promotion of *Jamón Serrano* and its marketing in the internal market, it currently reveals itself as a quality scheme that offers insufficient protection against the opening to new markets and the possibility that foreign operators produce a typically Spanish product. The losses that its lack of recognition in bilateral agreements make necessary, at least, the study on the feasibility of its registration as a protected geographical indication. Precisely, this is the main argument that *Jamón Serrano* producers hold for the change of quality scheme from the TSG to the PGI.

As for recognition as a PGI, it is true that it is not common for these intellectual property rights to comprise such a wide extension, but it is also true that, gradually, products have been registered with an equal or greater extension. However, the link with the environment is based on reputation and the specification offers multiple references in this regard.

In relation to the possible generic character of the term, the European Commission will assess, if applicable, the oppositions that may be raised in this regard, although factors such as the correct marketing of the product to date or the restrictive interpretation of the community jurisprudence on the matter suggest that the genericity of the term seems to be ruled out.

---

[27]This is the case, for example, of the PGI of meat products "*Beef from the Catalan Pyrenees/ Vedella dels Pirineus Catalans/Vedell des Pyrénées Catalanes*", and "*Rosée des Pyrénées Catalanes*" registered by Spain and France in 2016.

References

Corthésy NGS (2021) Country name designation and international IP protection of national competitive identities. J Intell Property Law Pract 16

Gallego Sánchez E, Fernández Pérez N (2019) Commercial law. First part. Tirant lo Blanch, Valencia

Gonzálvez Pérez JM (2022) Origen Geográfico y Calidad de los Productos Agroalimentarios: el caso del Jamón Serrano. Competencia, propiedad intelectual y tutela de consumidores en el sector agroalimentario. Tirant lo Blanch. Valencia

Guillem Carrau J (2021) Denominaciones geográficas de calidad. Tirant lo Blanch. Valencia

Huerta M (2021) El sinsentido de oponerse a la IGP del jamón serrano. Available at https://revistas.eleconomista.es/alimentacion/2021/marzo/el-sinsentido-de-oponerse-a-la-igp-del-jamon-serrano-IJ6803749

Montero García-Noblejas P (2016) Denominaciones de origen e indicaciones geográficas. Tirant lo Blanch. Valencia

Montero García-Noblejas P (2021) Towards a core unitary legal regime for geographical indications in the European Union digital market. J Intell Property Law Pract 16

Ribeiro De Almeida A, Carls S (2021) The criteria to qualify a geographical term as generic: are we moving from a European to a US perspective? Int Rev Intell Property Comp Law. Available at https://link.springer.com/article/10.1007/s40319-021-01045-x

Seville C (2009) EU intellectual property law and policy. Edward Elgar Publishing, London

Zappalaglio A (2022) Anatomy of traditional specialities guaranteed: analysis of the functioning, limitations and (possible) future of the forgotten EU quality scheme. GRUR Int 71(12): 1147–1161. Available at https://www.zappalaglio.com/publications

Chapter 6
Geographical Indications in South America: It's Not All About the Label. Cultural Factors and Networked Governance

Patricia Covarrubia and Kerry Purcell

Abbreviations

GI   Geographical Indication
IP   Intellectual Property
oriGIn   Organization for an International Geographical Indications Network
WIPO   World Intellectual Property Office
WTO   World Trade Organization Agreement

6.1   Introduction

The World Trade Organization Agreement (WTO) on Trade-Related Aspects of Intellectual Property Rights (TRIPS) is a multilateral agreement on Intellectual Property (IP) and thus, applicable to all WTO members. TRIPS came into force in 1995 and member states were obliged to protect GIs from unfair competition and imitation. In South America, countries started to regulate GI as *sui generis* right or as part of the intellectual property system (see Table 6.1). Therefore, there has been increased interest in the region to protect GIs even though the legislations and frameworks are relatively new in comparison with Europe. That said, the TRIPS does not have a regulatory framework, leaving it members to establish *ad hoc* procedural institutions which determine the scope of protection (Barjolle et al. 2017).

P. Covarrubia (✉) · K. Purcell
The University of Buckingham, Buckingham, UK
e-mail: patricia.covarrubia@buckingham.ac.uk; kerry.purcell@buckingham.ac.uk

© The Author(s) 2025
E. Vandecandelaere et al. (eds.), *Worldwide Perspectives on Geographical Indications*, https://doi.org/10.1007/978-3-031-71641-6_6

Table 6.1 List of South American countries and GIs framework (OriGIn and national IP offices)

Countries[a]	Argentina	Bolivia	Brazil	Chile	Colombia	Ecuador	Paraguay	Peru	Uruguay	Venezuela
Legislation	*sui generis*	*sui generis*	Main IP Law	Main IP Law	Main IP Law	Main IP Law	*sui generis*	Main IP law	Main Trade-mark Law	By Resolution
	1. Law No. 25.163 (1999)—General Rules for the Description and Presentation of Wines and Wine-based Spirits Law No. 25.380 (2000)—2. Legal Regime for Indications of Source and Appellations of Origin of Agricultural and Food Products	Law No. 1.334 (1992)—on Appellations of Origin	Law No. 9.279 (1996)—on Industrial Property *Title IV* Geographical Indication	Law No. 20.160 on Amendments to Law No. 19.039 on Industrial Property *Title IX* Geographical Indications and Denominations of Origin	Law Decision No. 486 Establishing the Common Industrial Property Regime CAN *Title XII* on Geographical Indications	Law Intellectual Property Law (Consolidation No. 2006-013) *Title XI* Geographical Indications	Law 4923/23 Geographical indications and denominations of origin	Legislative Decree No. 1075 Establishing the Common Industrial Property Regime CAN	Law No 17.011 (1998) Establishing Provisions o Trademarks—*Chapter XII* Geographical Indications	SAPI-RPI-AO No. 19
Registration body	1. Instituto Nacional Vitivinicultura (wines and spirits) 2. Secretariat of food, bioeconomy and regional development	SENAPI (national IP Office)	INPI (national IP Office)	INAPI (national IP Office)	SIC (national IP Office)	SENADI (national IP Office)	DINAPI (national IP Office)	INDECOPI (national IP Office)	DNPI (national IP Office) by seeking previous advice from any organisation which is identified with competences	SAPI (national IP Office)

	of the ministry of agriculture, livestock and fisheries								in the matter according to the nature of the product or service
Numbers of GIs Registered	116	5	102	140	30	7	7	10	0	6

[a]French Guiana, Guyana and Suriname are in South America. French Guiana is an overseas department of France; Guyana (formerly known as British Guiana) and Suriname (formerly known as Dutch Guiana) gained their independency recently. They are not considered in this research

For the last 20 years, the world economy has experienced significant growth, and although this mainly comes from the industrial sector, the world has seen diversification in exports including agriculture and natural resource-based sectors (Dodds 2003). This has permitted micro-enterprises and small communities or group of people, to benefit from the IP system and thus, providing smaller entities, including those who work through non-conventional channels, to benefit.

WIPO estimates 58,800 protected GIs in 2020 either as *sui generis* right, a legislative act, or the trade mark system, and/or labelling rules (WIPO 2021). The Organization for an International Geographical Indications Network (OriGIn) estimates that there are over 8000 protected GIs protected by *sui generis* right; the majority are in: Europe 3677; Asia 3463 and in third place, South America with 423 (OriGIn 2023). Generally, countries in South America have straightforward GI or IP legislation and processes to follow. Moreover, some countries have friendly campaigns run by their national IP offices, or Ministries but the main question remains: why have there not been a floodgate of registrations in countries rich on agricultural products and handicraft that are unique to the region? (see Table 6.1 with number of GIs registered). Indeed, there is vast literature covering the debate on how IP, and specifically the GI registration system helps in the economic development of a business and a country, including less-developed economies as it may boost rural development (Navarra and Thirion 2019). However, there is a clear absence of engagement with the GI registration process in South America.

This chapter begins by providing a background on GIs in South America. Within this, the legislative framework, and current status of GI registrations will be discussed. In this discussion, the fundamental issue of the absence of trust between citizens in the respective governmental body supported by a government indicator in respect of democracy is highlighted. The chapter then moves forward to examine the pragmatic recommendation of establishing a horizontal decentralised network hosted by WIPO to promote engagement with the GI registration process hindered by the absence of trusts between citizen and state. The recommendation of such a network compliments current literature, which argues that we perceive control at the "core of the GI system" highlighting the importance of managing GIs by the local stakeholders (Marie-Vivien 2020). Within the section, a horizontal trans governmental network hosted by WIPO is presented as a solution to supporting communities in the pre-application stage, with the overall aim of increasing the number of GIs registrations and potentially, boosting rural development.

6.2 Geographical Indication in South America—Laying the Ground

Distinctive signs help to develop the image of a product or service. A GI signals a link between a product and its specific place, a geographical link, its unique production methods, and distinguishing qualities. One of the most emblematic

South American examples are the cases of the Colombian coffee growers that effectively use the GI *Café de Colombia,* the first foreign Protected Geographical Identification registered in the European Union (Commission Regulation (EC) No 1050/2007 of 12 September 2007); spirit drinks such as Pisco (Peru, Resolution No 072087-DIPI of 12 December 1990; Chile, Decree 581 of 30 December 1999) and handicraft such as the Peruvian pottery, by the Consejo Regulador named *Chulucanas* (Resolution No 011517-2006/DSD-INDECOPI of 26 June 2006). Some of the emblematic products are part of the country's national heritage and a few are registered in the UNESCO Representative List of the Intangible Cultural Heritage of Humanity. Examples of the latter are, traditional weaving of the Ecuadorian toquilla straw hat (Inscription: 7.COM 11.12), registered as denomination of origin (*Sombrero de Montecristi*, Resolution No. 988698); and, traditional knowledge and techniques associated with Pasto Varnish mopa-mopa of Putumayo and Nariño (Inscription: 15.COM 8.a.1), registered as denomination of origin (*Mopa Mopa Barniz-Pasto*, Resolution 70002).

In the region, GI registrations are usually handled by their national IP office and others by the Minister of Agriculture; the terminology used are *appellation of origin/ denomination of origin* (defined as in the Lisbon Agreement), and *indication of source* and *geographical indication.* GIs are available to agricultural, foodstuffs, spirit drinks and wines, including handicraft. Brazil and Chile have widened the scope to include services.

Applications can be submitted by any person (natural or legal) that is directly engaged in the extraction, production, or processing (individually (except Chile, Venezuela, and the CAN countries) or collectively). Moreover, any organisation that are responsible for the promotion or protection of the producers' interests can also apply. Additionally, applications can also be submitted by national, regional, provincial, or communal authorities, located within the territories of their respective competences.

In the South America region, the state is the owner of the GI, and therefore the management is governed by the state that delegates the administration to private or public bodies (usually representing the actors within the chain of production); yet even in case of private bodies, the state has a say (Barjolle et al. 2017). Successful applicants obtain an authorisation of 'GI use' by the competent national office. The duration of the authorisation is unlimited, although in Paraguay and the CAN countries, authorisation is restricted to 10 years, however, it may be renewed.

GIs is a common property, and although the rights given are private, the entry is open to anyone that fits the criteria established (e.g., territory limits, etc.) as it is a *collective exclusive right* in the public domain (Réviron and Chappuis 2011). Such rights include how, when, and by whom, the GI is enjoyed (Kaul 2005). Therefore, trough the collective action and support contained within the proposed network recommended in this research, registration of GIs would enhance these rights within the network by addressing issues of conflict (Navarra and Thirion 2019). For instance, early conflict is based on representativeness, membership, and transparency (Marie-Vivien et al. 2019). Moreover, as noted by Hughes (2010), a growth threat is more apparent when countries "lack stable, transparent, [and] democratic institutions of government".

That said, in the pre-registration stages, there are numerous procedural steps and several phases that must be followed. For instance, the recognition of the product, which specifically identified and codifies the products' characteristics such as methods, techniques, raw material, etc.; the negotiations between producers in defining the product and geographical boundaries; and the internal and external quality control mechanisms, e.g., certification body; among others. Given the complexities establishing a consensus during pre-registration, it is bound to bring conflict among the group. Documented cases are noted in different jurisdiction, such as India (Manchikanti et al. 2022), France and Vietnam (Pick 2023). This is further heightened by the absence of trust between citizen and state. Therefore, there is a need to create a solid structure that would support it in the journey, from pre-registration to its future success (social, cultural, and economic). This structure of support would be facilitated by the proposed network.

Unfortunately, there is an absence of research examining why the producers, even though seeing GIs as a positive tool, are reluctant or weary to engage with the GI registration process. Yet, by noting the process and phases, one can see the intervention of public administrations and participation of public or private institutions (and privates which delegation is given somehow to the state) (Barjolle et al. 2017) which may be a reason. There exists an expectation that producers, and all involved in the value chain, entrust all institutions and consequently their roles, government and non-governmental organisations within the process and phases. However, it is questionable if these third parties are making the right decisions to benefit the GI's collective organisation and seeking a common good or to benefit themselves looking for an individual goal. We argued that both structures of governance, at product level and national level, impact the willingness of the group to participate, due to the exercise of powers by several state institutions. A GI structure includes private level (product level) governance, meaning that there must be a set of governance rules to produce the GI and the implementation of internal control mechanisms to ensure that the GI product consistently match the characteristics expected of it by consumers (International Trade Centre 2018); and the public level (national level) governance, meaning that there shall be a governing body, inspection body, regulatory institution, etc. For these phases to succeed, there must be a robust product level *governance structure* and indeed, a crucial need to lay the foundation for the people who produce, and or create the product (aside all in the value chain). For instance, in Europe collective organisations is a prerequisite where they adopt a code of practice which might become a 'masterpiece'.

While in South America several governmental bodies engage in GI campaigns e.g., *Sello de Origen* (Chile) aiming to promote the use of distinctive signs; free application fee in Ecuador for national products, etc., which are welcomed, the GI system remains under-utilised by many (see Table 6.1). The fact is that although there is a wealth of literature illustrating the importance of GI protection (Marie-Vivien and Biénabe 2017), not only as an economic objective but social too (Réviron and Chappuis 2011; Covarrubia 2016), there are still issues regarding the lack of

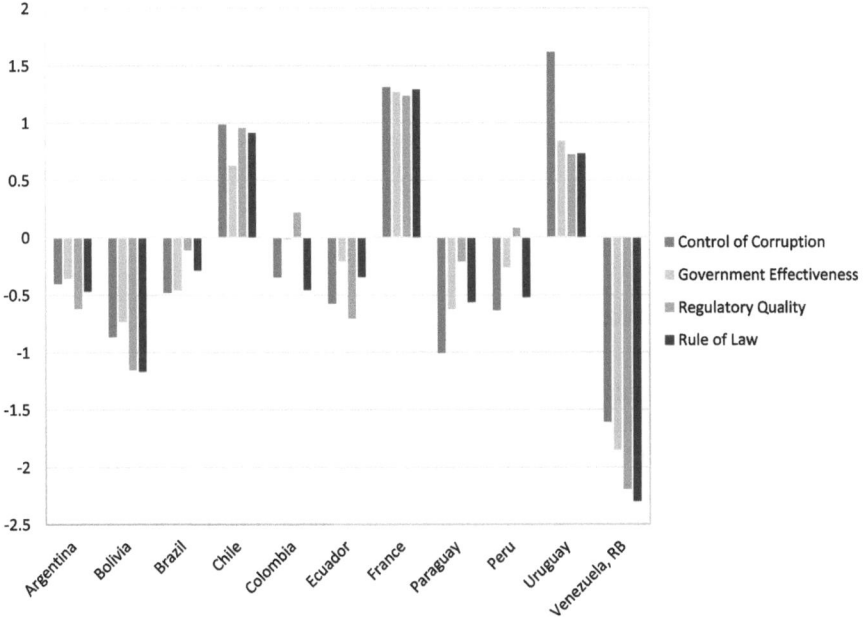

Fig. 6.1 South America Government Indicators 2021 (DataBank Worldwide Government Indicators). France is used for comparative reasons (CIRAD's location)

engagement with the GI registration process. These factors may vary e.g., costs, lack of infrastructure, lack of awareness, etc.; however, in this chapter we have focused on trust since many countries have lack of confidence between citizens and state.

In South America, trust is mainly based on high levels of corruption (Rodriguez 2019) affecting the producers to be empowered and deal with public administration, and bureaucracy even among themselves. Some people, organisations or the like, may favour short term remuneration and could be influenced by socio-political context (De Herde et al. 2022). Therefore, the role of the state and its interventionism influences the interest of actors to apply for a GI (Barjolle et al. 2017). Moreover, it is noticeable that most of the GI system in South America uses a top-down state driven approach. Therefore, it is expected that the state act in different functions, creating legislation, policing, and sanctioning; however, to what extend it is desirable for the state to intervene in different stages of the GI process when this is challenged by cases of corruption, be it petty or grand? and generally, by poor governance (see Fig. 6.1). Marie-Vivien et al. (2019) argued that while there is enough literature on the conflicts that arise during the drafting of the GI specifications, there is a gap on the conflicts regarding governance. Therefore, we are not proposing the *ideal* governance model as 'one size fits all', rather trying to contextualise and draft a pragmatic solution to promote engagement with the GI registration process.

Vast literature exist which assesses the value chain governance in GI, however, limited research exists on recontextualising the strategic relevance of governance networks in the GI process, specifically addressing the trust degree of GI interest groups. By building trust, potentially it may build a driver for long-term collective performance. Therefore, this research has employed general networked governance to offer a pragmatic solution that seeks to promote and harmonise GI registration. The rationale is that in principle, a GI needs organisation and cooperation among all actors involved in the GI product and its chain, and consequently, collective action is needed to address collective problems.

6.3   Network Creation

As discussed above, although there are clear benefits of GI registration, there is lack of engagement with the GI registration process in South America. This is due to several of the reasons already noted. In this chapter we look and the high levels of governmental corruption in South America (Rodriguez 2019) and submit that GI registrations suffer further issues of engagement with the registration process due to loss of confidence and trust between citizen and state. This chapter posits that to address issues of trust and confidence between citizen and state, for the governance of GIs, the establishment of a horizontal decentralised network hosted by WIPO is the most pragmatic solution.

A network is defined as a collective of persons, organisations, or institutions that are interconnected by a common objective or aim and engage in collaborative efforts and communication to attain this objective (Stocker 2006). Networks can exhibit various structural configurations, although regardless of their specific structure, the fundamental purpose of networks is to enable and enhance collective action. The founding purpose of networks is to harness their power termed 'network effect' to address and regulate matters pertaining to governance. Governance networks have been increasingly characterised as a mechanism for actors to exert influence on international outcomes (Hafner-Burton et al. 2009). The core motivation of the proposed network once established, would be to leverage actor knowledge, reputation, and recourses to promote engagement with the GI registration process.

Government networks have the capacity to function either within an international organisation or in conjunction with such entities. The establishment of governmental networks within a traditional international organisation such as WIPO has been characterised as 'breathing new life and power into the organisation itself' (Slaughter 2017). We argue that the proposed network should be established within WIPO as the network host. The rationale for this is one of pragmatism. WIPO as an intergovernmental organisation's, primary function is to foster collaboration and harmonisation among independent nation-states on a range of issues regarding IP. By hosting such a network, WIPO can at a higher level with policy makers by the fostering of relationships that drive cooperation, harmonisation, and best practice within international standards. Subsequently, demonstrating that transgovernmentalism and intergovernmentalism can work in a co-efficient manner.

Network design is imperative to overall functional success of the network. GIs reflect the socio-cultural context of the jurisdictions in which products are created. Consequently, any network established for the governance of GIs in terms of cooperation, harmonisation, and best practice, must be able to accommodate the diversity in socio-cultural attitudes. The two prevailing network topologies are commonly referred to as vertical networks and horizontal networks. International vertical governance networks are structured by a clear formalised hierarchy of authority and decision-making. Within this vertical structure WIPO would be at the apex of the structure with domestic governments delegating their autonomous governing authority. We argue that this vertical structure may prove ineffective, as it is questionable whether such delegation of autonomous governing authority would be agreed if there is governmental corruption at play.

The notion of horizontal government within the constitutional discourse encompasses a strong and cohesive political centre that directs and coordinates the actions of several stakeholders engaged in policy implementation (Torfing et al. 2012). Horizontal governmental networks consist of officials from the domestic government. The distinguishing characteristic of international horizontal networks is they unify international organisations, the varying domestic legislators, judges, and related domestic parties such as associations to collect and share information and distil best practices (Junki 2006). This recourse mechanism is vital for the proposed network when inviting jurisdictions to engage with the network. It is more likely to obtain domestic engagement if political actor autonomy (corrupted or not) is preserved.

Horizontal networks can adopt one of three infrastructure typologies: centralised, decentralised, and distributed. A centralised governance infrastructure is as a framework in which the power to make decisions and exercise control is consolidated within a singular central authority. This is reflective of the vertical structure discussed above. In adopting this centralised infrastructure, the horizontal network structure would position WIPO in the centre of the network. The actors would be seen as of equal status but would be positioned just below WIPO in terms of power and control. WIPO's role would be to set policies and to promote best practice regarding GI registration. This infrastructure could in theory be a workable solution, however we still have concerns regarding the level of engagement if political autonomy is impinged upon by WIPO. We have further concerns how this structure would function given that WIPO is not the governing body for GI registration.

Rather we recommend a decentralised infrastructure be adopted within the network. The concept of a decentralised governance network infrastructure refers to the dispersion of decision-making authority and control across multiple actors, as opposed to being concentrated inside a singular central institution (WIPO) (Gol et al. 2019). In a decentralised infrastructure, the decision-making processes commonly employ a consensus mechanism, wherein several actors engage in collaborative efforts to achieve mutual agreements and jointly make choices (Bevir 2006). This structure would place WIPO at the centre of the network, but not as a power source, rather as a support structure to enable actors to make decisions by virtue of a collective consensus mechanism. This structure would also offer opportunities to include external bodies e.g., Food and Agricultural Organization (FAO) within the support structure to promote GI registration.

Decentralised and distributed governance structures share numerous similarities as they both disperse decision making authority within the network. However, the distinction resides in the degree of dispersion and the mechanisms of consensus utilised. Distributed governance networks can be described as self-organising systems that facilitate innovation and spontaneity in subsequent processes (Duncan and Schoor 2015). We would not seek to recommend this structure as it would weaken the role of WIPO and the reginal bodies as the infrastructure and could potentially create a higher risk of ongoing corruption to continue. Therefore, our recommendation would be an international horizonal decentralised network hosted by WIPO to support the GI registration process.

6.4   Conventional Wisdom—Trust and Cooperation

As noted in Table 6.1, GIs are controlled by the government or owned by the state, consequently, the group, that is, the network governance has only the *right to use* a GI (Covarrubia 2016). The proposed horizontal decentralised network would address issues of perceived corruption in a two-fold strategic manner to address issues of trust and confidence between citizen and state for the governance of GIs. Firstly, by the exploitation of actor reputation influence, and secondly, establishing links between the respective jurisdictional associations and artisans, micro-enterprises, farmers, and indigenous peoples.

The first strategic element would be to utilise as social power within a networked context (Montgomery and Hafner-Bourton 2006). In short, the proposed network would seek to exploit actors with 'good' reputations in GI governance to employ their social power to influence not just South American jurisdictions, but all jurisdictions to implement and follow best practices in respect of GI registration. This power can also be harnessed in a coercively to deter corruptive behaviour (Cook et al. 2013).

In terms of the second strategic element, associations would be established within South American jurisdictions that would work with WIPO and the respective regulatory bodies to support artisans, micro-enterprises, farmers, and indigenous peoples in GI registration in a support structure devoid of corruption. Therefore, corrupted behaviour is targeted at heart of the network by actor influence (Cialdini 2007). Furthermore, the proposed horizontal decentralised structure offers additional support and reassurance to citizens who have issues of trust and confidence, by virtue of the role of the respective associations. One of the roles of the association could be to act as a third party on behalf of the citizen in the GI registration process. This 'middleman' approach removes the need for the citizen to interact with the national government directly as they would be supported and represented through the respective associations. Furthermore, the associations would be supported by both WIPO and the regional bodies if there were any conflict between the association and

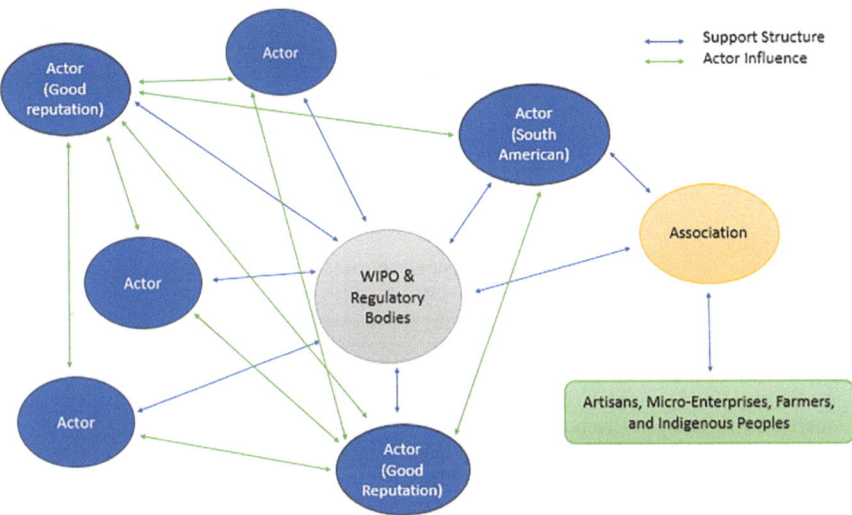

Fig. 6.2 Proposed two-fold strategic network mechanism to promote public actor engagement with the GI registration process

state. Associations would further receive indirect support in terms of conflict between them and the state in the employment of the first strategic element, utilising actor social power to influence national governments in terms of best practices in GI governance.

The Fig. 6.2 serves as a visual representation of the utilisation of social power within a network structure. The proposed network aims to leverage the influence of actors with strong reputations in GI governance to encourage all jurisdictions, including those in South America, to adopt and adhere to best practices in GI registration. This ability can also be utilised coercively to discourage corrupt behaviour.

6.5   Conclusion

The South America region shows its compliance with the TRIPS Agreement in regard the protection of GIs from unfair competition and imitation, be it by sui generis right or integrated into the IP system. While globally there is a trading increase in the agricultural and natural resource-based sector, which South American is rich on, the protection of agricultural products has not seen a flood gate of registrations.

The chapter evidence one of the potential main flaws, not of the system per se, but to the country, which is a region which is tainted with corruption. And while the debate and literature continue to show that the GI system may help in boosting rural development, the absence of trust between citizens and state, may be a big deterrent.

Therefore, this chapter provides a potential solution by establishing a horizontal decentralised network hosted by WIPO to promote engagement with the GI registration process.

References

Barjolle et al (2017) The role of the state for geographical indications of coffee: case studies from Colombia and Kenya. World Dev 98:105–119

Bevir M (2006) Democratic governance: systems and radical perspectives. Public Adm Rev 66(3): 426–436

Cialdini RB (2007) Influence: the psychology of persuasion. Collins, New York

Cook KS et al (2013) Social exchange theory. In: DeLamater J, Ward A (eds) Handbook of social psychology. Springer, Cham, pp 61–88

Covarrubia P (2016) Protection of non-agricultural GIs: a window on what is happening in Latin America. Eur Intellect Prop Rev 38(3):129–131

De Herde V et al (2022) Lock-ins to transition pathway anchored in contextualized cooperative dynamics: insights from the historical trajectories of the Walloon. J Rural Stud 94:161–176

Dodds J (2003) Intellectual property rights in agriculture. Biotechnol Agric Ser:149–160

Duncan CM, Schoor MA (2015) Talking across boundaries: a case study of distributed governance. Int Soc Third Sector Res 26:731–755

Gol ES, Stein M, Avital M (2019) Platform governance toward organizational value creation. J Strateg Inf Syst 28(2):175–195

Hafner-Burton EM, Kahler M, Montgomery AH (2009) Network analysis for international relations. Int Organ 63(3):559–592

Hughes J (2010) Coffee and chocolate—can we help developing country farmers through geographical indications? (September 29, 2010). Available via SSRN: https://doi.org/10.2139/ssrn.1684370. Accessed 18 Dec 2023

International Trade Centre (2018) Adding value to the origin of products through geographical indications (GIs). Material delivered by the SME Trade Academy

Junki K (2006) Networks, network governance and networked networks. Int Rev Public Adm 11(1):19–34

Kaul I (2005) Private provision and global public goods: do the two go together? Glob Social Policy 5(2):137–140

Manchikanti P, Datta S, Bandopadhyay TK (2022) Foodstuffs and geographical indications in India: an analysis. In: Bhattacharya NS (ed) Geographical indication protection in India. Springer, Singapore

Marie-Vivien D (2020) Protection of geographical indications in ASEAN countries: convergences and challenges to awakening sleeping geographical indications. J World Intel Property 23:328–349

Marie-Vivien D, Biénabe E (2017) The multifaceted role of the state in the protection of geographical indications: a worldwide review. World Dev 98:1–11

Marie-Vivien D et al (2019) Controversies around geographical indications: are democracy and representativeness the solution? Br Food J 121(12):2995–3010

Montgomery AH, Hafner-Bourton EM (2006) Power positions: international organisations, social networks and conflict. J Conflict Res 50(1):3–27

Navarra C, Thirion E (2019) Geographical indications for non-agricultural products. Cost of non-Europe. Report EPRS | European Parliamentary Research Service

Organization for an International Geographical Indications Network (OriGIn) (2023) GIs Worldwide Compilation. Available via https://www.origin-gi.com/worldwide-gi-compilation/. Accessed 18 Dec 2023

Pick B (2023) Intellectual property and development: geographical indications in practice. Routledge, Oxford

Réviron S, Chappuis JM (2011) Geographical indications: collective organization and management. In: Labels of origin for food: local development, global recognition. CAB International, Wallingford, pp 45–62

Rodriguez A (2019) Defining governance in Latin America. Public Organ Rev 19(1):5–19

Slaughter AM (2017) The chessboard and the web. Yale University Press

Stocker G (2006) Public value management: a new narrative for networked governance? The American Rev Public Adm 36(1):41–57

Torfing J, Peters GB, Pierre J, Sørensen E (2012) Horizontal, vertical, and diagonal governance. In: Interactive governance: advancing the paradigm. Oxford University Press, pp 86–103

World Intellectual Property Organization (WIPO) (2021) World Intellectual Property Indicators report: worldwide trademark filing soars in 2020 despite global pandemic. Geneva, November 8, 2021. PR/2021/883

Chapter 7
Analysis of Origin Labelling Schemes in the Southern Mediterranean Countries; The Case of Deglet Nour of Tolga in Algeria

Fadhila Bacha and Fatima El Hadad-Gauthier

Abbreviations

AO	Appellation of origin
ALGERAC	Algerian Accreditation Organisation
EU	European Union
DSA	Direction des services agricoles : agricultural services department
GI	Geographical Indication
IPR	Intellectual Property Right
MADR	Ministry of Agriculture and Rural Development, Algeria
OECD	Organisation for Economic Co-operation and Development
TRIPS	Aspects of Intellectual Property Rights related to Trade
WTO	World Trade Organisation
WIPO	World Intellectual Property Organisation

7.1 Introduction

The geographical indication (GI) is a denomination that allows the quality and reputation of a product to be linked to its place of origin (Amsallem and Rolland 2010). This quality sign highlights the particular characteristics of a product attributable to its terroir: know-how, tradition and local natural resources used

F. Bacha (✉)
Agro Institute of Montpellier, UMR MoISA, Montpellier, France
e-mail: fadhila.bacha@supagro.fr

F. El Hadad-Gauthier
Mediterranean Agronomic Institute of Montpellier (IAMM-CIHEAM), UMR MoISA, Montpellier, France

© The Author(s) 2025
E. Vandecandelaere et al. (eds.), *Worldwide Perspectives on Geographical Indications*, https://doi.org/10.1007/978-3-031-71641-6_7

(Prévost et al. 2014; Giovannucci et al. 2009; Barjolle et al. 1998). With the institutionalisation of the WTO, this quality sign benefits from international protection under the TRIPS Agreement, which allows the protection of products against imitation and illicit appropriation of their names.

Like the countries of the Southern Mediterranean, Algeria has undertaken a series of measures for the implementation of a labelling system by signs related to origin. Currently, three products have been labelled under GI: *Bouhezza cheese, Beni Maouche dried figs* and *Deglet Nour dates of Tolga*. This study analyses the implementation of the GI labelling process for *Deglet Nour dates* of *Tolga* in order to identify the key success factors or barriers to this process. Through semi-structured interviews with resource persons and the exploitation of existing literature, the analysis focuses on the examination of the specifications and the strategies of the actors.

7.2   Context: GI, History and Internationalisation

Since antiquity, GIs have been used to identify the place and particular characteristics of products related to their origin: wine from the island of Thasos in Macedonia, the brickworks of ancient Egypt, Corinthian bronze, Arabian perfumes or wines from Naxos (Le Goffic 2012). The origin of products was used as a reference for quality, which was the basis of commercial reputation (Sylvander et al. 2006; Rangnekar 2004). However, GIs appeared and gained their fame in Europe, and historically concerned wines and spirits, particularly in France, the cradle of quality signs, which established for the first time a law, the 1905 law, on Appellations of Origin. This law originally aimed to regulate viticulture in France. Subsequently in 1958, they were recognised as Intellectual Property Rights (IPR) through the Lisbon Agreement by the World Intellectual Property Organisation (WIPO) and in 1994, by the TRIPS Agreement of the WTO which allows the protection of products against imitation and illicit appropriation of their names.

The signing of the TRIPS agreement gave a real international impetus to the law of GIs since WTO member states are required to provide regulations allowing interested parties to request protection of GIs (Belletti et al. 2014; Arfini et al. 2011). However, this agreement does not set national legislation. Depending on economic circumstances and national contexts, their legal recognition has taken different paths (Sylvander et al. 2006). On the one hand, there is the very old European *"sui generis"* system, with laws protecting only GIs and on the other hand, collective trademarks and certification marks, laws against unfair competition, which are generally used in Anglo-Saxon countries (Giovannucci et al. 2009). However, some countries do not have specific regulations for GIs. In some cases, they are not officially registered but are used commercially.

The inclusion of GIs, as a new category of intellectual property rights, in the scope of the TRIPS Agreement, illustrates the success of the European Union (EU) negotiations as opposed to countries such as the United States, Australia and

Canada. Currently, the EU continues to defend its model so that it can be extended internationally. To this end, the defence of GIs is included in the negotiations of bilateral trade agreements it conducts with third countries, particularly developing countries (Bernault and Collart 2012). Furthermore, it supports the implementation of a system for recognising GIs in various countries. The aim is to assist these countries in building an institutional framework necessary to overcome certain difficulties such as the collective organisation of producers and the cost of legal protection. In France and Europe, GIs rely on a technical, institutional and financial mechanism through national and European public policies. Developing countries do not have such institutional and financial resources (Bérard and Marchenay 2009).

The legal frameworks governing the recognition and protection of GIs have multiplied around the world in a very divergent manner. This reflects the diversity of objectives attached to GIs. The purposes of GIs relate to market access or also aim at non-market dimensions, such as territorial development, preservation of cultural heritage and natural resources. In recent years, many developing countries show a strong interest in the protection of GIs. These countries mobilise GIs within the framework of several objectives such as the promotion of exports of specific products, a tool for rural development and safeguarding national heritage (Kalinda 2010; Fort 2014).

Although about 90 percent of GIs come from member countries of the Organisation for Economic Cooperation and Development (OECD), interest in GIs is growing in developing countries (Giovannucci et al. 2009; Le Goffic 2012; Bérard and Marchenay 2007). In the Mediterranean region, various initiatives are observed in the Southern Mediterranean countries for the protection and registration of local products (Lebanon, Tunisia, Morocco, Algeria). In 2014, the three Maghreb countries only registered one GI (Tyout Chiadma Olive Oil in Morocco) while the European Mediterranean concentrated 81% of non-wine appellations of the EU (Cheriet 2017). Currently, Tunisia has (05) labelled products under (GI) and (01) product under Appellation of Origin (AO) (Ministry of Agriculture Tunisia 2023), while Morocco has registered (73) (GI) and (AO) in a national register including (06) products (AO) (OMPIC 2023). In Algeria, three products have been labelled under (GI): *Bouhezza cheese, dried figs of Beni Maouche* and *Deglet Nour dates of Tolga*.

The geographical extension of GI initiatives in the various countries of the Southern Mediterranean has introduced new concerns and justifications around the implementation of this quality sign as a public policy instrument in areas such as rural development, food security and biodiversity. Therefore, these concerns are the subject of new aspects of the international debate on GIs currently under development (Allaire 2008).

7.3 Valorisation Through Quality Signs, Advantages and Impacts

Labelling through distinctive signs related to origin, though GI or AO, is a factor that allows the valorisation of local products. They can be considered as relevant levers for developing countries and territories (Fort 2014). Indeed, GI as a legal protection framework promotes both, the distinction of product quality and access to local and international markets (Amsallen and Rolland 2010). GIs are considered as offering numerous economic opportunities, which do not only benefit producers, but also actors throughout the supply chains (Bagal et al. 2011; Arety 2013; Vandecandelaere et al. 2018).

GIs can be a powerful tool in response to various sustainable development issues, in its three pillars: economic, environmental and social. When we talk about a historical, traditional, typical product, rooted in a place and in a society, surrounded by know-how and passed down over time, the GI takes on a dimension that can be described as heritage (Barjolle et al. 1998). GIs can be engines of rural transformation leading to more sustainable development, on the one hand, because economic sustainability is an important step towards environmental and social sustainability, and on the other hand, because specifications can directly influence environmental sustainability depending on the requirements considered (local species or breeds, specific agricultural practices, etc.) (Vandecandelaere et al. 2009) (see Fig. 7.1).

However, it should be emphasised that these effects are complex and uncertain: for example, it is not enough to register a GI in an intellectual property register for the product price to increase, biodiversity to be preserved and local knowledge to be valued. These positive impacts depend on the conditions of implementation and management (Vandecandelaere 2011; Vandecandelaere et al. 2018; Barjolle and Sylvander 2002). Geographical limitations and technical requirements can hinder the supply of raw materials (Arety 2013). In some cases, the distribution of added value, does not favour primary producers (Arety 2013; Vandecandelaere et al. 2018). Some authors question the implementation of GIs and consider them as collective monopoly rights (Rangnekar 2004; Herrmann 2011). In some contexts, the use of GIs is not appropriate, it can even present disadvantages, if it results from a poor design or a lack of governance structures. As an example, this is the case when management devices are defective, public policy incentives are limited or when there is a risk of domination by a limited number of companies (Giovannucci et al. 2009).

The valorisation of quality linked to origin can generate positive effects in economic, social and environmental terms. For this, the GI recognition system must meet certain conditions which are essential factors for the success of a GI: an effective institutional and legal framework; a robust specification; distinguishable quality; marketing efforts; a willingness of consumers to pay a high price; collective marketing strategies; good governance; support by public authorities (Vandecandelaere 2011; Vandecandelaere et al. 2018; Belletti et al. 2014; Giovannucci et al. 2009; Rangnekar 2004).

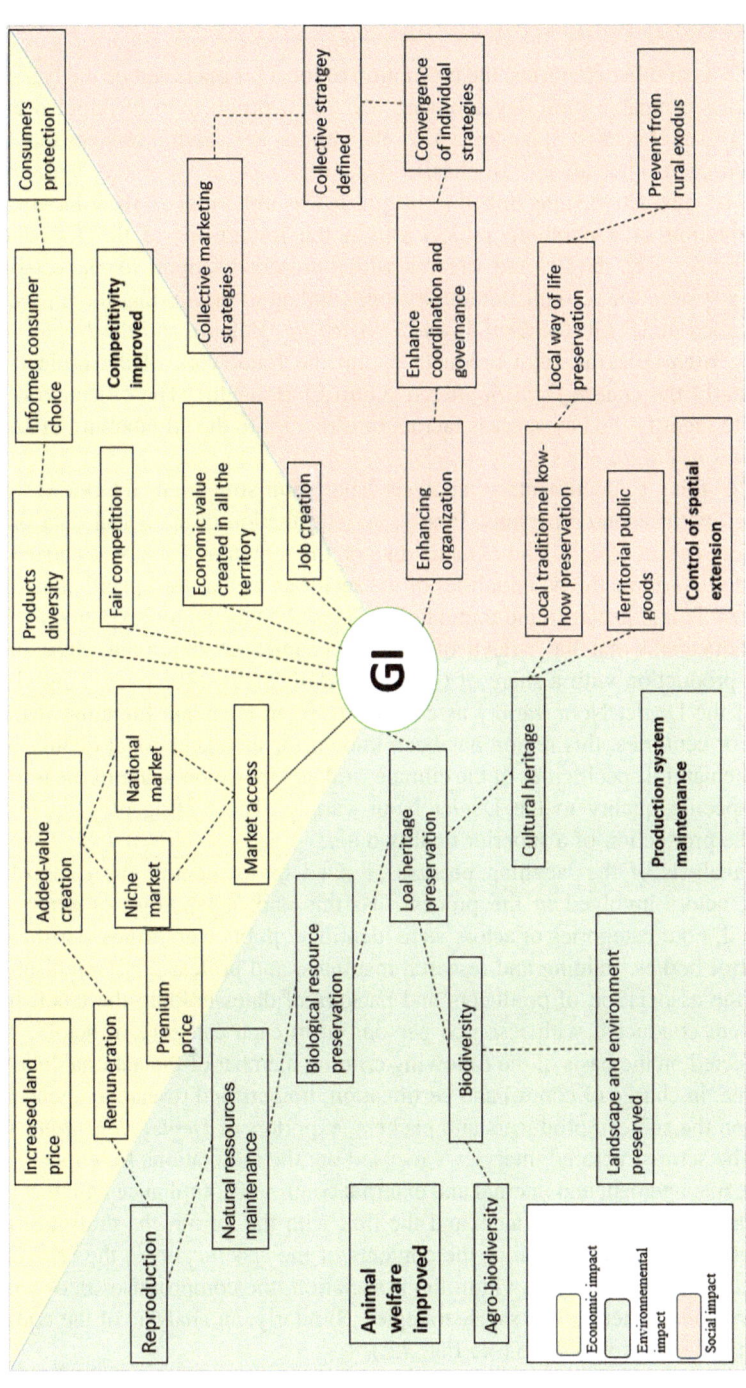

Fig. 7.1 GI relations to sustainable development. (Source: authors)

7.4   The GI Valorisation *of Deglet Nour Dates of Tolga*

In Algeria as in other countries, the promotion of local products can be the basis of a process of sustainable rural development, and allow products to distinguish themselves for market access. Since the 2000s, the various agricultural policies that have succeeded each other always emphasise the importance of valorising agricultural products by distinctive signs linked to origin. It was only after 2016, following the implementation of a twinning project within the framework of the Association Agreement between the EU and Algeria, which provided support for the establishment of a system for recognising quality signs linked to origin, that three products were labelled under GI: the *Beni Maouche dried fig* (Béjaia), and the *Deglet Nour dates* of *Tolga* (Biskra), pilot products within the framework of the project, and subsequently the *cheese from Bouhezza* (Oum El Bouaghi). The objective of the study is to identify the key success factors or obstacles to the GI labelling of *Deglet Nour dates of Tolga*.

To do this, a documentary analysis and semi-structured interviews were conducted with resource persons. Firstly, an exploitation of statistical data on the production, export of dates and climatic data on the region of Tolga was carried out, in order to determine the potentialities of the sector as well as the specificities of the territory of Tolga. It was found that dates are classified as the top exported agricultural product and that the wilaya of Biskra occupies an important place in the national production with a share of (42%) (MADR 2019). Moreover, Tolga is the cradle of the Deglet Nour variety as evidenced by an abundant literature (MADR 2019). For centuries, this region has been known for the exceptional quality of its dates. The natural specificities of the climate, soil and irrigation water of the territory give a specific quality to the Deglet Nour variety. The hydrogeological texture allows the production of a superior quality date.

The analysis of the labelling process allowed us to identify the roles of the different actors involved in the process. To this end, a typology of actors was developed. Four categories of actors were identified: public authorities, certification and control bodies, training and research institutes, and professional organisations, notably the association of producers and packers of dates holding the label. Interviews were conducted with resource persons from each category of actors. They were selected on the basis of the following criteria: member of the national labelling committee, in charge of control and certification, trainers and researchers who have worked on the subject, producers and packers, exporters of *Deglet Nour dates from Tolga*. The semi-structured interviews focused on: the motivations for choosing the label, the management and internal and external control of compliance with the label, the characteristics of the products and the link with the terroir, the drafting of the specifications, the justification of the contents of the specifications, the criteria for membership and governance within the association, the compromises between the producers, the marketing and sales strategies. Similarly, an analysis of the contents of the specifications was done (see Fig. 7.2).

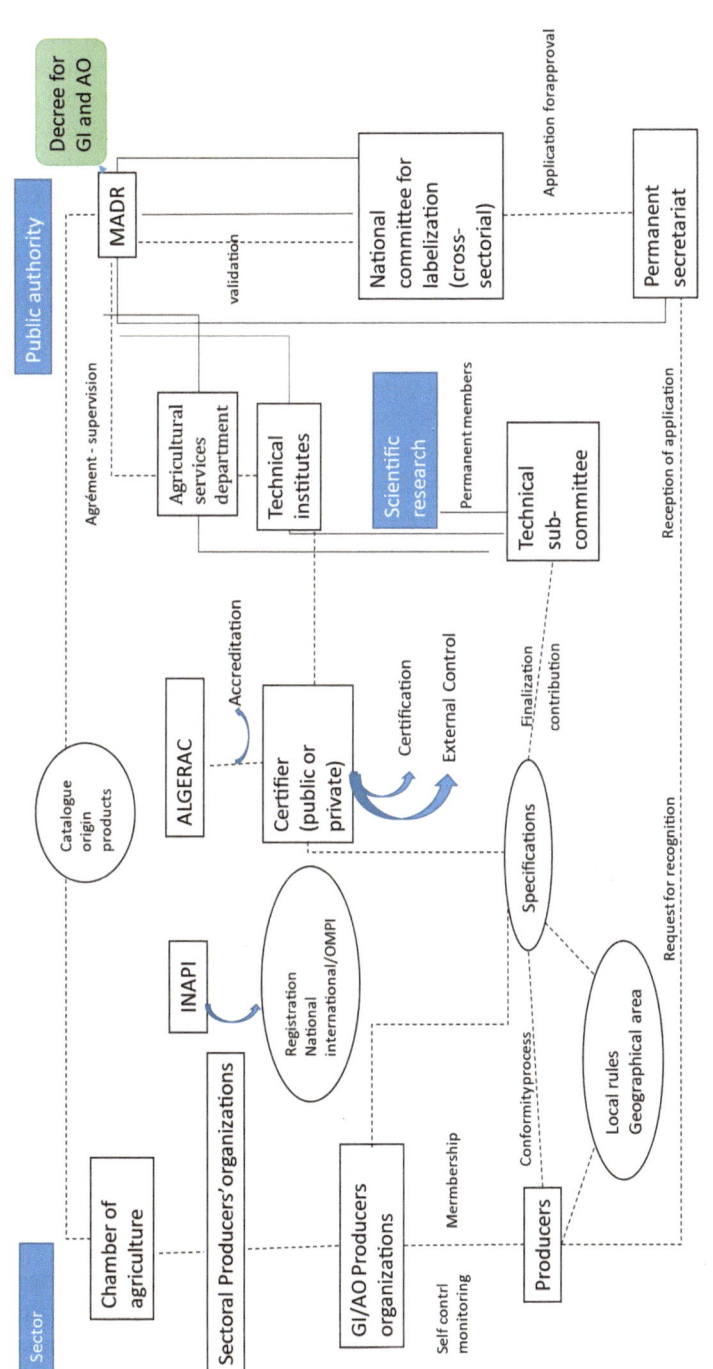

Fig. 7.2 Stakeholders and process for GI registration and protection. (Source: authors)

Subsequently, a documentary analysis of several studies as well as legal texts was conducted with the aim of understanding the choice of the GI label as well as the interest of public authorities and actors in the sector for this label. The analysis showed a strong willingness of public authorities to support the date sector through labelling for three main motivations: to facilitate access to the international market, to face the risks of usurpation, and to fight against the devaluation of products in the informal circuit. This labelling was put in place in a bottom-down approach with technical support from the EU. On the other hand, the date producers-exporters in Biskra saw the interest of labelling in order to position themselves on the European market. Aware of the specific quality of their products, they wish to distinguish their products by the recognition of the production terroir. The registration of the GI at the international level allows the reputation of the products to be spread beyond the local market, like other date producing-exporting countries such as Tunisia.

The way in which the specific quality is defined in the specifications depends on the type of product and the strategy of the producers: defensive or offensive strategy (Vandecandelaere et al. 2018). According to the results of the surveys with the resource persons, the labelling of the *Deglet Nour dates of Tolga*, falls within a rather defensive approach whose objective is to protect a solid reputation in order to safeguard the traditional practices used in the region. The specificities defined in the specifications essentially describe the existing practices in terms of production and harvesting. According to the producers, the sorting, preservation and presentation operations are also essential in preserving the quality. Producers who do not respect the practices described in the specifications are excluded. Regarding commercial valorisation, a higher gross margin is also the result of effective marketing and sales strategies and tools (Rangnekar 2004). In this area, the members of the GI *Deglet Nour of Tolga* association affirm the existence of negotiations between them concerning prices as well as the quantities produced, but they are often informal and do not only concern the members of the association. After the labelling under GI, this association has not been mobilised as a negotiation and consultation tool, particularly to build common marketing and sales strategies. The actors in the sector and the members of the GI *Deglet Nour of Tolga* association, have not engaged in a collective organisation dynamic. They have grouped together to request a collective label, while continuing to pursue individual strategies backed by relational networks (family relationships, neighbourhood, . . .), which leads to informal agreements for information sharing. These While these strategies are not without interest, the values of sharing and mutual aid strengthen the ties between producers, but they constitute an obstacle to a solid collective organisation, one of the necessary conditions for the success of valorisation through the GI.

Finally, the legal and institutional framework is recent, the labelling process under the GI/AO sign is operational, but it is not finalised. To date, the Algerian Accreditation Organisation (ALGERAC), is not yet accredited for agricultural products, resulting in the absence of accredited certification bodies. Thus, producers who want to export have faced the difficulty of certifying their product. This has led to a disinterest of the actors in the sector for the labelling of their product under GI and this is one of the reasons for the weakness of the adherence to the association in recent years.

7.5 Conclusion

Drawing inspiration from the success of the European model, a dynamic for the protection and registration of local products has been established in the countries of the South of the Mediterranean, a region characterised by the diversity of local resources, ecosystems, and traditional practices. In this context, Algeria has initiated a set of policies to establish a labelling system for geographical indications.

The study showed that the choice of the GI label for *Deglet Nour dates* of *Tolga* is based on a political will with adoption by the actors in the sector. The interest for the State is to protect the local variety of Deglet Nour dates in order to support the date sector which represents an export potential. For the producers of Tolga, they are aware that their territory has a reputation and they wish to enhance it and distinguish themselves on the markets, mainly internationally. To this end, the specificities of the GI described in the specifications are the same traditional practices generally used in the region.

This political will does not seem sufficient to meet the challenges of labelling, particularly commercial valorisation. The results show that the factors that hinder the success of the labelling process are mainly institutional and organisational. On the one hand, the management of the label is characterised by a weak collective appropriation of the GI by the actors. On the other hand, the legal and institutional frameworks are finalised but show shortcomings, particularly in the accreditation process of certification bodies. The absence of accredited certification bodies hinders the certification process. As a result, producers cannot export under GI. This has led to a low incentive to engage in a collective approach, a necessary condition for the success of a labelling process.

References

Allaire G (2008) Diversity of Geographical Indications and positioning in the new international trade regime. International Seminar on Terroir Products, Geographical Indications, and Sustainable Local Development in Mediterranean Countries. Antalya, Turkey

Amsallem I., Rolland E. (2010). *Geographical indications: product quality, environment and cultures.* In: *Common knowledge,* 9, AFD, FFEM.

Arety (2013) Study on assessing the added value of pdo/pgiproducts. Européenne Commission

Arfini F, Albisu LM, Giacomini C (2011) Current situation and potential development of geographical indications in Europe. In: Barham E, Sylvander B (eds) Labels of origin for food. Local development, global recognition. CABI International, Cambridge, pp 29–44

Bagal M, Barjolle D, Vandecandelaere E, Tartanac F, Kaamon Kpohomou H, Chabrol D (2011) Quality linked to geographical origin and geographical indications in West and Central Africa. FAO-OAPI

Barjolle D, Boisseaux S, Dufour M (1998) The link to the terroir. Review of research work. Institute of Rural Economy, ETHZ

Barjolle D, Sylvander B (2002) Some success factors of "origin products" in European agro-food sectors: markets, internal resources, and institutions. Econ Soc 36(9)

Belletti G, Brazzini A, Marescotti A (2014) To use or not to use protected geographical indications? An analysis of firms' strategic behaviour in Tuscany. In: Thévenod-Mottet E, Tregear A (2006) Qualité, origine et globalisation : justifications générales et contextes locaux des Indications Géographiques. In: Casabianca F, Sylvander B (eds) Produits agricoles de qualité : spécificités culturelles et enjeux de globalisation. Sciences en Partage, INRA, Paris, pp 19–38

Bérard L, Marchenay P (2007) Terroir products: understanding and acting. CNRS – Terroir resources. CNRS – Ressources des terroirs - Cultures, usages, sociétés

Bérard L, Marchenay P (2009) Places, cultures, and diversities: an anthropological view on localized productions. In: Tekelioglu Y, Ilbert H, Tozanli S (eds) Terroir products, geographical indications, and sustainable local development in Mediterranean countries. CIHEAM, pp 31–37. (Options Méditerranéennes, n° 89)

Bernault C, Collart C (2012) Articulation of national, community, and international law on quality signs: the European approach. Paper presented at the Lascaux Program Workshop on the Valorization of Agricultural Products: legal approach. San Jose (Costa Rica)

Cheriet F (2017) Valorization of terroir products in Algeria: ongoing initiatives, institutional constraints, and perspectives. Working Paper (2). UMR MOISA

Fort F (2014) Terroirs in the Mediterranean: concepts, theories, practices, and research perspectives. In: Doctoral Seminar "Terroirs in the Mediterranean". UMR MoISA, Montpellier (France)

Giovannucci D, Josling T, Kerr W, O'Connor B, Yeung MT (2009) Guide to geographical indications: linking products to their origins. International Trade Centre, Geneva

Herrmann R (2011) The socio-economics of geographical indications. In: Worldwide Symposium on Geographical Indications. WIPO, pp 39–53

Kalinda FX (2010) The protection of geographical indications and its relevance for developing countries. Ph.D. dissertation in Law, University of Strasbourg

Le Goffic C (2012) Champagne from California? American Chablis? Protection of geographical indications in Europe and the United States. Territories of Wine, 4

MADR (2019) Statistical series. Ministry of Agriculture and Rural Development, Algeria

Prévost P, Capitaine M, Gautier-Pelissier F, Michelin Y, Jeanneaux P, Fort F, Javelle A, Moïti-Maïzi P, Lériche F, Brunschwig G, Fournier S, Lapeyronie P, Josien É (2014) Terroir, a concept for action in the development of territories. VertigO 14(1)

Rangnekar D (2004) The socio-economics of geographical indications. UNCTAD-ICTSD Project on IPRs and Sustainable Development. Issue Paper, 8

Sylvander B, Allaire G, Belletti G, Marescotti A, Barjolle D, Thévenod-Mottet E, Tregear A (2006) Quality, origin, and globalization: general justifications and national contexts, the case of geographical indications. Can J Reg Sci 29(1):43

Vandecandelaere E (2011) Socio-economic reasoning underlying the development of ge1ographical indications: combining economic and public good dimensions to contribute to the sustainable development of territories. In: WIPO, Worldwide Symposium on Geographical Indications, pp 1–14

Vandecandelaere E, Arfini F, Belletti G, Marescotti A (2009) Linking people, places, and products: a guide for promoting quality linked to geographical origin and sustainable geographical indications, 2nd edn. FAO, Rome

Vandecandelaere E, Teyssier C, Barjolle D, Jeanneaux P, Fournier S, Beucherie O (2018) Strengthening sustainable food systems through geographical indications: an analysis of economic impacts. FAO, Rome

Webography

Ministry of Agriculture, Hydraulic Resources and Fishing Tunisia (2023) Consulted on 11 March 2024. http://www.aoc-ip.tn/index.php/carte-aoc-ip
OMPIC (2023) National register of Geographical Indications and Appellations of Origin. Consulted on 11 March 2024. http://www.ompic.ma/sites/default/files/Registre%20National%20 IG-AO.pdf

Chapter 8
Silent Registered EU GIs: What Is at Stake?

Andrea Zappalaglio, Giovanni Belletti, and Andrea Marescotti

Abbreviations

GI	Geographical Indications
PDO	Protected Designations of Origin
PGI	Protected Geographical Indication
EU	European Union
IPRs	Intellectual Property Rights
ISMEA	Istituto di Servizi per il Mercato Agro-alimentare

8.1   Introduction

This contribution presents the results of a research on the topic of 'silent Geographical Indications' (GIs) in the EU. These are defined as registered Protected Designations of Origin (PDO) or Protected Geographical Indications (PGI) which, for a relevant period of time, have not been employed or employed considerably beyond their expected potential, with little to no evidence of administrative activity from the relevant producers' associations and no related activities such as marketing, online promotion and alike.

The under-utilisation of registered GIs is a phenomenon evident worldwide and in the EU itself (European Commission 2021; Marie-Vivien 2020), despite the fact that the European regulatory framework in place ensures a bottom-up registration

A. Zappalaglio (✉)
School of Law, University of Leeds, Leeds, UK
e-mail: a.zappalaglio@leeds.ac.uk

G. Belletti · A. Marescotti
Department of Economics and Management, University of Florence, Florence, Italy

© The Author(s) 2025   93
E. Vandecandelaere et al. (eds.), *Worldwide Perspectives on Geographical Indications*, https://doi.org/10.1007/978-3-031-71641-6_8

process, which requires the active participation of producers for its activation. Very often, in fact, protected GIs remain empty boxes due to the fact that the recognition process is implemented top-down, without any involvement of the producers who then have to use them, or the registration results useless or ineffective due to changed market conditions or too high implementation costs.

The matter of under-utilisation of registered GIs is also extremely relevant from a theoretical point of view, as it is related to a number of fundamental questions concerning the role, nature, and functioning of GIs, as well as to the justifications of GI Law and its relationship with other Intellectual Property Rights, such as trademarks. However, this topic is still deeply under-researched in both economic and legal literature, with no substantive analysis published yet. In fact, the issue of silent GIs can be tackled from two diverging perspectives: one considers silent GIs as detrimental to the GI system as a whole, while the second argues for the need to maintain their protection in view of, among others, the prevention of genericisation of GIs, the need to protect intellectual property rights to avoid misleading indications and misuses, and the indirect economic, social, and patrimonial effects they can still play.

Hence, this paper contributes to the debate by conducting a first analysis on a subset of registered geographical indications, the Italian PDOs and PGIs that appear in the EU register under Class 1.6 (Fruit, vegetable, and cereals fresh or processed). Following the application of a mixed economical/legal methodology, the work carries out an assessment of this class of products, identifying the potential 'silent GIs' and investigating their peculiarities. A complete presentation of the methodology adopted in this study is provided below.

This contribution is structured as follows. Section 8.2 introduces the state of the art and the legal framework concerning the cancellation of GIs and concludes that the formal analysis of the applicable rules is insufficient to conduct an in-depth assessment of the topic of silent GIs; Sect. 8.3 presents the methodology applied to the present analysis; Sect. 8.4 illustrates the results of the conducted assessment and, finally, Sect. 8.5 discusses the meaning of such findings and draws some conclusions.

8.2   Legal Framework and State of the Art

Regulation (EU) 2024/1143 on geographical indications for wine, spirit drinks and agricultural products, as well as traditional specialties guaranteed and optional quality terms (Regulation 2024/1143) takes into consideration the possibility for registered GIs to fall into disuse, by featuring provisions on their cancellation. In particular, art 25(1) reads:

> 1. *The Commission may, on its own initiative or on a duly substantiated request by a Member State, a third country or any natural or legal person having a legitimate interest and established or resident in a third country, by means of implementing acts, cancel the registration of a geographical indication in the following cases:*

(a) where compliance with the requirements for the product specification can no longer be ensured; or

(b) where no product has been placed on the market under the geographical indication for at least the preceding seven consecutive years.

This provision largely reflects the previous rules applicable at the time where the research presented in the present contribution was conducted, thus confirming the validity of the argument presented here (cf. art 54(1) Regulation 1151/2012).

The national legislation of Member States also includes provisions on GI cancellation, which usually replicate the text of the above-mentioned rules. This is the case, for instance, of article 14 of the Italian *Decreto Ministeriale* 14 October 2013.

The cancellation of a GI, however, is a rare occurrence. Indeed, according to the EU legal register of protected GIs, 'eAmbrosia', to date, only 4 agricultural products have been cancelled. These are:

- *'Salaisons Fumée, marque nationale Grand-Duché de Luxembourg'* and *'Viande de porc, marque nationale Grand-Duché de Luxembourg'*. These Luxembourgish goods were registered in 1996 following the 'Simplified Procedure', i.e. art 17 of the old Regulation 2081/1992. They were cancelled on 18 February 2022 due to lack of use after having remained substantively inactive on the register, without amendments or notifications of any kind.
- The German PGI *'Holsteiner Karpfen'* was cancelled on 9 February 2022 due to the deterioration of the conditions of production that made it impossible.
- The French PGI *'Volailles de Loué'* was cancelled on 3 February 2022 due to the decreasing volume of products marketed under the registered name and the decision of the producers to promote the sales under a different name, focusing on the promotion of a trademark instead of the PGI label.

This is because of various nonexclusive factors. For instance, in some countries, such as Germany, the National Competent Authority is considered lacking the 'legitimate interest' to take action (Guerrieri 2022, p. 72). Moreover, practice suggests that the presentation of a cancellation request usually falls within the essential roles of producer associations rather than National Competent Authorities. This is also what the new Regulation 2024/1143 suggests in Recital 41 of its Preamble.[1]

However, both scholars (Galtier et al. 2013; Carbone et al. 2014; Belletti et al. 2014a, b; Cardoso et al. 2022; Belletti and Marescotti 2021) and practice suggest that GIs can become inactive or severely underused for a number of reasons, among other things: (1) the raw materials/methods of production related to the GI good can become very rare and/or expensive; (2) the GI is the result of a top-down registration

---

[1] 'Producer groups play an essential role in the application process for the registration of geographical indications and in the management of their geographical indications. Producer groups may be assisted in the preparation of their application by interested parties such as regional and local authorities. Producer groups should be equipped with the means to better identify and market the specific characteristics of their products. The role of the producer group should therefore be clarified.

processes conducted by local public authorities without proper level of involvement of the local communities of producers; (3) there is a disproportionate imbalance between the costs of the use of GIs as certification tools and the obtained benefits that make the use of the GI inconvenient. Hence, a formal analysis of the legal framework is insufficient to evaluate the extent of the 'silent GIs' phenomenon.

Therefore, based on this background, the present contribution hypothesizes that the phenomenon of inactive or underused protected GIs must be more widespread than what the EU register may lead to believe. This theory is tested based on the results of an empirical analysis, the methodology of which is presented in detail in the next section.

8.3   The Methodology Adopted for the Identification and Analysis of 'Silent GIs'

The present analysis tracks Italian GIs that, albeit registered, appear to be silent. This was accomplished by implementing a three-step methodology:

1. Identification of PDO-PGI products for which there are no or very small certified values and volumes in 2018–2021, according to the ISMEA-Qualidò databank, in order to build a first list of potential silent GIs. Every year, Ismea carries out a survey of Control Bodies, PDO-PGI Consortia, Producers' Associations and any other subjects, aimed at detecting for each registered PDO-PGI product the number of companies belonging to the PDO-PGI chain, the certified production of each company, the ex-farm, wholesale and consumer prices, and other data on the destination of sales. In this way, the turnover at origin and consumption of each PDO and PGI is also estimated, using both the prices recorded by ISMEA itself both at origin and consumption, and the prices provided by the Consortia or Associations.[2]

   The methodology was operationalised by identifying a composite set of turnover threshold values, aimed at identifying potential silent PDOs-PGIs. The turnover is measured at origin, i.e. at the time of sale of the product from the farm that applied for certification, depending on the product so at the gate of the farm or processor. Based on the distribution of the annual turnover values of the universe of the 114 products surveyed in the database, we identified three thresholds: 20,000, 30,000 and 50,000 euro. For each threshold, products with an average of 4 years below the threshold were identified, and then identified those with at least 1 year above the threshold; this is to take into account years of non-production caused by abnormal adverse weather or pathogenic phenomena,

---

[2]More details about the ISMEA-Qualidò methodology of data collection and elaboration are available at: https://www.ismeamercati.it/flex/cm/pages/ServeAttachment.php/L/IT/D/5%252Fe%252F5%252FD.b9e56f4fb1d37702772d/P/BLOB%3AID%3D2934/E/pdf

which are frequent in the case of fruit and vegetable production[3]. Subsequently, we checked for each product whether the lack of turnover in 1 or more years was due to a lack of prices, but nevertheless quantities of marketed product were present, in order to exclude it from the list of silent GIs. Finally, for products identified as silent or suspect, we checked the year of registration in order to exclude those products registered very recently (after 2016).

2. Identification of PDO-PGI products in this first list whose specifications have never undergone amendments, even minor, according to the *eAmbrosia* EU database[4] as well as to the Italian ministerial documentation.[5]

3. The GIs identified following the application of the above-mentioned criteria will then be subject to additional desk research aimed at gathering more information on the real status of the PDO-PGI products identified in the previous steps. The desk analysis followed a predetermined grid and was carried out by drawing information from the web, in particular from the Qualigeo website[6] (https://www.qualigeo.eu/) and from the Google search engine, including the 'news' and 'shopping' sections. The analysis focused in particular on the existence and level of activity of a Consorzio di tutela or Association of producers and of an updated website and/or Facebook page related to the GI product; on the presence of firms marketing the GI product even without the PDO-PGI label, and on local fairs and festivals centred on the GI product; and other signs of life of the protected product.

This empirical approach will make it possible to assess the current situation concerning 'silent GIs' registered in PDO-PGI EU register. This will open the doors to a debate on critical issues such as the reliability of the register as well as the legal and economic nature of GIs.

8.4   Presentation of the Results

Figure 8.1 represents the distribution ordered according to the average turnover at origin in the 4-year period 2018–2021 of the 114 PDO-IGP products registered in Italy under Class 1.6 (Fruit, vegetable, and cereals fresh or processed) and included in the Ismea database. It is evident from the Pareto diagram the enormous concentration of turnover, in fact the most important product (*Mela dell'Alto Adige*) represents 31.3% of the total PDO-IGP turnover of the Class, the first three products 56.9% and the first ten 76.5%. At the other extreme, the 60 smaller products all together account for 1% of the total turnover of the Class. The situation is very

---

[3] An emblematic case is that of chestnut cultivation, which in Italy in recent years has suffered attacks from a parasite, the pine beetle, which has reduced or cancelled production in some areas.

[4] https://ec.europa.eu/agriculture/eambrosia/geographical-indications-register/

[5] https://www.politicheagricole.it/flex/cm/pages/ServeBLOB.php/L/IT/IDPagina/396

[6] https://www.qualigeo.eu/

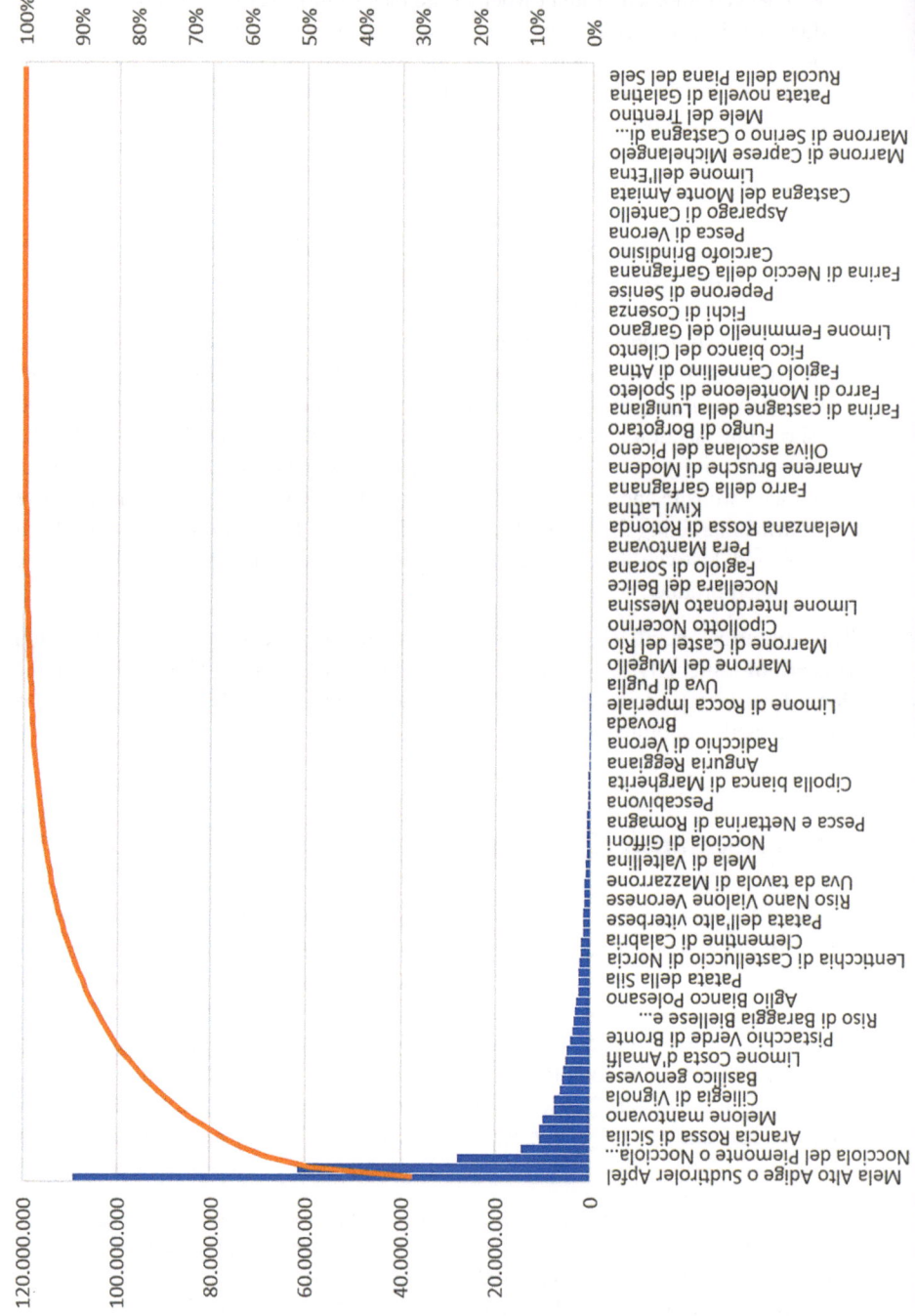

Fig. 8.1 Turnover at the producer's gate for Class 1.6 PDO-PGI products, Italy
Caption: the figure shows the value of turnover at origin of each PDO-PGI product and the cumulated total of turnover at origin for all the products registered under Class 1.6. (Source: our elaborations on ISMEA data)

similar if, instead of looking at the average turnover, one analyses the turnover of the best year: the share of the 60 smallest products (52.6% of the total number of registered GI in the class) rises to 1.3% while that of the top three falls to 54.4%.

The high level of concentration of turnover should not be surprising, since alongside some GIs with a very large extension (sometimes an entire province or region, such as the *South Tyrolean Apple*, the *Piedmont Hazelnut* or the *Sicilian Orange*) and therefore a very high potential number of producers, there are GIs linked to small products with a very small territory delimited by specifications (the Sorana Bean, in Tuscany, is an example). However, even if these figures are influenced by the fact that, for some products, the sales prices and thus the turnover have not been communicated, although quantities are present, albeit in many cases modest, the presence of numerous products with little or very little certified production is evident.

Step 1 of the methodology presented in the previous section identified, for each of the three thresholds, those products with an average turnover value at origin of the 4 years below the threshold, and those with no value above the threshold in any year. Considering for example the highest threshold of €50,000, the potentially silent PDO-PGIs are 31 (Table 8.1, Section A).

Subsequently, for each product below the threshold, we verified the certified quantities and excluded all products with non-negligible certified quantities, and for which therefore the lack of turnover was caused by the lack of price indication. This check identified 8 products (out of the 31 with a 50k threshold) with a certified production volume but no price indication. These products were excluded from the list of potentially silent GIs (Table 8.1, Section B).

Finally, PDO-PGIs registered after 2016 (see step 2 of the methodology) were excluded from the list of silent suspect products on the assumption that a period of several months or even a few years may elapse from the moment of registration to the actual start of certification. In fact, the preparation of the control plan and its approval by the competent authorities, the registration of companies, the activation of traceability procedures, etc. are necessary. In the list of silent products, there were four PDO-PGI products registered in the year 2020, which have been removed.

At the end of the procedure, the total number of registered GIs to be considered as suspected silent GIs is 22 if the most inclusive criterion is applied, and 14 if the most selective criterion is applied. This is a not insignificant number, respectively 19% and 12% of the total universe observed (Table 8.1, Section C). The value resulting from the application of the most selective criterion (14 PDO-PGI silent) seems to us the most correct, since there are some registered products of very limited size, involving a small number of enterprises and characterised by strong seasonality; in these cases, therefore, achieving a turnover of 50,000 € can already be considered a relative commercial success.

We applied the next steps of the methodology (step 3) on the larger group resulting from step 1, i.e. 22 PDO-PGI products (threshold 50k€ on average in years 2018–2021). Steps 2 and 3 were implemented jointly. Table 8.2 summarises the results of this analysis. Orange coloured cells indicate the presence of criticality. In the last column we express a summary judgement on the level of 'silence' on a

Table 8.1 Potentially silent PDO-PGI based on the application of different criteria based on average and yearly turnover, quantities certified, and year of registration

	Threshold 20k €			Threshold 30k €			Threshold 50k €		
	Average 2018–2021 < 20k (a)	Peak >20k€ in at least 1 year (b)	(a)–(b)	Average 2018–2021 < 30k (a)	Peak > 30k€ in at least 1 year (b)	(a)–(b)	Average 2018–2021 < 50k (a)	Peak > 50k€ in at least 1 year (b)	(a)–(b)
Section A) All the products based on turnover (average and peak)									
Number	31	5	26	32	3	29	34	3	31
Section B) All the products based on turnover (average and peak) excluded products with a relevant quantity									
Number	23	5	18	24	3	21	26	3	23
Section C) All the products based on turnover (average and peak) excluded products with a relevant quantity and products registered after 2016									
Number	19	5	14	20	3	17	**22**	3	19

Caption: The table shows the potential silent GIs according to the different criteria presented in the text
Source: our elaborations on ISMEA data

Table 8.2 Results from steps 2 and 3 and identification of Silent PDO-PGIs

	Year of registration	Producers' organization	Website	Facebook profile	Festival/Fair	Shop on line	Amendment PS 2018–2023	Silent (from 0 to 4)
Arancia del Gargano	2007	Yes	Yes, active	Yes, active	Yes	Yes	No	**0**
Asparago di Cantello	2016	Yes	No	No	Yes	No	Yes (2023)	2
Carciofo Brindisino	2011	Yes	No	Yes, but not updated	No	Limited	No	4
Castagna Cuneo	2007	Yes	Yes, partially active	No	Yes	Yes	Yes (2023)	1
Castagna del Monte Amiata	2000	Yes	Yes, but not updated	Yes, active	Yes	No	No	2
Ciliegia dell'Etna	2011	Yes	Yes, but not updated	No	Yes	Limited	No	3
Fagioli Bianchi di Rotonda	2011	Yes	Yes, but not updated	Yes, but not updated	Yes	Limited	No	3
Fagiolo Cannellino di Atina	2010	Yes	Yes, but not updated	No	No	Limited	No	4
Fagiolo di Sarconi	1996	Yes	No	Yes, but not updated	Yes	Yes	No	3
Farina di Neccio della Garfagnana	2004	Yes	Yes, but not updated	Yes, active	Yes	No	Yes (2022)	2
Farro di Monteleone di Spoleto	2010	No	No	No	Yes	Yes	Yes (2021)	3
Fichi di Cosenza	2011	Yes	Yes, active	Yes, active	Yes	Yes	No	**0**
Fico bianco del Cilento	2006	Yes	No	No	Yes	Yes	No	3
Ficodindia di San Cono	2013	Yes	No	No	Yes	No	No	3
Limone Femminello del Gargano	2007	Yes	Yes, active	Yes, active	Yes	Yes	Yes (2018)	**0**
Marrone di Caprese Michelangelo	2009	Yes	No	Yes, but only dedicated to the Festival	Yes	No	No	3

Nocciola Romana	2009	Yes	No	No	Yes	Yes	Yes (2022)	2
Patata novella di Galatina	2015	Yes	No	Yes, but not updated	No	No	No	4
Peperone di Pontecorvo	2010	Yes	No	Yes, but only dedicated to the Festival	Yes	Yes	No	2
Peperone di Senise	1996	Yes	Yes, partially active	No	Yes	Yes	Yes (2020)	1
Pesca di Verona	2010	Yes	Yes, but not updated	No	No	No	No	4
Scalogno di Romagna	1997	Yes	Yes, active	Yes, active	Yes	Yes	Yes (2022)	0

Caption: The table shows the PDO-PGIs that according to the application of criteria 2 and 3 presented in the text have characteristics of silent PDO-PGIs of greater or lesser intensity, measured on a scale of 0 (less silent) to 4 (more silent)

Source: our elaborations on data from various sources

scale from 0 (unsilent) to 4 (fully silent) that takes into account the various elements in the table. Obviously, the summary assessment is not the mere sum of the criticalities, as it takes into account their specific relevance to the product[7], as well as other aspects not in the table such as the presence of research activities on the product itself.

The assessment led to the identification of four registered PDOs/PGIs which can be considered fully silent. Particularly, these share some common characteristics, such as: (1) no recent online presence, or no online presence at all, including online shopping; (2) no social activities, such as participation in fairs or events; and (3) no signs of amendments and/or other administrative activities.

Other 7 PDO-PGIs are identified as likely silent (score = 3). In these cases, there are some signs of online sales of the product, but above all, there are yearly festivals or fairs centred on the product of origin, sometimes even explicitly referring to the PDO or PGI.

At the other extreme, 4 of the 22 products on the short list were identified as active: in fact, there is a Consortium or Association with a website and an up-to-date Facebook profile, a product festival held regularly at least once a year, usually in the harvest season, widespread presence of online sales of the product with clear reference to the PDO-PGI.

8.5 Discussion and Conclusions

The paper makes a contribution to the existing literature on GIs by exploring the under-researched topic of 'silent-GIs' in the EU. These are defined as the registered PDOs/PGIs which, for a relevant period of time, have not been employed or employed considerably beyond their expected potential, with little to no evidence of administrative activity from the relevant producers' associations and no relevant marketing activities, real-life presence and alike.

Identifying cases of truly 'silent' GIs is a complex task that requires the combination of various criteria. Indeed, elements such as the turnover and the certified production volume cannot be decisive when taken in isolation. In fact, the conditions underpinning the production of a GI good can be extremely diverse, depending among other on the typology of the product and its seasonality, the number of active producers, the extension of the area of production and more. In other words, what a PDO/PGI may consider a 'modest' turnover, can be a successful result for another. Similarly, although the absence of administrative activity for a significant period of time, i.e. 15 or more years, can be considered suspect, it is not enough to determine

---

[7]For instance, the absence of online sales is considered in our analysis less indicative for fresh and very perishable products (e.g. cherries), while it is very indicative for storable products (e.g. spelt or dried beans). Similarly, the lack of amendments is less alarming for products whose recognition is more recent.

the 'silent' status of a GI. This is because the producers' associations are not required to regularly exchange administrative information with their national competent authorities or the European Commission, unless they do not deem it necessary and/or are in the position of doing so.

This is why the present contribution has adopted a mixed economic and legal methodology, assessing the vitality of all the Italian PDOs/PGIs registered under Class 1.6 (Fruit, Vegetables and Cereals, fresh or processed) through a combination of elements applied progressively in three steps: (1) PDOs/PGIs for which there are no or very small certified values and volumes in 2018–2021 have been identified, thus leading to a list of 'suspect' silent GIs. From this starting point, the analysis has turned to (2) the administrative activity of the relevant producers' associations, such as the presence of amendments and other communications to the competent national and EU authorities. Finally, (3) all these findings have been combined into a table that also included the results of desk research aimed at identifying online activities as well as the existence of social events related to the identified products, such as fairs and competitions.

This mixed methodology has led to the identification of a short list of 22 suspected PDO-PGIs products, among these only 11 products are classified on the basis of our parameters as fully or likely silent, representing the 9.6% of the Italian PDOs/PGIs registered in Class 1.6. Therefore, although many products have a low or very fluctuating volume and turnover certified and sold as PDO-PGI, in their territories, these products exist and are well represented in local traditions, as witnessed by festivals, fairs, and events. Thus, in a significant number of cases where the certified value is low, according to our analysis the GI product nevertheless exists and is marketed, albeit on a presumably local and/or more or less informal market, and exists an active production system. However, producers do not see the need to use PDO-PGI in marketing, as they do not perceive a cost-benefit advantage. In some cases, this can be caused by a Product specification containing rules perceived as too restrictive by producers; and in a couple of cases, we have verified that a request has recently been made to amend the Product specification in order to make the rules easier for users to comply with. It is also worth noting that there are some cases in which after long periods of under-utilisation/non-utilisation of the PDO-PGI there is a resumption, or attempted resumption, of activities. In the cases of manifest silence, there is always the absence of consortia/associations, or their *de facto* inactivity, which indicates how the presence of collective forms of organisation is a factor facilitating the effective use of the PDO-PGI.

However, on the basis of the analyses conducted in this study, it is not possible to state whether, despite a limited level of use by the companies, PDO-PGI plays a relevant role in the valorisation of the product and the maintenance of its production system and patrimonial value. The fact that in some small communities there are festivals and fairs centred on the product, even with explicit reference to the PDO-PGI, suggests that the EU formal recognition reinforces the product's reputation and the producers' sense of belonging to a local tradition. On the other hand, it is not known whether in these cases, the protection guaranteed by the PDO-PGI prevents misuse of the geographical name by companies outside the territory, while still favouring local companies and maintaining the patrimonial value of the product of origin.

In conclusion, this paper represents the first attempt at a systematic analysis of the topic of 'silent GIs'. At the same time, it aims to provide the stimulus for further research on this topic, especially on profiles that exceed the scope of the present investigation. These may include, among others, the assessment of the practical reasons for the non-use or under-use of registered GIs, as well as the role of other intellectual property instruments, such as trademarks, as possible competitors of GIs in this scenario. The implications in terms of public policies in support of GIs are also relevant, especially in those cases (particularly frequent in the Global South) where the state promotes the registration of GIs. A policy for the effective development of GIs should in fact also appropriately consider support for the take-off of GI initiatives, in terms of implementation of post-registration procedures, capacity-building of actors and strengthening of production chains.

Acknowledgements The authors would like to thank ISMEA (Istituto Servizi per il Mercato Agricolo e Alimentare) and in particular Mrs Antonella Giuliano for providing most of the raw data employed in the present research, and the participants to the Conference "Worldwide perspectives on geographical indications" (Montpellier, July 2022) for the useful comments and suggestions.

References

Belletti G, Marescotti A (2021) Evaluating Geographical Indications. A Guide to tailor evaluations for the development and improvement of geographical indications. FAO—University of Firenze DISEI. Available at: http://www.fao.org/3/cb6511en/cb6511en.pdf

Belletti G, Brazzini A, Marescotti A (2014a) To use or not to use protected geographical indications? An analysis of firms' strategic behavior in Tuscany. In: Paper presented at AIEAA Congress, Alghero, 26–27 June

Belletti G, Brazzini A, Marescotti A (2014b) Collective rules and the use of protected geographical indications by firms. Int Agric Policy 1:11–20

Carbone A, Caswell J, Galli F, Sorrentino A (2014) The performance of protected designations of origin: an ex post multi-criteria assessment of the Italian cheese and olive oil sectors. J Agric Food Ind Organ 12(1):121–140

Cardoso VA, Lourenzani AEBS, Caldas MM, Bernardo CHC, Bernardo R (2022) The benefits and barriers of geographical indications to producers: a review. Renew Agric Food Syst:1–13

European Commission, Directorate-General for Agriculture and Rural Development (2021) *Evaluation support study on geographical indications and traditional specialities guaranteed protected in the EU—Final report*. Publications Office. https://data.europa.eu/doi/10.2762/891024

European Commission. Regulation 2024/1143 Regulation (EU) 2024/1143 of the European Parliament and of the Council of 11 April 2024 on geographical indications for wine, spirit drinks and agricultural products, as well as traditional specialities guaranteed and optional quality terms for agricultural products [2024]L series (23 April 2024)

Galtier F, Belletti G, Marescotti A (2013) Factors constraining building effective and fair geographical indications for coffee: insights from a Dominican case study. Dev Policy Rev 31(5):597–615

Guerrieri F (2022) Cross-national comparative analysis of procedural laws and practices in the EU
    Member States. In: Zappalaglio and others (eds) Study on the functioning of the EU GI system.
    Max Planck Institute for Innovation and Competition
Marie-Vivien D (2020) Protection of geographical indications in ASEAN countries: convergences
    and challenges to awakening sleeping geographical indications. J World Intel Property 23:328–
    349

Chapter 9
Domain Name Protection for Geographical Indications: A European Gamechanger

Latha R. Nair

Abbreviations

ADR	Alternative Dispute Resolution
ccTLD	Country Code Top-Level Domains
CI	Craft and industrial
DNS	Domain Name System
EU	European Commission
EUIPO	European Union Intellectual Property Office
FAO	Food and Agriculture Organisation
GI	Geographical Indication
gTLDs	Generic Top-Level Domain
ICANN	Internet Corporation for Assigned Names and Numbers
INTA	International Trademark Association
IPR	Intellectual Property Rights
oriGIn	Organisation for an International Network for Geographical Indications
TRIPs	Agreement on Trade Related Aspects of Intellectual Property Rights
UDRP	Uniform Domain Name Dispute Resolution Policy
WIPO	World Intellectual Property Organization

L. R. Nair (✉)
K&S Partners, New Delhi, India
e-mail: Latha@knspartners.com

© The Author(s) 2025
E. Vandecandelaere et al. (eds.), *Worldwide Perspectives on Geographical Indications*, https://doi.org/10.1007/978-3-031-71641-6_9

9.1 Introduction

Geographical Indications (GIs) are prominently recognized as one of the intellectual property rights (IPR) under Section 3 of the Agreement on Trade Related Aspects of Intellectual Property Rights (the TRIPs Agreement). GIs and trademarks are equally effective branding tools for the relevant right holders, despite the different modes of ownership of these two rights. In those goods branded using GIs, both these IPRs complement such goods in that the GI would indicate the geographical origin, provenance, and uniqueness of the goods whereas the trademark would indicate the specific producer being the commercial source of the goods sold thereunder. Trademarks are private monopoly rights of a single proprietor, whereas GIs are collective community rights belonging to multiple legitimate users. Yet, today, these two rights are protected and enforced using similar legal principles before intellectual property offices and courts in most countries. Besides the possibility of registration of both these rights with the relevant authorities, in common law countries, it is also possible to rely on the unregistered rights, acquired through use, in these two IPRs.

Both GIs and trademarks are used for branding purposes. However, there is a significant difference in the way each of these rights come into existence. While trademarks can be adopted and used overnight, GIs emerge as IPRs only after a duration of use, acquisition of reputation, and consequent recognition. Due to this difference, a trademark is capable of being visible to the public almost instantly upon adoption whereas a GI is a late bloomer as an IPR and is visible to the public only after several years. This certainly places trademark owners in an advantageous position in comparison to bodies administering GI rights and their stakeholders. For instance, soon after adoption, trademark owners can secure a registered right for a trademark and domain name. On the other hand, a GI body would take years to be formed, let alone the time taken for deciding to register the right as a GI or a domain name. Also, because of the advantage of instantaneous adoption and the head start in getting a registration, trademark owners can prevent third parties from misappropriating their rights as a trademark or a domain name. Whereas a GI body would embark on those steps only after its formation, by when misappropriation of the GI as a domain may have already occurred, unless preventive steps were astutely initiated by the right holders (which is rarely the case).

This paper examines this specific disadvantage of bodies administering GIs and their stakeholders in the context of the Report of the Second WIPO Internet Domain Process published in 2001. Specifically, this paper focuses on how the prejudicial observations in the Second Report continue to haunt the domain name protection efforts of GI bodies and their stakeholders to date and examines some of the promising developments in Europe that could act as a trailblazer in bringing in a global shift to this situation.

9.2 Report of the Second WIPO Internet Domain Name Process, 2001

In September 2001, WIPO released the Report of the Second WIPO Internet Domain Name Process[1] ("the Second Report"). As the name suggests, there was a first process too. The first process investigated the interface between trademarks and internet domain names and recommended the establishment of a Uniform Domain Name Dispute Resolution Policy (UDRP). UDRP is the legal framework for the resolution of disputes between a domain name registrant and an unauthorized third party over the abusive registration and use of an Internet domain name in the generic top-level domains (gTLDs) such as .com, .net, .org, .biz, .info, .mobi, and .name. Thereafter, in 1999, the Internet Corporation for Assigned Names and Numbers (ICANN) adopted UDRP as an international mechanism to respond to complaints related to the domain name system (DNS). Several countries have also adopted UDRP on a voluntary basis to deal with similar complaints with respect to country code top-level domains (ccTLDs) such as .in (for India), .us (for the United States), .fr (for France), and .uk (for the United Kingdom).

As per a report available on the website of WIPO,[2] the WIPO Arbitration and Mediation Center has administered over 67,000 complaints under the UDRP and related policies. Together, these proceedings have involved parties from 185 countries and over 120,000 Internet domain names.

The Second Report discussed several identifiers apart from trademarks as a potential right, based on which complaints could be filed under the UDRP. One such identifier was the rights based on GIs. Unfortunately, the Second Report excluded GIs as a recognized prior right forming the basis for a complaint under the UDRP. There were various justifications given for this exclusion. These are:

The (then) existing international legal framework for the protection of GIs was developed for, and applied to, trade in goods. The Paris Convention, the Madrid (Indications of Source) Agreement, and the TRIPs Agreement all deal with the misuse of geographical identifiers in relation to goods. There is, thus, not a ready and easy fit between these rules and the predatory and parasitic practices of the misuse of GIs in the DNS.

The Second Report noted that the mere registration of a GI as a domain name by someone with no connection whatsoever with the geographical locality in question, however cheap and tawdry a practice, did not appear to be, on its own, a violation of the then existing international legal rules with respect to GIs. Such a registration may violate existing standards if it were associated with conduct relating to the relevant

---

[1] Please see the link as accessed on February 21, 2024. https://www.wipo.int/amc/en/processes/process2/report/html/report.html

[2] Please see the link as accessed on February 21, 2024. https://www.wipo.int/amc/en/center/caseload.html#:~:text=WIPO%20Center%20has%20administered%20over,over%20110%2C000%20Internet%20domain%20names

goods. However, there are many circumstances in which a domain name registration, even though constituting a false or unauthorized use of a GI, may not constitute a violation of existing international rules because there is no relationship between the domain name and goods. Existing rules, therefore, would offer only a partial solution to the problem of what is perceived to be the misuse of GIs in the DNS.

There is a major problem with respect to the applicable law because of the different systems that are used at national levels to protect GIs. This problem highlighted the lack of a multilateral system for the recognition of GIs. The (then) existing international framework for GI protection would only partially answer the perceived problems of false indications of source and GIs within the DNS. The Second Report noted that, because of the need to resort to a choice of applicable law to resolve the issue of the recognition of the existence of a GI, very complicated questions would be involved in the application of the UDRP in this area and that the international opinion on these questions was far from settled.

The Second Report concluded by recognizing the widespread dissatisfaction with the use of GIs as domain names by unconnected persons. It noted that this problem could be addressed only by creating a new law in view of the inadequacies of the existing law. It exhorted the international community to advance multilateral discussions on (i) the definition of the circumstances in which the registration and use of GIs as domain names should be proscribed; and (ii) the establishment of a multilaterally agreed list of GIs or other means to satisfactorily deal with the interaction of differing systems and levels of protection for GIs.

As this paper is being written, the Second Report completed 23 years since its publication. In these 23 years, there were multiple unsuccessful representations to get GIs onto the high table of identifiers before the ICANN. Importantly, in these 23 years, GIs have also emerged as a prominent IPR in their own stance across continents of the world and have come to the forefront as an important and universally recognized IPR. Whether the findings in the Second Report, pertaining to the eligibility of GI rights as a valid basis for domain name complaints, are still valid is a rhetorical question today.

9.3   The Emergence of GIs as a Dominant IPR

With the advent of the TRIPs Agreement in 1995, GI rights that were dormant for centuries were woken up across the globe. While Europe was always a haven for origin-guaranteed goods protected under the GI banner, the obligation, among others, under the TRIPs Agreement to enact laws to protect GIs got many non-European nations to also sit up and notice these hitherto dormant and lesser-known products. The TRIPs agreement and the European examples also brought forth the positive transformative potential of GIs for the development of the communities owning these products. Consequently, there was noticeable progress in the awareness, recognition, protection, and collective policy development of GIs. This has resulted in the metamorphosis of GIs from a lesser-known IPR to a powerful tool with the potential for social and economic development. To appreciate the current stark irrelevance of the findings of the Second Report summarised above, it is pertinent to trace some of these developments that led to the emergence of GIs in the last 23 years as a prominent IPR.

An Increased Pro-GI Stand in Protection Activities Around the Globe
One of the noticeable changes was that several nations started to increasingly and consciously protect GIs. Since there are multiple tools to protect GIs, the legal means adopted by these countries varied from certification marks, collective marks, and sui generis protection to passing off and unfair competition laws.[3] The website of the Organisation for an International Network for Geographical Indications (oriGIn) has a compilation of the details of global GI protection.[4] This includes not only details of the GIs of each country and the different world regions but also the type of products that are protected or recognized as GIs. Currently, the statistics available on oriGIn's website show that there are over 110 countries that protect GIs through various legal instruments. The statistics pertaining to GI protection in the different world regions are also interesting and sharply indicative of the ever-increasing awareness among Asian and African countries of the potential of GIs as an IPR tool for development. For the purpose of providing the statistics pertaining to GIs, the said website categorizes the world regions as Africa, Asia, Europe, Oceania, North America, South America, and Central America. A piechart generated from the statistics available on the said website of the total number of GIs from each region (at the date of writing this paper) is as follows (see Fig. 9.1):

---

[3] Paragraph 234 of the Second Report analyses this and states as follows: "Commentators who oppose the introduction of protection for geographical indications in the DNS recognize that there exist uniform rules governing the subject matter, but believe that the level of harmonization achieved by those rules is insufficient to constitute an adequate basis for the protection proposed in the Interim Report. In particular, they highlight three specific areas where sufficient uniformity is lacking. First, they argue that the harmonized rules at issue (in particular the TRIPS Agreement) incorporate several qualifications and exceptions to the protection for geographical indications, striking a delicate balance of interests, which would be difficult, if not impossible, to reflect faithfully in the UDRP.[232] Secondly, they state that the legal and administrative mechanisms giving effect at the national level to the internationally harmonized norms vary widely, ranging, as noted above, from sui generis registration systems, certification or collective marks, the law on passing-off, unfair competition law and consumer protection legislation.[233] Thirdly, and perhaps most importantly, they claim that there exists no uniform view at the international level of what is to be deemed a protectable geographical indication, and that, consequently, terms which are protectable in some jurisdictions, are freely available in others".

[4] Please see the link as accessed on June 28, 2024. Worldwide gi compilation - oriGIn l Organization for an International Geographical Indications Network (origin-gi.com) The data on this link is dynamic as oriGIn keeps updating the same.

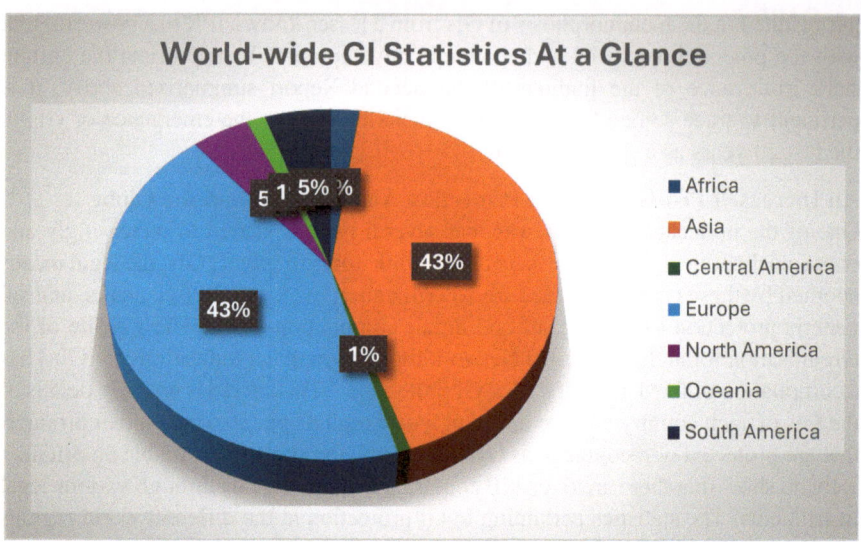

Fig. 9.1 Percentage of GI in different regions of the world

It can be gleaned from the above chart that Asian countries are equally interested and invested in the protection of GIs as European countries. This was not the case two decades ago. The statistics on the website indicate that China leads the pack in Asia with a total of 1784 GIs and that Japan and India are right behind China with respective total numbers of 569 and 388.[5]

Unlike Europe which is dominated by GIs for wines and spirits, Asia and Africa have more agricultural products, food stuffs, and craft products that are GIs. While Article 22.2 of the TRIPs Agreement mandated members to provide legal means to protect all GIs against misleading uses, Article 23.1 thereof stipulates an "additional protection" for GIs relating to wines and spirits, where it is not necessary to demonstrate that the unauthorized use of the relevant GI by a third party is misleading. This issue of preferential treatment to wines and spirits was controversial immediately after the World Trade Organisation was established and it continues to be an eyesore in the TRIPs Agreement.[6]

The website of oriGIn categorizes the products identified as GIs according to their type and has provided some statistics indicating 29 product types and one type categorized as 'services'. In total, there are 8293 products listed as of the date of writing this paper. While wine and spirit product names are a total of 2949 (alcoholic

---

[5]The Indian numbers given on the website of oriGIn do not appear to be updated. India's GI Registry indicates 643 registered GIs on its website as on June 28, 2024. See the link below. https://ipindia.gov.in/IPIndiaAdmin/writereaddata/Portal/Images/pdf/Year_wise_GI_Application_Register_-_26-04-2024.pdf

[6]Please see link as accessed on May 29, 2024 https://www.wipo.int/edocs/mdocs/geoind/en/wipo_geo_mvd_01/wipo_geo_mvd_01_2.doc

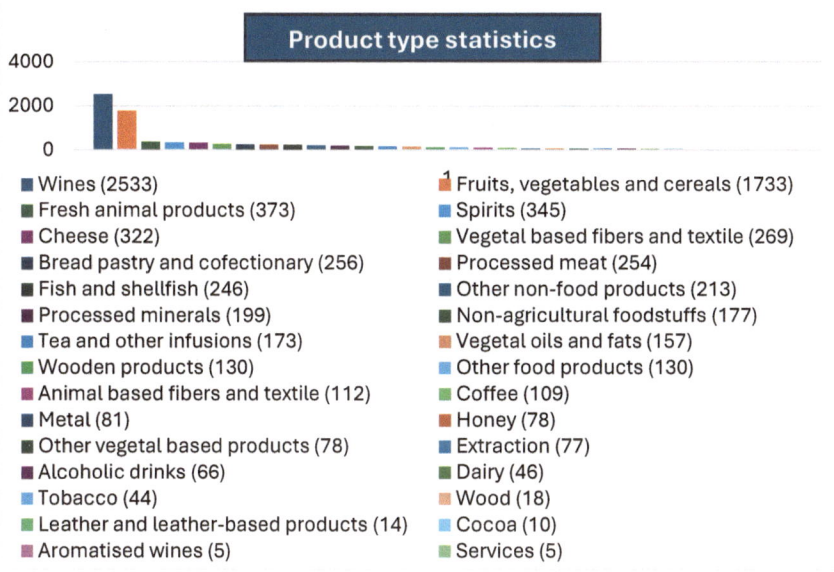

Product type statistics

4000

2000

0

■ Wines (2533)
■ Fresh animal products (373)
■ Cheese (322)
■ Bread pastry and cofectionary (256)
■ Fish and shellfish (246)
■ Processed minerals (199)
■ Tea and other infusions (173)
■ Wooden products (130)
■ Animal based fibers and textile (112)
■ Metal (81)
■ Other vegetal based products (78)
■ Alcoholic drinks (66)
■ Tobacco (44)
■ Leather and leather-based products (14)
■ Aromatised wines (5)

■ Fruits, vegetables and cereals (1733)
■ Spirits (345)
■ Vegetal based fibers and textile (269)
■ Processed meat (254)
■ Other non-food products (213)
■ Non-agricultural foodstuffs (177)
■ Vegetal oils and fats (157)
■ Other food products (130)
■ Coffee (109)
■ Honey (78)
■ Extraction (77)
■ Dairy (46)
■ Wood (18)
■ Cocoa (10)
■ Services (5)

Fig. 9.2 Share of GI products by category

products, wines, aromatized wines, and spirits), fruits, vegetables, and cereals grabbed the second position with a total number of 1733 names. Also, it is interesting to note that the statistics extracted in the chart below for wine and spirit product names are starkly outnumbered in percentage (35%) by all the remaining product names (total of 65%). The Fig. 9.2 below is a depiction of the share of each of these products based on the numbers provided on oriGIn's website[7] as of the date of writing this paper:

GI awareness and GI protection activities have, therefore, become highly prevalent among developed and developing nations, and GIs, irrespective of the category of goods they indicate, stand shoulder to shoulder with any other IPR today.

Emerging Harmonization and Uniformity in GI Protection
In 2001, when the Second Report was published, GIs were not one of the top IPRs of interest to nations. The Second Report also was probably right when it pointed out the lack of uniformity or harmonization among nations of the world in the methods adopted for GI protection when it declined to identify GI rights as a basis for the transfer of domain names in the UDRP complaints. The data available on the website of oriGIn indicates that the situation has shifted dramatically since 2001. It appears that of the 9332 GIs identified and listed there (at the date of writing this paper) and indicated as protected globally, 94% are protected using sui generis means. This

---

[7]Worldwide gi compilation - oriGIn | Organization for an International Geographical Indications Network (origin-gi.com)

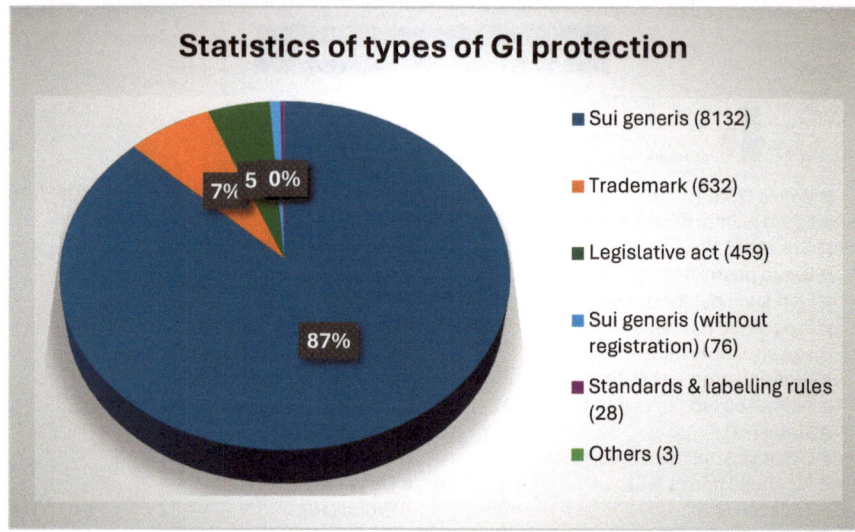

Fig. 9.3 Share of GI protection type

94% is a combined percentage of sui generis protection offered through registration (87%) and that offered without registration (7%) by various countries. There is only a minuscule of GIs protected using methods such as trademark law, legislative acts, and labeling regulations. If the statistics available on the website of oriGIn were to be relied on, the Second Report's finding that the methods employed for GI protection around the world are not uniform or that they lack harmonization would appear to be an outdated finding.

Below (see Fig. 9.3) is a pie chart generated from the relevant statistics on the type of GI protection found around the world as per the details found on the website of oriGIn.

The above statistics also compel us to reexamine the findings in the second report regarding the harmonization of the laws relating to GIs. With an increased level of protection for GIs globally, and the above statistics pointing to a largely accepted means of protection for GIs, namely, sui generis laws, the findings in the Second Report appear disjuncted today.

Organizational and Institutional Support for GIs as a Branding Tool

In the last two decades, GI bodies and stakeholders have been relentlessly working to unite their cause for the protection of GIs. This is evident from various developments such as the emergence of GI rights-based organizations such as oriGIn, the inclusion of GI bodies and advocacy of their rights among brand associations, and GIs being tested as tools for enhancing socio-economic development.

Founded in 2003, oriGIn calls itself a "truly global alliance of GIs from a large variety of sectors, representing some 576 associations of producers and other

GI-related institutions from 40 countries".[8] The first goal of oriGIn is stated as campaigning for the effective legal protection and enforcement of GIs at the national, regional, and international levels through campaigns aimed at decision-makers, media, and the public at large. Its second goal is stated as promoting GIs as a sustainable development tool for producers and communities. Since its inception, oriGIn has been actively involved in policy and advocacy matters relating to GIs and has been regularly holding events for stakeholders of GIs around the world. Its efforts in the last two decades have contributed in a meaningful way to bring GIs and their stakeholders from the fringes to the forefront of IPR discussions.

It is significant that certification and collective marks, two different types of trademarks, are used by some countries such as the United States, Australia, New Zealand, and Japan as tools for the protection of GIs. Associations of brand owners such as the International Trademark Association (INTA) and MARQUES have included GIs in their activities over the past decade. The members of these associations primarily consist of trademark owners, including owners of certification marks and collective marks as well as trademark lawyers. Both these organizations have dedicated committees[9] and teams[10] that delve into issues pertaining to GIs and evolve policies that positively impact GIs and their stakeholders.

The activities of organizations like INTA and MARQUES towards evolving policies for GIs, supporting stakeholders with their publications,[11] and disseminating information about GIs have helped GIs to move towards gaining an equal footing as brands along with trademarks. Additionally, these are testimonies of the acceptance and inclusion of GIs as powerful brands by these organizations.

The power of GIs as a tool for social and economic development is also being explored by the Food and Agriculture Organisation (FAO) of the United Nations. FAO believes that origin-linked products can become the pivotal point of an origin-linked virtuous circle, through a territorial strategy of promotion, whose effects are reinforced over time.[12] The promotion and preservation of origin-based quality can contribute to rural development, food diversity, and consumer choice. Accordingly, in 2007, FAO launched a program on origin-linked quality to contribute to rural development by assisting member countries and stakeholders in the implementation of origin-based quality schemes, both at institutional and producer levels that are

---

[8] Please see the link as accessed on June 28, 2024. Worldwide gi compilation - oriGIn | Organization for an International Geographical Indications Network (origin-gi.com).

[9] Please see the link as accessed on February 21, 2024. https://www.inta.org/committees/geographical-indications-committee/

[10] Please see the link as accessed on February 21, 2024. https://www.marques.org/teams/teammembers.asp?TeamCode=GeInTeam

[11] For example, at a 2015 European conference held in Rome on "GIs, Trademarks & Domain Names" INTA launched an online searchable publication on GIs, Certification Marks, and Collective Marks. Please see the link as accessed on February 21, 2024. https://www.inta.org/events/2015-europe-conference/

[12] Please see the link as accessed on February 21, 2024. https://www.fao.org/geographical-indications/our-approach/en

tailored to individual economic, social, and cultural contexts. Thereafter, it launched a guide in 2009 for promoting quality linked to geographical origin and sustainable GIs.[13] In 2018, FAO released another study on the economic impact of strengthening sustainable food systems through GIs.[14] The main objective of the study, on nine operational GIs,[15] was to provide additional evidence regarding the economic impacts of GIs on value chains and producers. In collaboration with universities, data pertaining to these GIs were collected and analyzed using quantitative and qualitative methods. The study details the analysis of these cases and points to the evidence of the positive economic impact GIs have on price, production volumes and market access. In the study, FAO states that the promotion of linkages between local producers, their local areas, and their food products through GIs is recognized as a pathway to nutritious food systems and sustainable development for rural communities throughout the world. FAO's interest in GIs demonstrates that GIs can not only be powerful brands but also agents of economic change.

Significant Legislative Change: Geneva Act, 2015
In May 2015, the Geneva Act of the Lisbon Agreement on Appellations of Origin and Geographical Indications was adopted under the auspices of WIPO. While the Lisbon Agreement applies only to appellations of origin, a special kind of GIs for products that have a particularly strong link with their place of origin, the Geneva Act extends that protection to all GIs covered by the TRIPS definition as well. This change accommodates the existing national or regional systems for the protection of distinctive designations in respect of origin-based quality products since the Act and the Regulations thereunder accommodate the different national/regional GI systems that exist around the world.

The Act is a significant milestone in the debates surrounding GIs and their inclusion in the DNS for two reasons. First, because it recognizes the wider term "geographical indications[16]" and has serious potential to put at rest the concerns raised in the Second Report regarding the lack of uniformity in the protection granted and the inability of the existing systems to fit in violations of GIs in the domain name world. One of the issues raised by the Second Report to deny an entry as a basis for domain name cancellation was that GIs enjoyed no uniform means of protection. By including the expression "geographical indications" in Article 2, the Geneva Act is bringing in more uniformity to the protection systems and is aligning itself with the wider expression 'geographical indications" used in the TRIPs Agreement. The difficulties and inadequacies of uneven protection systems offered to GIs by the

---

[13] Please see the link as accessed on February 21, 2024. http://www.fao.org/3/i1760e/i1760e.pdf

[14] Please see the link as accessed on February 21, 2024. http://www.fao.org/3/a-i8737en.pdf

[15] *Colombian coffee, Darjeeling tea* (India), *Futog cabbage* (Serbia), *Kona coffee* (United States), *Manchego cheese* (Spain), *Penja pepper* (Cameroon), *Taliouine saffron* (Morocco), *Tête de Moine cheese* (Switzerland) and *Vale dos Vinhedos wine* (Brazil).

[16] Please see the link as accessed on February 21, 2024. https://www.wipo.int/edocs/pubdocs/en/wipo_pub_239.pdf. Article 2 of the Geneva Act deals with the subject matter of the treaty as "Appellations of Origin and Geographical Indications".

nations of the world is mitigated to a great level by this inclusion. By becoming a member of Lisbon agreement, the unequal treatment would be addressed largely for those wishing to protect or enforce their rights in a foreign country. Secondly, Article 11[17] of the Act recognizes that though GIs are in respect of goods, their use in respect of services could, in some circumstances, violate GI rights, including situations involving possible dilution of GI rights.

9.4   Trailblazer Laws in Europe and Rays of Hope on the Horizon

Unlike trademarks that enrich corporations and private individuals, GIs, when protected well, could enrich an entire community in ways that can lead to their social and economic development. It is this power of GIs that is helping them emerge from the fringes towards the mainstream. The developments that have been taking

---

[17] Article 11 Protection in Respect of Registered Appellations of Origin and Geographical Indications:

(1) [Content of Protection] Subject to the provisions of this Act, in respect of a registered appellation of origin or a registered geographical indication, each Contracting Party shall provide the legal means to prevent:

(a) use of the appellation of origin or the geographical indication

(i) in respect of goods of the same kind as those to which the appellation of origin or the geographical indication applies, not originating in the geographical area of origin or not complying with any other applicable requirements for using the appellation of origin or the geographical indication;

(ii) in respect of goods that are not of the same kind as those to which the appellation of origin or geographical indication applies or services, if such use would indicate or suggest a connection between those goods or services and the beneficiaries of the appellation of origin or the geographical indication, and would be likely to damage their interests, or, where applicable, because of the reputation of the appellation of origin or geographical indication in the Contracting Party concerned, such use would be likely to impair or dilute in an unfair manner, or take unfair advantage of, that reputation;

(b) any other practice liable to mislead consumers as to the true origin, provenance or nature of the goods.

(2) [Content of Protection in Respect of Certain Uses] Paragraph (1)(a) shall also apply to use of the appellation of origin or geographical indication amounting to its imitation, even if the true origin of the goods is indicated, or if the appellation of origin or the geographical indication is used in translated form or is accompanied by terms such as "style", "kind", "type", "make", "imitation", "method", "as produced in", "like", "similar".

(3) [Use in a Trademark] Without prejudice to Article 13(1), a Contracting Party shall, ex officio if its legislation so permits or at the request of an interested party, refuse or invalidate the registration of a later trademark if use of the trademark would result in one of the situations covered by paragraph (1).

place in the GI space since the Second Report are hard to ignore. Yet, in a world that is rapidly turning digital and online, GIs continue to be kept out of the DNS.

One of the conclusions in the Second Report was that the problem of the use of GIs as domain names by unconnected persons could be addressed only by creating a new law in view of the inadequacies of the existing law. It appears that the recent developments in Europe have taken a cue from the said conclusion. The discussion below pertains to two recently enacted European laws. The author regards these two laws as trailblazers that would go on to strengthen the position of GIs as a basis for claims for domain name cancellations under UDRP. Let us look at how these laws seek to remedy the situation faced by GIs and how the rest of the world can emulate these laws to offer better protection to GIs in an increasingly digital world.

EC Regulation No. 2024/1143[18]

On April 11, 2024, the European Parliament notified Regulation (EU) 2024/1143 Of The European Parliament and of the Council on geographical indications for wine, spirit drinks and agricultural products, as well as traditional specialties guaranteed and optional quality terms for agricultural products ('the Regulation 2024/1143'). The Regulation 2024/1143 which came into effect from May 13, 2024, amends Regulations (EU) No 1308/2013, (EU) 2019/787 and (EU) 2019/1753 and repeals Regulation (EU) No 1151/2012. Among others, Regulation 2024/1143 has specific provisions regarding the use of GIs as domain names and stipulates safeguards for the prevention of misuse of GIs through domain name registrations. Some of the salient features of the Regulation 2024/1143 in this context are listed below:

- Recital 33 says that protection should be granted to names entered in the Union Register of GIs with the aim of ensuring that they are used fairly and to prevent practices liable to mislead consumers. To strengthen the protection of GIs and to combat counterfeiting more effectively, the said recital states that the protection of GIs should also apply to all domain names that are accessible in the Union, irrespective of the place of establishment of the relevant registries.
- Recital 45 says that the relationship between internet domain names and the protection of GIs should be clarified as regards the scope of the application of the remedy measures, the recognition of GIs in dispute resolution, and the fair use of domain names. It further states that alternative dispute resolution systems of country-code top-level domain name registries throughout the Union should acknowledge GIs as a right to be invoked during such disputes.
- Recital 55 says that, given the increased use of online intermediary services, the enforcement of the protection of GIs against domains names that contravene that protection deserves particular attention. The recital recognizes that it is necessary to hence equip the competent national authorities with the tools to react properly to a violation of the protection of a GI by a registered domain name.
- Article 26.1 states that GIs are protected against, (i) any direct or indirect commercial use on products not covered by the registration, but comparable to

---

[18] https://eur-lex.europa.eu/eli/reg/2024/1143/oj

the products registered or where use of that GI for any product or any service exploits, weakens, dilutes, or is detrimental to the reputation of, the GI, including when used as an ingredient; (ii) any misuse, imitation, or evocation, even where the true origin of the goods are indicated or if the protected name is translated, transcribed or transliterated or accompanied by expressions such as 'style', 'type', 'method', 'as produced in', 'imitation', 'flavour', 'like' or similar, including when those products are used as an ingredient; (iii) any false and misleading usage as to the provenance, origin, nature or essential qualities of the product that is used on the inner or outer packaging, advertising material, in documents or information provided online relating to the product or packaging that is liable to convey a false impression as to its origin; and (iv) any other misleading practice liable to create a false impression as to origin. Further, Article 26.2 states that Article 26.1 would apply to all domain names accessible in the European Union. Article 42.4 supplements this by stating that Member States must take appropriate administrative and judicial steps to disable access to domain names that contravene Article 26(2) from their territory.

• Article 35 deals with the protection of GI rights in domain names. Article 35.1 states that ccTLD registries in the EU must ensure that alternate dispute resolution (ADR) procedures for domain names recognize registered GIs as a right that can be invoked in those procedures. Further, Article 35.2 empowers the European Commission (EC) to lay down provisions entrusting the European Union Intellectual Property Office (EUIPO) to establish and manage a domain name information and alert system—upon the submission of an application for a GI by an applicant, such a system would provide the applicant with information about the availability of the GI as a domain name and, on an optional basis, about the registration of a domain name identical to their GI. ccTLD registries established in the Union are required to provide the EUIPO with the relevant information and data, on a voluntary basis.

EC Regulation No. 2023/2411[19]

EC Regulation 2023/2411 on the protection of geographical indications for craft and industrial (CI) products came into effect on November 16, 2023. This Regulation deals with GIs relating to crafts and industrial (CI) products ('the CI Regulation').

The CI Regulation was enacted by the EU to ensure uniform recognition and protection throughout the Union for GIs for CI products, which the Regulation states to be a priority for the Union. In its recitals, the CI Regulation also refers to the Geneva Act which offers protection for GIs regardless of the nature of the goods to which they apply, and states that to fully comply with the international obligations under the Geneva Act, the CI Regulation is enacted. The CI Regulation is again a significant milestone in the recognizing the rights of GIs as a basis for protection against domain name violations. Some of the provisions of the CI Regulation relevant to such basis are listed below:

---

[19] https://eur-lex.europa.eu/eli/reg/2023/2411

- Recital 38 states that to strengthen the protection of GIs for CI products and to combat counterfeiting effectively, the protection under the CI Regulation should also apply to domain names on the internet.
- Recital 44 states that ccTLD registries established in the Union offering ADR procedures to settle disputes relating to the registration of domain names should ensure that such procedures also cover GIs. The recital elaborates that, where a ccTLD registration: (i) contravenes the protection of a GI, (ii) is being used in bad faith, or (iii) has been made by its holder without a right to or a legitimate interest in the GI, it should be possible for the relevant ccTLD registries established in the EU to revoke or transfer such registration to the relevant producer group by following an appropriate ADR or judicial procedure.
- Recital 45 states that the EC must evaluate the feasibility of establishing an information and alert system against the abusive use of GIs for CI products within the domain name system and submit to the European Parliament and to the Council a report. It further states that based on the outcome of that evaluation, the EC should, where necessary, present a legislative proposal to establish such a system.
- Recital 64 states that to strengthen the protection of GIs for CI products and to combat counterfeiting more effectively, the protection under the CI Regulation should apply to both the offline and online environment, including domain names on the internet.
- Article 40 deals with the protection of GIs. Article 40.1 states that GIs are protected against, (i) any direct or indirect commercial use of the GI on products not covered by the registration, but are comparable to the products registered or where the use of the name exploits, weakens, dilutes, or is detrimental to, the reputation of the protected GI; (ii) any misuse, imitation, or evocation, even where the true origin of the goods or services are indicated or if the protected name is translated, transcribed or transliterated or accompanied by expressions such as 'style', 'type', 'method', 'as produced in', 'imitation', 'flavour', 'like' or similar, including when those products are used as an ingredient; (iii) any false and misleading usage as to the provenance, origin, nature or essential qualities of the product that is used on the inner or outer packaging, advertising material, in documents or information provided online relating to the product or packaging that is liable to convey a false impression as to its origin; and (iv) any other misleading practice liable to create a false impression as to origin. Further, Article 40.3 states that the protection of GIs would also apply to any use of a domain name that is contrary to paragraph 40.1.
- Article 46 stipulates that ccTLD registries in the European Union shall ensure that any alternative dispute resolution procedures for domain names recognise registered GIs as a right that can be invoked in those procedures.

Effect of These Laws on Domain Name Protection
While some countries of the EU already provide for GI rights as a basis for ADRs pertaining to domain name disputes, these two Regulations elevate the protection of GIs in the context of domain name disputes to another level. This is a highly

welcome move in the context of the current remedies available to GIs in respect of instances of domain name misappropriation. In an increasingly digital world, it is important for laws protecting GIs to take note of the implications of online trade. This is truer in the post COVID-19 era where the need for and acceptance of digital presence in business is taken for granted. The domain name alert system contemplated under both these Regulations is surely a safeguard for GI owners against this background.

Additionally, extending the protection of GIs to any use of a domain name that is contrary to levels of protection provided under these Regulations is a game changer in GI protection in the DNS. The provisions cover direct or indirect commercial use on dissimilar goods bound to dilute the GI or use that can evoke a GI. These provisions go even beyond mere recognition of GIs as a right based on which domain name transfers could be demanded. The inclusion of such provisions in laws that are specifically meant for GI protection sends a stronger message to the IP community about the need for inclusion of rights still perceived by some as those of a lesser nature in the context of domain name protection.

9.5   Conclusion: Winds of Change?

Europe has always been the torch bearer in GI protection. Though the developments and advancements surrounding GIs and related policies are expanding, the problem of the misuse of GIs as domain names by unconnected persons has remained a constant due to the inflexible policies of ICANN towards GIs. Thanks to COVID-19, online trading has become the norm for most businesses. For a GI producer body, the use of the GI as part of the domain name that resolves to a website can provide direct market access. Besides, it also lends credibility to its customers of the source, quality, and authenticity of the product. Thus, more than ever before, the denial of a place for GIs in DNS is hugely disadvantageous to right holders and consumers. It is hence important for producers, practitioners, academics, and other stakeholders of GIs to start influencing the policymakers in their respective countries to support this process.

While more than two decades have passed post the Second Report denying GIs a place in the DNS, GIs have only thrived and emerged as a significant IPR. These latest developments from Europe in the context of protection of GIs against domain name misappropriation are a game changer and a tool that has the real potential to blunt the effects of the Second Report for many reasons.

For one, it acts as a model law to numerous developing countries in Asia and Africa to make policy shifts in their domain name and GI protection laws by providing for similar legal provisions. Such policy shifts will support GI communities in such countries to reclaim their usurped domains and conduct trade from internet portals that helps consumers identify the products of such communities.

Cybersquatting is still a business model for many. Often GI right holders are unable to get relief against such cyber squatters even though these are bad faith registrations. Enactment of similar laws globally can act as a deterrent to cyber squatters who choose to bargain over domains adopted based on GIs.

The domain name alert system contemplated in these Regulations is yet another model for nations of the world to emulate. Unlike trademark owners with deep pockets who can afford watch services, GI communities are disadvantaged to afford such services. Assuming that these services, when implemented, would not be as costly as a private watch service, it would further lend support to GI communities to prevent misuse of GIs as domain names by third parties.

The Regulations assume significance in the DNS where consistent blind eyes and deaf ears have been meeting the pleas of GI bodies and stakeholders to recognize their rights as a basis for domain name transfers in the UDRP framework. The European Union[20] and some other countries[21] have also tried to remedy this situation by incorporating GI-based rights as a basis for the complaints under ccTLD Alternative Dispute Resolution (ADR) policies. However, this significant step by Europe to address this issue by incorporating provisions in the GI laws of the Union is an answer to problematic conclusion in the Second Report that the issue of the use of GIs as domain names by unconnected persons could be addressed only by creating a new law in view of the inadequacies of the existing law.

Once enacted into the GI laws of the Union, Europe can also influence its bilateral partners during free trade negotiations to enact similar provisions into their ccTLD policies. As GI rights are becoming more and more equalized as an IPR by global nations, these Regulations are just what is needed to give an impetus to the emerging self-sufficient image of GIs.

---

[20] Please see link as accessed on October 24, 2023 for European Union's ADR policy for domain names. https://eurid.eu/d/24830115/alternative-dispute-resolution_en.pdf . Relevant clause is B(1)(b)(9).

[21] Please see link as accessed on October 24, 2023 for Mexico's ADR policy for domain names. https://www.dominios.mx/policies/ . Relevant provision is listed under the section "Intellectual Property Dispute" under clause 5. Also see link as accessed on October 24, 2023 for Czech Republic's ADR policy for domain names. https://www.nic.cz/files/nic/PravidlaCZAJ.pdf . Relevant provisions are found under paragraph 2.2.9 of the Rules, and Paragraph 2.2, 3.1 and 3.1.1 of Annexure 3 to the Rules.

The opinions expressed in this chapter are those of the author(s) and do not necessarily reflect the views of the [NameOfOrganization], its Board of Directors, or the countries they represent.

Part II
GI Legal Systems and Roles of Public Actors

Chapter 10
Study on the Protection System and Economic Impact of GIs in China

Hui Xu, Xuan Yang, and Xin Gu

Abbreviations

GI	Geographical Indication
EU	European Union
CNIPA	China National Intellectual Property Administration
AQSIQ	General Administration of Quality Supervision, Inspection and Quarantine
IP	Intellectual Property
TSG	Traditional Specialties Guaranteed

10.1 The GIs Protection System in China

China's vast land and abundant resources, more and more geographical indications containing economic value, social value and cultural value have registered. The Protection System of geographical indications in China is still being perfected. At present, there are overlapping and parallel in management of the protection system.

At the present stage, there are mainly two modes for the protection of geographical indications in China: the first is the certification trademark and collective trademark protection system established by the legal norms dominated by the trademark law. And The Trademark Office of China National Intellectual Property Administration (CNIPA) shall be responsible for the administration, that is registration and examination and approval of trademarks.

H. Xu (✉)
China Center for International Economic Exchange, Beijing, China

X. Yang · X. Gu
Intellectual Property Development and Research Center of CNIPA, Beijing, China

© The Author(s) 2025
E. Vandecandelaere et al. (eds.), *Worldwide Perspectives on Geographical Indications*, https://doi.org/10.1007/978-3-031-71641-6_10

The second is the protection mode of special law (*sui generis* institutional forms), One way is the system of geographical indication protection products that the CNIPA is responsible for providing protection and management; The other is the protection system of geographical indications of agricultural products, which is managed by the Ministry of Rural Agriculture (see Table 10.1).

The two GI protection modes are now presented.

10.1.1 The Private Law Protection System: The Trademark Law

China has established the system of protection of registered geographical indications through the Trademark Law. Under the Trademark Protection System, the protection of geographical indications can be divided into two aspects: positive protection and negative protection. The former refers to the special provisions for the registration of Geographical Indications as certification marks and collective marks, the latter refers to the Provisions on the Prohibition of Geographical Names in trademark registration.

According to paragraph 2 of Article 16 of the Trademark Law of the people's Republic of China, a geographical indication refers to a sign indicating that a commodity comes from a certain region with the specific quality, reputation or other characteristics being mainly determined by the natural or human factors of the region, which is consistent with the definition of geographical indications stipulated in paragraph 1 of Article 22 of the TRIPS Agreement.

"The Measures for the registration and administration of collective trademarks and certification trademarks" issued and implemented in 2003 have made more detailed provisions on the registration of geographical indications as collective trademarks and certification trademarks, and related matters of management.

"The trademark examination and adjudication standard" formulated in December 2016 and "Guidelines for Trademark Examination and Trial" implemented in January 2022 specifically provides for the "examination of GI collective trademarks and certification trademarks",[1] including the examination of the applicant's subject qualification and the specific quality of the goods used in collective trademarks and certification trademarks as geographical indications, review of the relationship

---

[1] According to Chapter IX "Examination and Trial of Collective Trademarks and Certification Marks" of the "Guidelines for Trademark Examination and Trial" Part 2 (Guidelines enacted according to Articles 10, 11, 30 and 31 of Trademark Law. See also the Regulations for the Implementation of Trademark Law, Article 4, Paragraph 1 "A geographical indication provided for in Article 16 of the Trademark Law may be registered as a certification mark or a collective trademark in accordance with the provisions of the Trademark Law and these Regulations.; Geographical indication trademarks are composed of signs specified in Article 8 of the Measures for the Registration and Administration of Collective Marks and Certification Marks").

Table 10.1 Different GIs protection system in China

Name of the legal tool for this chapter	GI trademark	GI products	GI agricultural products
Management authority	The Trademark Office of CNIPA	China National Intellectual Property Administration	The Ministry of Rural Agriculture
Legal basis	Trademark Law, Regulations for the Implementation of the Trademark Law, Measures for the Registration and Administration of Collective Marks and Certification Marks[a]	Provisions on the Protection of Geographical Indication Products[b]	Measures for the Administration of Geographical Indications of Agricultural Products[c]
Legislative level	The Trademark Law is enacted by the National People's Congress	Departmental regulations	Departmental regulations
Objects of protection	Agricultural products, handicrafts, Processed products and other products	Agricultural products, handicrafts, Processed products and other products	Mainly primary agricultural products
Applicant	from a group, association or other organization within the area indicated by the geographical indication	Application institutions for the protection of geographical indication products designated by the people's governments at or above the county level or recognized associations and enterprises	Farmers' professional cooperative economic organizations, industry associations, etc., determined by the people's governments at or above the county level
Protection and oversight sector	Local intellectual property offices, market supervision and administration bureaus, holders of geographical indication certification trademarks and collective trademarks	Local intellectual property offices and entry-exit inspection and quarantine departments	Local agricultural administrative departments and holders of geographical indication registration certificates
Protection system	Geographical Indication Trademark Protection	Geographical Indication Products protection	Geographical indications for agricultural products protection
Can be transferred	Yes	no	no
Term of protection	10 years renewed	Once approved, the conditions for protection are met and protection is granted in perpetuity	Once approved, the conditions for protection are met and protection is granted in perpetuity

[a]Trademark Law (Revised in 2019); Regulations for the Implementation of the Trademark Law (Revised in 2014); Measures for the Registration and Administration of Collective Marks and Certification Marks (2023), Implemented on February 1, 2024
[b]Provisions on the Protection of Geographical Indication Products (2023), Implemented on February 1, 2024
[c]Measures for the Administration of Geographical Indications of Agricultural Products (Revised in 2019), Implemented on April 25, 2019

between reputation or other characteristics and the natural or human factors of the area identified by geographical indication.

10.1.2 The Special Law Protection System (or Sui Generis Institutional Forms)

The Special Law Protection System, at present refers to the protection of geographical indication products established by the "Provisions on the protection of geographical indication products", and by the "Measures for the administration of Geographical Indications of agricultural products" (Agricultural Law) is the main legal norms established for the protection of geographical indications of agricultural products.

Before being under CNIPA, the "Provisions on the protection of geographical indication products" were first issued by AQSIQ (General Administration of Quality Supervision, Inspection and Quarantine) in July 2005 and stipulate that the application acceptance, examination and approval of geographical indication products, the registration, supervision and administration of special geographical indication signs, etc.

As for geographical indications of agricultural products, the Agricultural Law was revised in 2002, and stipulates that "agricultural products that meet the requirements of the specified place of origin and production specifications can apply for the use of geographical indications of agricultural products in accordance with the provisions of relevant laws or administrative regulations."

In 2008, the Ministry of agriculture promulgated the "Measures for the administration of geographical indications of agricultural products", and subsequently published a number of supporting specifications, stipulating the application review procedures and quality control technical specifications of geographical indications of agricultural products.

10.2 Registration Statistics of geographical indications in China

Under the present GIs protection system, there are three types of categories: the GI trademarks, the GI products and the GI agricultural products. By the end of 2019, China had registered 5324 geographical indication trademarks (there is no statistics on the split between collective and certification trademarks) and 2385 geographical indication products, there were 8484 approved enterprises for the use of special marks [based on the 2019 annual report of the State Intellectual Property Office], and 2778 registered geographical indications of agricultural products.

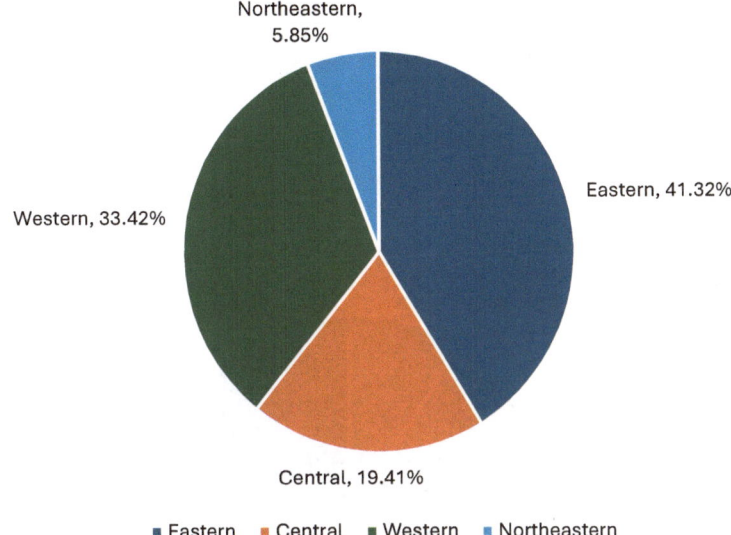

Fig. 10.1 Distribution of GI trademarks in economic regions. (Source: CNIPA 2019b. Evaluation Report on the Status of China IP Development in 2019.Intellectual Property Development and Research Center)

10.2.1   Geographical Distribution of GIs in the Three Categories

Midwestern of China is rich in geographical indications resources, the number of registered GIs more than half.

GI Trademarks
According to the distribution of geographical indications trademarks in the four major economic regions of China, in 2019, nearly three-quarters of the registered geographical indications trademarks are concentrated in the eastern region (41.32%) and the western region (33.42%), the central region (19.41%) and the northeast region (5.85%) had a relatively low proportion. The top five provinces were Shandong (720), Fujian (492), Hubei (437), Sichuan (382) and Jiangsu (323) (see Fig. 10.1). By the end of 2019, the first place of geographical indication trademark is the French Bordeaux Wine Industry Joint Committee, with 133 geographical indication trademarks in China.

GI Products
In 2019, CNIPA approved five geographical indications under the Provisions on the Protection of Geographical Indication Products, and 301 enterprises were approved for the use of those special geographical indications products. By the end of 2019, 2385 products for geographical indications had been approved, and 8484 enterprises were approved for the use of special geographical indications (see Table 10.2) (CNIPA 2019a, b).

Table 10.2 The five GI products approved in 2019

1.	Tangyin Beiai
2.	Pupu Yao tea
3.	Miluo zongzi
4.	Xincheng glutinous corn
5.	Lianjiang kelp

Source: Announcement of CNIPA on the Implementation of the Protection of Geographical Indication Products (No. 331)

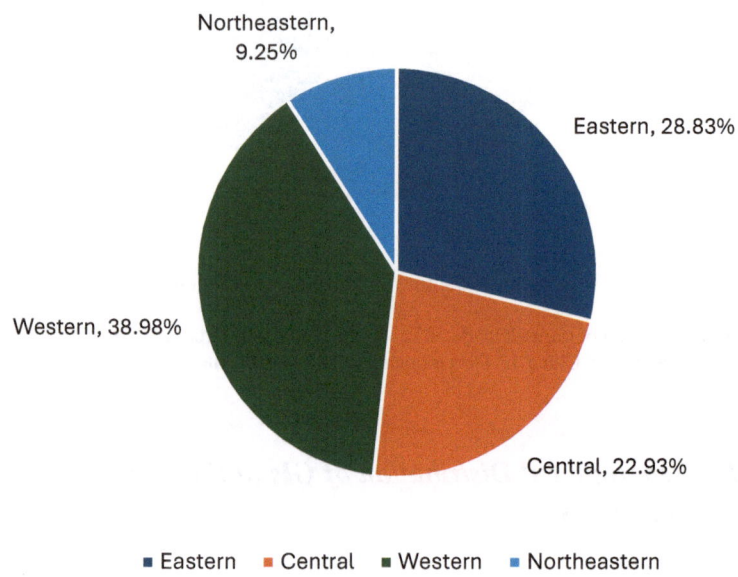

Fig. 10.2 Distribution of GI products in economic regions 2019. (Source: CNIPA 2019b. Evaluation Report on the Status of China IP Development in 2019. Intellectual Property Development and Research Center)

According to the distribution of geographical indications products in China's four major economic regions, the western region (38.98%) has the largest number of approved geographical indications products, accounting for nearly 40% of the national total, the eastern region (28.83%) and the central region (22.93%) were followed by the northeast region (9.25%). The top five provinces were Sichuan (295), Hubei (165), Guangdong (158), Guizhou (146) and Henan (116) (Fig. 10.2).

GI Agricultural Products

From the distribution of geographical indication agricultural products in China's four major economic regions, the western region (40.71%) has the largest number of registered geographical indication agricultural products, accounting for more than 40% of the national total, the eastern region (25.85%) was closer to the central region (24.15%), while the northeastern region (9.29%) was less numerous. The top five provinces were Shandong (330), Sichuan (176), Shanxi (155), Hubei (154) and Heilongjiang (140) (see Fig. 10.3).

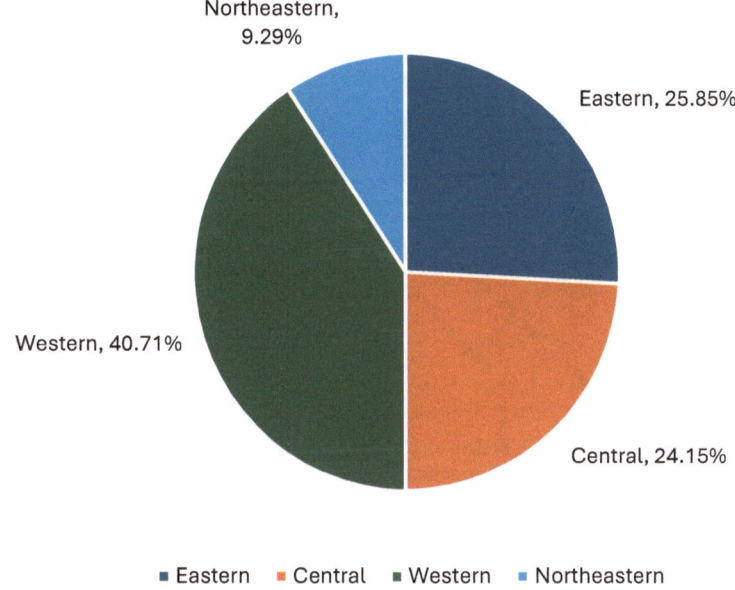

Fig. 10.3 The Distribution of GI agricultural products in economic regions. (Source: CNIPA 2019b. Evaluation Report on the Status of China IP Development in 2019.Intellectual Property Development and Research Center)

10.2.2   Main Types of GI Products in China

The main types of GIs are fruits and vegetables, tea, medicinal materials, ceramics and wine.

GI Trademarks
From the registration of various types of geographical indication trademarks, they are mainly concentrated in fruits and vegetables (1877 pieces), aquatic products, domestic livestock and poultry (1267 pieces), etc. (see Fig. 10.4).

GI Products
From the registration of various types of GI products, they are mainly concentrated in fruits and vegetables (672 pieces), other types (443 pieces) and tea medicinal materials (416), these three types of geographical indication products account for more than 60% of the geographical indication products under quasi protection (see Fig. 10.5).

GI Agricultural Products
From the registration of various types of geographical indications agricultural products, the type with the highest registration volume is fruits and vegetables (1224), accounting for nearly half of China's GI agricultural products, followed by aquatic livestock and poultry (702) (Fig. 10.6).

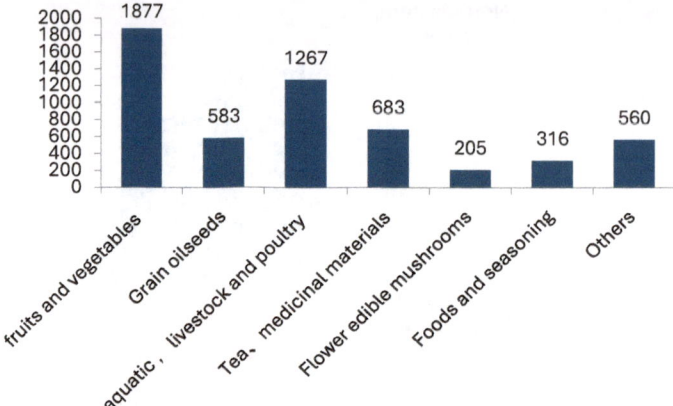

Fig. 10.4 Various types of GI trademarks. (Data sources: Zhinong361 database of the China Center for Intellectual Property in Agriculture (CCIPA))

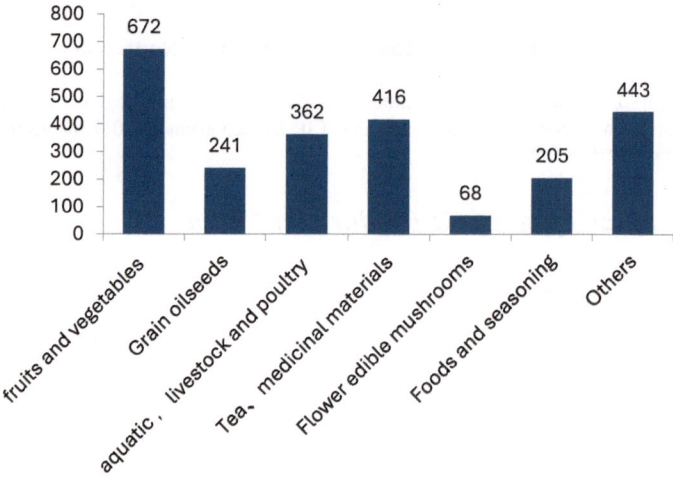

Fig. 10.5 Various types of GI products. (Data sources: Zhinong361 database of the China Center for Intellectual Property in Agriculture (CCIPA))

10.2.3 *Annual Cumulative Registrations in the Three Categories*

The annual cumulative registrations of the three GI categories show different trends. Among the three GI systems, the total amount of geographical indication trademarks is the largest, and the growth rate of geographical indication products is relatively stable, while the growth of geographical indication agricultural products is mainly concentrated in the decade from 2010 to now.

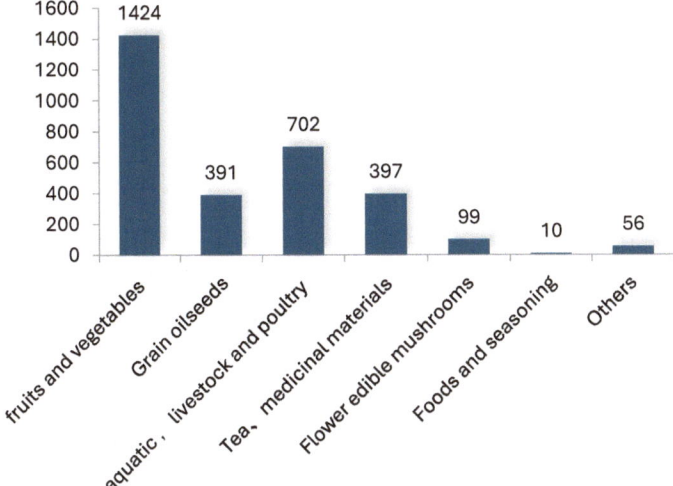

Fig. 10.6 Various types of GI agricultural products. (Data sources: Zhinong361 database of the China Center for Intellectual Property in Agriculture (CCIPA))

GI Trademark Increase Rapidly

In the 8 years after 2012, the registered volume of geographical indication trademark has increased significantly, and by the end of 2019, it has increased to nearly twice the cumulative registered volume of geographical indication agricultural products and geographical indication products (see Fig. 10.7).

Protection Under Multiple Categories

Although China adopts three different categories in its protection system to ensure the development and operation of rich GI resources, the three protection systems are not completely complementary. Therefore, the applicant will have to choose the protection mode in combination with the product characteristics and register it at the corresponding competent management Authority. Due to the diversity of products and markets, many GI products are managed and protected by two or even three systems at the same time. According to statistics, 87.36% of geographical indication products in China have been registered in only one Authority, 11.30% of geographical indication products have been registered at two Authorities, and 1.34% of geographical indications are registered at three Authorities at the same time (see Fig. 10.8).

10.2.4   Foreign GI in China and Chinese GI in EU

At present, the GIs applied by foreign countries in China, including the GIs from the EU are mainly GI related to products based on wine (89.27%), followed by food seasonings based on milk and cheese (8.29%). Three types of products such as

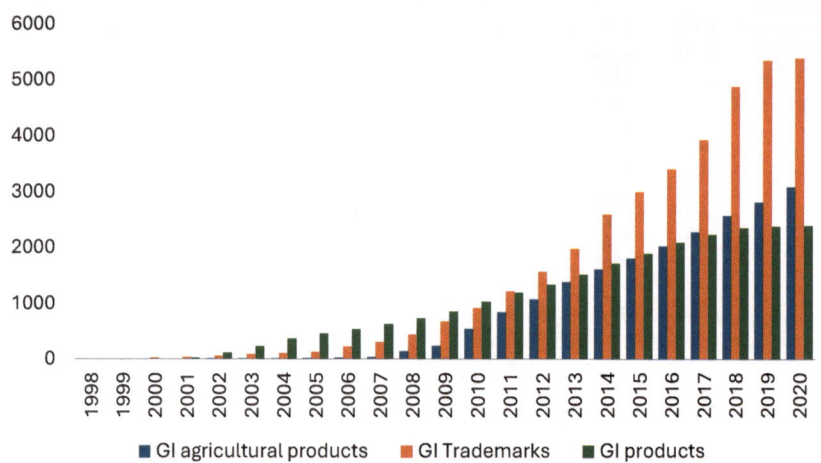

Fig. 10.7 Registered volume of GI. (Source: authors own elaboration, based on data published by CNIPA and the Ministry of Agriculture)

Fig. 10.8 GIs registered in three authorities at the same time. (Data sources: Zhinong361 database of the China Center for Intellectual Property in Agriculture (CCIPA))

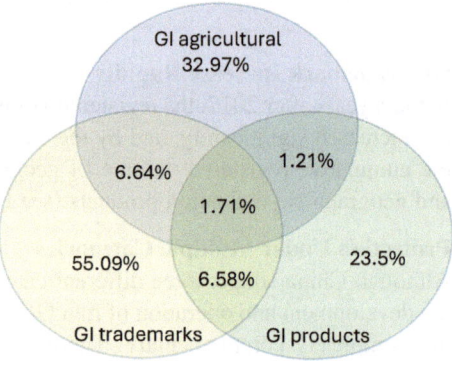

aquatic animal husbandry, tea and medicinal materials and flower edible fungi have not yet entered China. According to the "100 + 175" protection list of the China-EU agreement on geographical indications, China's protected geographical indication products in the EU are widely distributed, with fruits and vegetables (30.63%), tea and medicinal materials (27.93%) are the most, followed by other landmarks dominated by ceramics and wine (12.61%), while aquatic livestock and poultry (8.11%), flower edible fungi (8.11%), grain and oil (6.31%) and food seasoning (6.31%) are relatively evenly distributed.

10.3   The Study on Economic Impact of GIs in China

Geographical indications' comprehensive values such as economic value, social value and cultural value have also been recognized by people. The protection and development of geographical indications has become a new driving force for regional economic development. Geographical indication has its unique quality, which is formed by the joint action of local natural factors and human factors. In regional economic development, we should make full use of the factor endowment of geographical indications, realizing the role of geographical indication resources in promoting regional economic development.

According to statistics, the sales of geographical indication products in 28 EU member States reached 74.8 billion euros in 2017, including 39 billion euros for alcohol, accounting for 51%, and 27 billion euros for agricultural products, accounting for 35%; The import and export volume of geographical indication products reached 32.1 billion euros, accounting for 42% of the total import and export volume in 2017 (EUROPEAN COMMISSION 2019)

With regard to the statistics of economic value of geographical indications, the EU has accumulated many years of empirical research experience and has also established an increasingly perfect economic statistical database and empirical research methods of geographical indications, such as study on volume and value of sales, imports and exports, price and premium of geographical indication products.[2]

10.3.1   Methodology of the Study

In order to count the economic value of geographical indication products in China, a sample of geographical indications is taken to investigate the sales price and sales volume of geographical indications protection products in China.

Three criteria have been used for sample selection:

- First, selection of geographical indications with strong recognizability and high market awareness from the list of geographical indications included in the China EU agreement on geographical indications signed in September 2020;
- Second, select the same type of geographical indications shared by the eastern and western regions to reflect the premium and market value of similar geographical indication products in different regions;

---

[2] $Value\ premium = \sum(GI\ volume \times GI\ price) - \sum(GI\ volume \times non\ GI\ price)$

$Value\ premium\ rate = \sum(GI\ volume \times GI\ price) / \sum(GI\ volume \times non\ GI\ price)$

- Third, the selected samples include both primary products with low economic value-added and processed products with high economic value-added (see Table 10.3).

According to the above standards, 15 geographical indication samples were selected for in-depth survey. From the perspective of regional distribution, there are 10 in the eastern region, 3 in the western region, and one in the northeast and central regions. They belong to vegetable and fruit products, tea and food seasoning respectively.

The selling price information of these 15 geographical indication products comes from Taobao and Jingdong e-commerce platforms. The price is the average unit price of the GI commodity.

The criteria for judging whether the product searched by the e-commerce platform is a geographical indication product are:

- the special logo of geographical indication product (see Fig. 10.9) is printed on the product package;
- the special logo of geographical indication product is printed on the product information introduction page;
- and the text mark of the trademark is printed on the product package with the registered collective GI certification trademark.

In addition to the above, the similar without GI is searched and compared to the GI one to analyze the difference of premium effect.

10.3.2   Results

The analysis of the economic values of GI products in China, calculated in the sample (see Table 10.3), shows the following:

The premium of GI products compared to the non-geographical indication protection products is significant, especially *Zhenjiang vinegar* has the highest price premium rate among the samples, reaching up to 450.68%, the premium rate of *Shanxi aged vinegar* is also as high as 288.24%,but *Jinxiang garlic* did not show a high premium;[3]

There are significant differences in the premium effect of GI products of different categories. Among the GI products with a premium rate of more than twice, the majority are processed products such as vinegar and tea, especially *Zhenjiang fragrant vinegar* and *Shanxi aged vinegar*, which have outstanding premium effects. The premium rates of primary goods such as *Jinxiang garlic*, *Yantai apples*, *Dongning black fungus* and *Cangshan garlic* is relatively low. It can be seen that GI has greater added value effect for processed products.

---

[3] Price Premium rate=$\frac{GIprice - NonGIprice}{NonGIprice} * 100\%$

Table 10.3 Characteristics of GI products sample and results on premium

Name	Place of origin (Province)	Type	Whether to register GI agricultural products	Whether to register GI Trade Mark	Whether to register GI products	Whether to enter the EU protection list	The price premium rate of GI products
Wuyi Rock Tea	Fu Jian	TEA	Y	N	Y	Y	163.64%
Wuyishan Dahongpao	Fu Jian	TEA	N	Y	N	Y	95.24%
Kangxian black fungus	Gan Su	Vegetable and fruit	N	Y	Y	N	108.33%
Minle Purple Garlic	Gan Su	Vegetable and fruit	N	Y	Y	N	43.34%
Qiongzhong green orange	Hai Nan	Vegetable and fruit	Y	Y	Y	N	113.18%
Chengmai Qiaotou sweet potato	Hai Nan	Vegetable and fruit	N	Y	N	Y	53.85%
Dongning black fungus	Hei Longjiang	Vegetable and fruit	Y	Y	Y	Y	14.29%
Pizhou garlic	Jiang Su	Vegetable and fruit	N	Y	N	Y	–
Zhenjiang vinegar	Jiang Su	Food seasoning	N	Y	Y	Y	450.68%
Jinxiang Garlic	Shan Dong	Vegetable and fruit	Y	Y	Y	Y	-9.65%
Cangshan Garlic	Shan Dong	Vegetable and fruit	Y	Y	Y	Y	33.20%
Yantai Apple	Shan Dong	Vegetable and fruit	Y	Y	Y	Y	3.38%
Longkou Vermicelli	Shan Dong	Food seasoning	N	N	Y	Y	42.86%
Shanxi aged vinegar	Shan Xi	Food seasoning	N	Y	Y	Y	288.24%
Anji white tea	Zhe Jiang	TEA	Y	Y	Y	Y	85.00%

Fig. 10.9 The special logo
of GI product

The impact of regional differences on the premium effect of GI products is relatively weak. By comparing the premiums of similar products in different economic regions, it is found that the correlation between the premium of geographical indications and regional distribution is not high. Such as *Zhenjiang fragrant vinegar* (450.68%) and *Shanxi aged vinegar* (288.24%), *Wuyishan Dahongpao* (95.24%) and *Anji white tea* (85.00%), *Cangshan Garlic* (33.20%) and *Minle purple skin garlic* (43.34%), etc. Premium rates for similar GI products in different regions are equivalent.

The use of the special logo of GI products affects the premium of GI products on e-commerce platforms. Such as *Yantai Apple* has almost no stores that have GIs' special logo on their product packaging on the two major e-commerce platforms. However, due to the registration of the GI collective certification trademark, most of the products in the store are labeled with the "*Yantai Apple*" trademark text on the packaging. But the prices of GI *Yantai Apple* and non GI products are basically the same, the premium rate is only 3.38%. The use of GI special logo can increase the recognition of GI products, the effect of the combination of the special logo of GI product and the Trade mark is remarkable;

The quality level determines the type of market, and lack of quality supervision leads to the withdrawal of geographical indication products from the market. Such as the *Hainan Lingshui cherry tomatoes*, GI product, according to field research there were no cherry tomatoes grown in Lingshui, just because the quality of cherry tomatoes has declined without quality control, and market demand has declined too.

More and more enterprises are willing to join in the production of GI products according to CNIPA statistics of enterprises approved to use the special logo of GI (see Table 10.4).

Table 10.4 Number of enterprises approved to use the special logo of GI

Year (1–12 month)	Numbers of enterprises
2019	301
2020	1052
2021	7677
2022	6373
2023 (1–11 month)	5662

Source: CNIPA statistics

10.4   Conclusion

In conclusion, building on the statistics and analysis of the economic study of GI in China, the following recommendations can be formulated:

1. **Increase the standardized use of geographical indications rules to improve the premium effect of geographical indications on commodities.**

The GI is conducive to improving the identification and recognition of quality products, and is conducive to improving the premium effect on commodities in the Internet economy. It is necessary that GI identification system is promoted in an orderly manner, so as to support a comparable quality level in the market channels and for promotion, unified standards for GI request examination, and unified supervision for protection, through a harmonized coordination mechanism between the three GI categories.

2. **Coordinate the development of GIs and brands, and guide enterprises to use GIs to enhance their goodwill.**

It is important to take full advantage of the premium function of geographical indications, to guide enterprises, especially unknown enterprises, to use GI to enhance product premium capacity and market influence, while ensuring quality production, strive to cultivate their own brands, and obtain brand premiums again through differentiated pricing, so that enterprises can obtain long-term economic benefits and ensure the coordinated use and development of geographical indications and brands.

3. **Play the positive role of geographical indications in regional economic development.**

Geographical indications can contribute to promoting the development of a region. Regions with GI resources should incorporate GI strategies into the comprehensive strategy of regional economic development and adopt "package" development strategies, such as GI protection, quality standards, government support to associations, industrial chain agglomeration, etc.

4. **Strengthen the protection of geographical indications, supervised by the government.**

The government should strengthen publicity and improve relevant legislation; the competent functional department should establish unified quality standards and supervise the quality of the entire industrial chain of geographical indications; finally, efforts should be made to develop intermediary organizations and industry associations, fully leveraging their leadership, quality, technical control, sales promotion, and bargaining power in the production of geographical indications.

References

AND-International (2019) Study on economic value of EU quality schemes, geographical indications (GIs) and traditional specialties guaranteed (TSGs). October 2019. EUROPEAN COMMISSION

CNIPA (2019a) Annual report

CNIPA (2019b) Evaluation report on the status of China IP development in 2019. Edited by Intellectual Property Development and Research Center, CNIPA

Chapter 11
Geographical Indication Regulations and Practices in Türkiye

Özden İlhan, Neşe Altıntaş, Nazlı Şimşek, and Sertaç Dokuzlu

Abbreviations

BEBKA	Bursa, Bilecik Eskişehir Kalkınma Ajansı (Bursa, Bilecik, Eskişehir Development Agency)
EBRD	European Bank for Reconstruction and Development
ePATS	Turkish Patent and Trademark Office Electronic Application System
EU	European Union
FAO	Food and Agriculture Organization of the United Nations
GI	Geographical Indication
GTB	Gemlik Ticaret Borsası (Gemlik Commodity Exchange)
HORECA	Hotel, Restaurant and Catering
IP	Intellectual Property
ISO	International Organization for Standardization
MAF	Ministry of Agriculture and Forestry
NGOs	Non-Governmental Organizations
PDO	Protected Designation of Origin
PGI	Protected Geographical Indication
PO	Producer Organization
QR	Quick Response Code

Ö. İlhan
Turkish Patent and Trademark Office, Ankara, Türkiye
e-mail: ozden.ilhan@turkpatent.gov.tr

N. Altıntaş · N. Şimşek
Ministry of Agriculture and Forestry of Turkey, Ankara, Türkiye
e-mail: nese.altintas@tarimorman.gov.tr

S. Dokuzlu (✉)
Bursa Uludag University, Bursa, Türkiye
e-mail: sdokuzlu@uludag.edu.tr

E. Vandecandelaere et al. (eds.), *Worldwide Perspectives on Geographical Indications*, https://doi.org/10.1007/978-3-031-71641-6_11

TAIEX   Technical Assistance and Information Exchange
TOBB   The Union of Chambers and Commodity Exchanges of Türkiye
TRIPS   Trade-Related Aspects of Intellectual Property Rights Agreement
Türkpatent   Turkish Patent and Trademark Office
Yörex   Local Products Fair

11.1 Introduction

Products of protected origin often possess characteristics which enable them to be distinguished in the market and compete effectively with comparable products (Belletti et al. 2017). In recent years, due to the negative effects of the Covid-19 pandemic, ongoing conflicts and the challenges posed by climate change, the issues of competition, rural development and sustainability have become strategically important considerations for many countries. The concept of geographical indications (GIs) emerges as an intersection of these strategic elements. GIs, if implemented effectively, have potential to increase the competitiveness of products in the markets and can contribute to sustainability and rural development. Consequently, several countries, recognizing the importance and potential benefits of GIs have started their efforts in 90s and, while others are embarking on similar endeavours more recently. Different legal frameworks regarding geographical indications have been put into force in many countries worldwide. Türkiye is one of these countries and published its first legal regulation on GIs in 1995.

Türkiye is located on an area where Europe meets Asia, creating a gateway between these two continents. It is surrounded by seas on three sides and has many different climate zones and rich soils that agriculture one of the main contributors to the economy of the country. Türkiye is the world's leading producer of hardshell nuts, figs and apricot and among the leading manufacturers of cotton and textiles (FAO 2021). Fertile lands, long history and diverse culture of the Anatolian peninsula, which has hosted many great civilizations, contributes to the abundance of traditional and local products.

This chapter examines the current GI protection mechanisms in Türkiye and highlights the roles of relevant institutions. It also underlines the importance of collective action in realizing the full potential of geographical indications and uses case studies to illustrate this point. The aim is to provide a detailed understanding of the GI protection system in Türkiye and its current outlook. The chapter concludes with strategic recommendations for current challenges and the future.

11.2 History of Geographical Indication Protection in Türkiye

First attempts of Türkiye to protect the origin of goods began during the Ottoman period with *"Trademark Ordinance"* dated 1871, which was based on the French *Registered Trademarks Law* of 1857. The law of 1871, however, could not be adapted to Turkish economic life and was repealed in 1965 with the *Law No. 551* (Official Gazette 1920) *"Trademark Law"* (Candan 2017). In 1930, the *Law No.1705* "The Law on the Prevention of Counterfeiting in Trade and the Surveillance and Protection of Exports" was adopted and the Ministry of Industry and Trade was given the authority to prevent the sale of any goods which may be misleading to consumers (Official Gazette 1930). Until 1995, only indirect protection was provided for geographical indications. During this period, unfair competition provisions of the *Turkish Commercial Code* and the *Law No. 551* provided protection for geographical indications.

By 1995, the protection of intellectual property rights in Türkiye started to develop rapidly with Türkiye's attempts to join the European Union (EU) and the subsequent harmonization process of national legislative framework.[1] On 31 December 1995 a set of statutes to adapt EU Directives to national legislation entered into force, which included a decision to update current provisions for intellectual property rights to achieve the same level of protection available in EU. Within this framework, a new decree law was prepared in 1995 and geographical indications were given direct protection. *Decree Law No. 555 on Protection of Geographical Indications* came into force on 27 June 1995 (Official Gazette 1995). During the time, geographical indications of agriculture and food products in EU were protected by *European Council Regulation No. 2081/1992* and *Decree Law No. 555* was based on this regulation.[2] Within the scope of this legislation a *sui generis* type of protection was given to geographical indications and former Turkish Patent Institute, became the competent authority for all registration processes.

For the purposes of this Decree-Law, GI was defined as the indicators of a product that possesses a specific quality, reputation, or other characteristics attributable to the area, region, or country where it originates. Similar to the European legislation *EC 2081/92*, geographical indications were divided into protected designation of origin (PDO) *"menşe adı"* and protected geographical indication (PGI) *"mahreç işareti"*. However, unlike EU legislation, protection covered a broader range of product categories and was provided to natural, agricultural, mining, industrial products, and handicrafts. It was also not mandatory to be a producer group to file GI applications and all producers of the product, consumer associations and public institutions concerned with the product, or the geographical region were given the

---

[1] See https://neighbourhood-enlargement.ec.europa.eu/enlargement-policy/turkiye_en

[2] Council Regulation (EEC) No 2081/92 on the protection of geographical indications and designations of origin for agricultural products and foodstuffs available at https://eur-lex.europa.eu/legal-content/EN/TXT/?uri=celex%3A31992R2081

right to apply. Applications were examined by the Trademark Department in the former Turkish Patent Institute. A commonly criticized provision of *Decree Law No. 555* was the publication of applications in the Official Gazette, in two national newspapers with the widest circulation and in one local newspaper before registration. The high cost of publications led to many applications being delayed and abandoned before registration. Even though the scope of protection and relation to trademarks was very similar to EC 2081/92 and in line with the TRIPS agreement, control modalities were not defined in detail and only the reports of the inspection to be carried out by the registrants were requested to be submitted to the Institute every 10 years. Furthermore, in 2009, several clauses of the law which defined the criminal and administrative penalties for the acts of infringement were cancelled by the supreme court on the grounds that penalties shall be prescribed only by law (Official Gazette 2009).

During this time more emphasis was started to be given to GIs in Türkiye and GIs were included in the *"National Rural Development Strategy"* published in 2006. Subsequently on 16 April 2008 intellectual property law chapter for Türkiye's EU accession negotiations was opened. To develop strategies in the intellectual property area and to ensure coordination and cooperation among related institutions, "Intellectual and Industrial Property Rights Coordination Board" was established on 21 May 2008. During the first meeting of this council in 2009 it was decided to commence studies to set out a national policy for geographical indications. On 2015 Turkish Ministry of Development's Higher Planning Council adopted the *"2015–2018 Action and Strategy Plan for Geographical indications"*.[3] A new intellectual property (IP) law was proposed to the national assembly in 2016 and the new *Industrial Property Law No. 6769* has come into effect on 10 January 2017, repealing the *Decree Law No. 555* (Official Gazette 2017).

11.3   Legal Framework and Role of Public Authorities in Türkiye

11.3.1   Legislation and Role of Türkpatent

Industrial Property Law No. 6769 (IP Law) dated 2016 and its implementing regulation is the current legal instrument which establishes the protection of geographical indications in Türkiye. Turkish Patent and Trademark Office (Türkpatent) remained the main competent authority for all registration processes including receiving, examining and registering applications. Post registrations processes including validation of yearly control reports, amendments and cancellations are also carried out by Türkpatent.

---

[3] See https://webim.turkpatent.gov.tr/file/486351e7-a30e-4d51-9d76-d3f470216cc0?download

Protection of Foreign GIs in Türkiye
For a foreign GI to be protected in Türkiye the geographical indication or traditional product name subject to the application must be protected by the country of origin, control requirements specified in Article 49 shall be met in the country of origin and country of origin shall provide equal rights of protection for Turkish GI application. In addition to these provisions, IP Law also states that the persons whose domiciles are situated abroad shall only be represented by trademark attorneys for matters relating to geographical indications. Therefore, GI applications from foreign countries must be filed through trademark attorneys which are designated by Turkpatent.
Source: (Mevzuat 2016).

Under IP Law; food, agricultural, mining and industrial products, and handicrafts can be registered as GIs. Like repealed Decree Law No. 555, geographical indications are divided into PDO "menşe adı" and PGI "mahreç işareti". IP Law also introduces protection for "geleneksel ürün adı /traditional product name" which are not considered as GIs but serve to protect names that have been used at least 30 years on the related market and originate from traditional production or processing technique, traditional composition or produced from traditional raw materials or ingredients.

Producer groups, public institutions related to the product or geographical area of the product, associations, foundations, and cooperatives operating for public interest which are related to the product and a single natural or legal person (proven that they are the only producer of the product) can file applications. The producer groups mentioned here is defined as associations, irrespective of its legal formation or composition, composed of producers of the same products.

All applications for registration are submitted directly to Türkpatent through the Turkish Patent and Trademark Office electronic application system (ePATS).[4] In addition to applications, all transactions related to GIs, like notification of deficiencies and objections, submission of audit reports, etc. are also carried out online through the ePATS system.

The application is examined by experts in Türkpatent Department of Geographical Indications. After initial examination, comments may be requested by Türkpatent from relevant institutions and organizations[5] to evaluate the technical information. Experts of the Department of Geographical Indications in Türkpatent work with the applicants to complete the deficiencies in the geographical indication files and to prepare the applications for publication. Applications examined and

---

[4] https://epats.turkpatent.gov.tr/run/TP/EDEVLET/giris
[5] For food and agricultural products technical opinions are requested from Ministry of Agriculture and Forestry while for mining products technical opinions are requested from General Directorate of Mineral Research and Exploration, for handicrafts technical opinions are requested from Ministry of Culture and Tourism.

found acceptable are published in the Bulletin.[6] Third parties may file an opposition within 3 months of the publication of the application. If no opposition is filed or all oppositions are rejected/resolved by Türkpatent, the application is registered, provided that the registration fee has been paid by the applicant.

Protection is granted to GIs and traditional product names only after registration. All users of GIs are obliged to comply with the specifications provided in the register. There is no requirement for a renewal of the registration and once registered they are protected indefinitely.

Official Logos
GIs must be used in combination with the official logos which were presented in the regulation which came into force on 10 January 2018. The logo shall either be found on the product itself, its packaging or be displayed in a visible manner at businesses.

Mense Adı   Mahreç İşareti   Geleneksel Ürün Adı

Source: Regulation on Geographical Indication and Traditional Product Name Logos (Mevzuat 2017).

IP Law also corrects an important point which was missing in the *Decree-Law No. 555* pertaining to amendment to specifications. Article 42 of IP Law of allows parties with legitimate interest to request amendments for a registered geographical indication or traditional product name.

11.3.2   The Control System of Geographical Indications

The current control system of GIs in Türkiye can be separated into three parts. The first is the controls under the 'sui generis' system of IP Law, which gives the main responsibility of controls to the registrants of GI. The registrant is the entity who has filed the application for the registered GI. Controls under IP Law cover all activities related to the compliance of the GI to the product specifications during the

---

[6]Bulletin "Bülten" is the relevant publication in which the matters specified in IP Law are published, regardless of the type of publication medium. Since 2017 the Bulletin has been published on the Türkpatent website twice a month for geographical indications related publications.

production, marketing or distribution stages, or the use or misuse of the GI or the official logos in the market.

The second is the ex-officio controls carried out by Ministry of Agriculture and Forestry (MAF) for food and agricultural products which are explained further under topic the role of Ministry of Agriculture and Forestry.

The third is the mainly complaint-based market controls carried out by Ministry of Trade. The Advertising Board under Directorate General for Consumer Protection and Market Surveillance is mandated to safeguard consumers against unfair commercial practices. The Advertising Board can receive complaints regarding misleading commercial advertisements related to GIs from consumers, public institutions, or competing businesses. Following investigations, the Board may halt non-compliant advertisements, mandate corrections, and impose monetary penalties or provisional suspensions of up to 3 months.

According to the IP Law, all GI applications must include provisions for control after registration including a "control committee" who will be carrying out all controls specified under IP Law after the registration. There is no restriction to composition of the control committees; however, they are required to be objective, have competent and sufficient number of staff and possess necessary resources and equipment for the efficient control of the product. The competence of the control committee is evaluated by Türkpatent during the application procedures. After registration, any amendment to the control committees is also subject to the approval of Türkpatent. The control committees are often made up of three or four representatives from public institutions which are present in the geographical area such as producer groups, chambers of commerce, universities, or agricultural directorates. Currently, it is not required for the control committees to be accredited by ISO 17065 but it is possible to identify private companies which specializes in the certification of products as the control committee. A model "control record" is also submitted during application which includes the specific control points defined in the GI specification.

Within the scope of IP Law, the registrant is required to keep a list of the producers of the registered geographical indication. The users of the registered GI are also required to notify the registrant of their involvement in the production and marketing activities. The list of GI users is then used to create a control plan by the control committee. Controls are carried out according to this plan by the control committee and a control record is maintained for each control activity. Control records are compiled to form a control report which must be submitted to Türkpatent at least once every year after registration. If it is determined by Türkpatent that controls are not duly performed by the registrant, a new registrant is determined by publishing an amendment request in the Official Bulletin. If no requests are made by suitable applicants to become the registrant of the GI in question, the GI is cancelled.

Infringement of GI Rights

The registrant or the rightful users of the GI may file an action before the court (Courts of Intellectual and Industrial Property Rights) to prevent infringing acts referred to in Article 53(1) of the IP Law. These acts include:

(a) any direct or indirect commercial use of geographical indications or official logos on similar or comparable products which do not carry the properties specified in the register, to benefit from the product's reputation.

(b) any deceptive use of the place of origin or its translation or expressions such as 'style', 'type', 'method', 'as produced in' or other similar descriptions on similar or comparable products which do not carry the properties specified in the register

(e) any use of false or misleading information regarding the origin of a product on its inner or outer packaging, advertising material or documents relating to the product.

(d) any misleading usage of the official logos.

The claims that can be brought before the court for infringement of GI rights by the registrant or the rightful users of the GI in case of infringement are defined in Article 149 of the IP Law. The claims that can be summarised as follows:

(a) Determination of whether the act constitutes an infringement.
(b) Prevention of potential infringement.
(c) Cessation of infringement actions.
(d) Removal of the infringement and compensation for material damages.
(e) Confiscation of products that constitute the infringement or warrant penalty, as well as tools such as devices or machines exclusively used in their production, without hindering the production of other products.
(f) Recognition of ownership rights over the products, devices, and machines confiscated under clause (d).
(g) Measures to prevent the continuation of the infringement, especially the alteration of shapes, deletion of trademarks on the confiscated products, devices, and machinery under clause (d), or their destruction if inevitable to prevent infringement of industrial property rights.
(h) If there is a valid reason or benefit, the final decision's publication in daily newspapers or similar means, in full or in summary, or its notification to the interested parties.

Source: (Mevzuat 2016).

Foreign registrants are also requested to file control reports every year. However, it is sufficient to only include the records of products that are imported to Türkiye and the evidence that these products are controlled in the origin country.

11.3.3 Role of the Ministry of Agriculture and Forestry

The Ministry of Agriculture and Forestry takes an active role in the examination, control and promotion of food and agricultural GIs. Department of Agricultural Marketing is primarily responsible for issues relating to GIs under MAF and works in close cooperation with Türkpatent. The main functions of the Ministry of Agriculture and Forestry regarding geographical indications are summarized below.

Technical Support to Applicants
MAF has a broad structure with provincial and district directorates in all provinces throughout Türkiye that provide support to GI producers. Applicants who want to prepare a geographical indication file for a food or agricultural product in any province or district can receive technical support from MAF's provincial and district organizations during the preparation of the specifications of their products. For example, MAF made significant contributions to the preparation of the application for the Türk Çam Balı (Turkish Pine Honey) geographical indication application with the project titled "Supporting the Sustainable Honey Value Chain in Türkiye through Geographical Indications" prepared and implemented by the Food and Agriculture Organization of the United Nations (FAO) and the European Bank for Reconstruction and Development (EBRD). During the project, agricultural engineers, food engineers and experts who are working at MAF provided technical support to the project team and shared the results of the project previously prepared on Turkish Pine Honey.

MAF also plays a pivotal role in establishing a sound legal framework to enhance the effective implementation of geographical indications of registered products. For example, the famous *Maraş Dondurması* (ice cream) is produced with a special production technique. This production method was called as "Maraş Style" in the Turkish Food Codex Ice Cream Communiqué. However, when *Maraş Dondurması* was registered as a geographical indication, the expression "Maraş style" was removed from the Turkish Food Codex by MAF. Now it can be used only when it is produced in accordance with geographical indication registration. In this context, MAF has facilitated the negotiations between Kahramanmaraş Chamber of Commerce and Industry (the GI registrant) and the ice cream sector representatives regarding the industrial production.

Technical Opinion on the Application Files of Agricultural and Food Products
The Ministry of Agriculture and Forestry also supports the examination process of the applications relating to food and agricultural GIs with Türkpatent by providing technical opinions and if necessary, opposition to the applications published in the Official Bulletin.

GI applications for food and agricultural products submitted to Türkpatent are first examined by Türkpatent experts and then forwarded to MAF for technical examination. MAF experts examine the applications technically, identify deficiencies or information that needs to be added, and report them to Türkpatent. Türkpatent forwards the examination result to the applicant in an official letter together with Türkpatent's opinions.

Control of Geographical Indications by MAF

Control of agricultural and food products geographical indications is among the duties of the MAF, in addition to the control duties of registrants. According to the Veterinary Services, Plant Health, Food and Feed Law No. 5996 dated 11/6/2010; "The Ministry inspects the compliance of the use of geographical indications or traditional product names related to agriculture and food with the characteristics specified in the registration". In this context, MAF's provincial and district organizations can carry out controls by themselves, and also accompany registrants who will carry out controls upon request.

Also for the establishment of an effective official control mechanism in geographical indications in Türkiye "Regulation on the Inspection of the Compliance of Geographical Indications and Traditional Product Names Related to Agriculture and Food with the Features Specified in the Registration" has been prepared by MAF. The said regulation is under evaluation at the moment and work continues on the draft.

Conducting Awareness Raising Studies

As an important market tool, GI could serve as a key to a strong and sustainable rural development in Türkiye. The importance of the GI as a legal tool is to protect traditional knowledge of the country. In this context, activities related to information on GI and promotion of registered GI products are carried out.

The target groups of awareness-raising activities are producers, producer organizations and consumers. During these activities, the importance of the participation of all actors in the value chain is specifically emphasized.

In this context, MAF has organised regional events. However, due to the pandemic for the 2020–2021 period, these meetings have been held by video conference method. These activities are planned to cover 81 provinces of Türkiye. The activities were attended by the administrators and technical personnel of the provincial and district directorates of MAF, GI applicants and registrants, relevant public institutions and organizations, municipalities, non-governmental organizations' (NGOs) representatives, and producer organizations. The main purpose of these activities is to contribute to the development of the geographical indications from the region to the point where it can develop, by emphasizing the geographical area and regional development.

In addition, trainings on GIs were provided within the scope of other rural development projects carried out by MAF.

In order to explain, promote and ensure the visibility of GIs, MAF promotes the use of media communication tools:

• GIs are included as a discipline in the Journal of Agriculture and Forestry; texts and images are published under the title of "There is News About Geographical Indications" since the beginning of 2021.

- MAF has supported the preparation of a short video for the promotion of Turkish GIs.
- MAF funded the development of short videos and posters for *Milas Zeytinyağı* (olive oil), *Bayramiç Beyazı* (nectarine), *Taşköprü Sarımsağı* (garlic) and *Giresun Tombul Fındığı* (hazelnut) which are registered in the EU. In the future other GIs of Türkiye that registered in the EU will benefit from similar short videos and posters.
- A book was written about "GIs of Türkiye". The design and printing work of the book is about to be completed.
- A public advertisement on GIs was prepared and broadcast.
- MAF conducted surveys to better understand consumer awareness and attitudes towards GIs and determine the value of GI products.

In collaboration with the European Union (EU), various initiatives have been undertaken, including Technical Assistance and Information Exchange (TAIEX) workshops and expert missions, which have incorporated field activities. Drawing upon the insights acquired through these activities, dedicated initiatives are being undertaken to identify and mitigate infringement of rights for national GI registrations.

11.3.4   Role of Other Public Institutions

GIs are included in the strategic plans of many public institutions in Türkiye. Cooperation of public institutions in different disciplines is required for the development of the GI system.

Presidency of the Republic of Türkiye; the Presidency's Directorate of Communications of Türkiye is actively engaged in branding of Turkish products abroad. In this context, promotional materials are prepared in English to tell the stories of GI products registered in the EU (SBB 2023).

Ministry of Tourism and Culture; Türkiye Culture Portal was established by the Ministry of Tourism and Culture in 2006 (T.C Kültür ve Turizm Bakanlığı 2023). Under this initiative, an inventory of origin-linked products of all 81 provinces of Türkiye was prepared and published online under Türkiye Culture Portal website. After the success of the Portal, governorships started to prepare more detailed information about local products, including registered GIs in the area and published on their websites.

The Ministry of Commerce; supports the market surveillance of geographical indications. In 2018, a cooperation protocol was signed between the Ministry of Commerce and Türkpatent to ensure coordination between two institutions in terms of geographical indications and traditional product names (Ticaret Bakanlığı 2018). The Ministry of Commerce has also enacted a regulation related to the sale of GI products. According to the Law on the Regulation of Retail Trade of Türkiye, large department stores and chain stores must allocate at least 1% of their shelf space to the sale of GI registered and/or local products. However, these GI or local products must be produced in the region where the store is located. At the same time, provincial trade directorate officials decide which GI/local products will be allocated for shelves (Mevzuat 2015).

Regional Development Agencies; that are operating under the Ministry of Industry and Technology also work on GIs and local products of their responsible region. Many Development Agencies prepared and published detailed books, brochures, or catalogues about registered or potential GIs of their regions. These Agencies work on a project basis and give priority to investments related to local products and also carry out technical support programs for the preparation of product specifications and GI files (Kalkınma Ajansları 2023).

Professional Organizations at Public Status; The Union of Chambers and Commodity Exchanges of Türkiye (TOBB) and its affiliates play a key role in organizing the traditional Local Products Fair (Yörex). This fair provides a platform for participants from various provinces in Türkiye to showcase their local products, including GIs. Yörex also hosts panels and seminars that facilitate the transfer of knowledge within the sector. These panels often feature industry experts, business leaders, and academicians who share insights and expertise on various aspects of protection, marketing, and distribution of GIs and other local products (Yörex 2023).

11.4 Geographical Indications in Türkiye with Figures

From 1996 to 2022, following the establishment of the *sui generis* system for GI protection in Türkiye in 1995, there was a marked escalation both in GI applications and registrations. Detailed below is a statistical overview that elucidates the trajectory of GI applications and registrations in Türkiye (Fig. 11.1).

By September 2023, there are 1466 registrations and 635 ongoing applications for GIs and traditional product names in Türkiye. Of these registrations, 14 have been registered, 5 are published, and 39 are ongoing applications under European Union

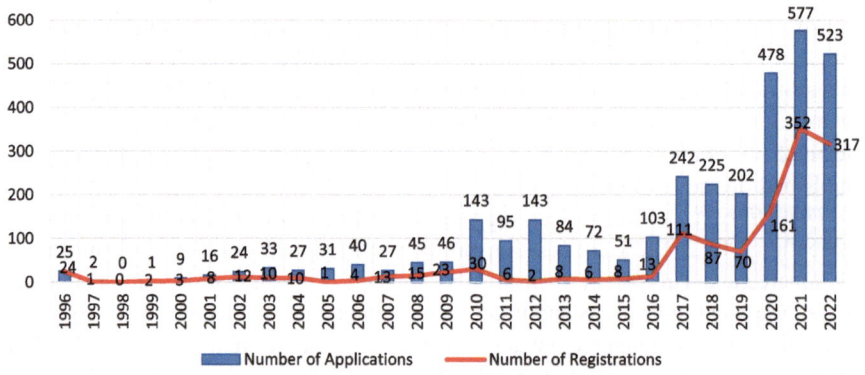

Fig. 11.1 Annual trends in gi applications and registrations in Türkiye (1996–2022). (*Source:* Compiled by the authors from GI registry website (Türkpatent 2023a, b))

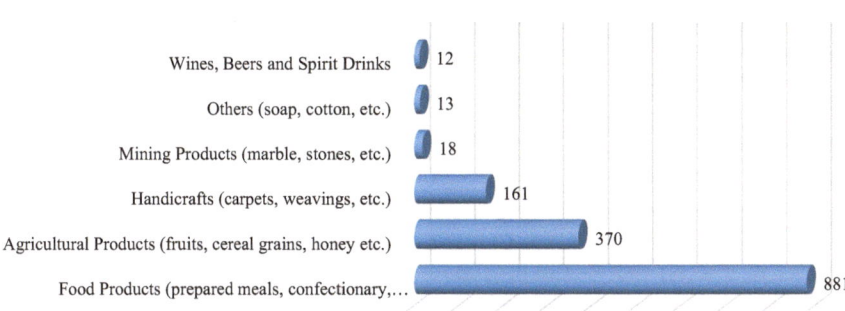

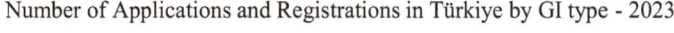

Fig. 11.2 Number of registrations according to product categories in Türkiye. (*Source:* Compiled by the authors from GI registry website (Türkpatent 2023a, b))

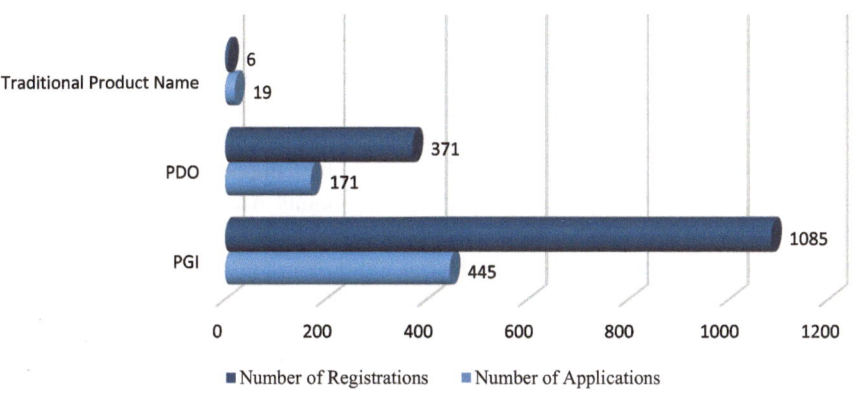

Fig. 11.3 Number of applications and registrations according to GI types. (*Source:* Compiled by the authors from GI registry website (Türkpatent, 2023a, b))

quality schemes (e-Ambrosia 2023a, b). Türkiye also protects 3 GIs under bilateral agreements with South Korea and Peru.

The number of GI registrations by product categories and for PDO, PGI and traditional products names are shown in Figs. 11.2 and 11.3 respectively.

The distribution of GI registrations by registrants is presented in Table 11.1. Municipalities and chambers of commerce/industry constitute the biggest portion of the GI registrants, while producer groups (associations, cooperatives or unions) represent less than 11% of all registrants.

Chambers (commerce/industry) that rank second in terms of number of registrations are public professional organizations with legal entity status established with an aim to meeting the common needs of their members and facilitating their

Table 11.1 Distribution of geographical indication (gi) registrants in Türkiye (2023)

Types of registrants	Number of registrants	Ratio (%)
Municipalities	532	36.29
Chambers of Commerce/Industry	432	29.47
Governorships	161	10.98
Producer Groups	152	10.37
Commodity Exchanges	134	9.14
Others (Development Agencies, Universities, Ministries)	55	3.75

Source: Compiled by the authors from GI registry website (Türkpatent 2023a, b)

professional activities (TOBB 2024). The members of the chambers are companies and traders.

11.5   Current Challenges of GI System in Türkiye

Successful implementation of the GI system requires awareness and participation of all stakeholders in the value chain, from producers to retailers. Despite the steady rise in GI applications and registrations in Türkiye, several challenges can be observed. One of the major problem is directly related to the distribution of geographical indications registrants. Producer groups, despite their importance, still represent a smaller fraction of the registrants (10.37%) which points to a lack of ownership from the producer's side and the centralized approach to GIs. While the state can provide much-needed resources, it can introduce a top-down approach that may underlie political, economic, and cultural interests which might affect objectivity. One prominent issue that arises when public entities are acting as primary applicants, is the possible exclusion of producers during the application phase. Some public institutions, especially municipalities and chambers, apply for geographical indication in order to gain the appreciation of local people and/or members with a populist approach. In this case, their focus is usually just on getting GI registration and implementation is disregarded. The primary duties of most of these institutions (especially municipalities) is also not directly related to GIs. This situation often leads to a weak link between the registrant and the producers, especially during the implementation of the GI. In such cases, the involvement of GI product producers in the process is either very low or non-existent, as a result, many producers remain unaware of the registration. The current top-down approach also creates challenges in the case of proactive litigation against non-conformities A similar negative situation can also occur when any producer organization applies for GI. There may be a risk of "exclusion" of some producers, especially when there is more than one producer organization working on the same product in the region or when some producers are not members of the applicant producer organization. However, despite all these risks, when producer organizations apply for GI, the number of producers who are aware about GI application process or GI registration is higher when it compares the public applicants.

Geographical Indications have the potential to benefit rural producers; however, their effectiveness is contingent upon the governance mechanisms implemented by local actors (Nizam and Tatari 2022). According to local GI stakeholders, one of the main challenges today is the difficulty in initiating collective action. Many aspects contribute to the reluctance of producers to engage in a GI system: the production scale differences between local producers, political pressures and concerns about losing favour of different fractions of NGOs, and the different costs to be considered (control, promotion, marketing activities).

Another challenge of the current system is the concerning trajectory of applications which are filed for products that lack specific characteristics. This trajectory potentially undermines the value of unique GI products, thereby attenuating the credibility of the GI framework. It's disconcerting to observe a significant portion of registered products either conspicuously missing from the market or not promoted as GIs but only as commodities (Denk and Bilici Sanalan 2023).

The incomplete or inadequately prepared product specifications of previously registered GIs has also become a challenge in recent years. These are especially prominent for GIs registered before 2010. The specifications of these GIs do not fully reflect the product features and cause difficulties during control processes. Since the product criteria specified in incorrect or incomplete registration contents do not reflect the actual product specifications, thus, it is not possible to maintain product quality with incorrect/missing criteria during the control phase. Similarly, some GI registrations may have product criteria that are difficult, expensive or time-consuming to analyse. Long-time-consuming analyses make control difficult, especially for products with short shelf lives. Analyses performed with special equipment in specific laboratories are generally expensive, and some registrants or rights holders have difficulty in covering these costs. All these factors damage the control effectiveness. Contrary to the repealed decree law, since Law No. 6769 allows the revision of registrations, deficiencies in some registrations have been corrected and updates have been made. However, many registrations still need to be revised.

Consumer awareness on GIs is another challenge. In Türkiye, consumers' interest in local products is quite high and many consumers are ready to pay more for these products between 5% and 30% (Meral and Şahin 2013; Kadanalı and Dağdemir 2016; Toklu 2016; Çakaloğlu and Çağatay 2017; Dokuzlu et al. 2019a, b; Çukur et al. 2020; Kan et al. 2021). However, consumer awareness of geographical indications and recognition of GI logos are low (Kalekahyası and Göktaş 2022; Meral 2023). This is one of the obstacles to the commercialization of GI products.

11.6 Success Factors for GI Implementation Learned from the Turkish Experience

Geographical indications have many potential benefits if they are managed and applied following certain principles. Some of these benefits are: premium price, increase in market share, fairer distribution in value chain, new market opportunities, preservation of the specific quality of the product, protection of the reputation of the product, better farm management, supporting rural development and rural tourism (Belletti et al. 2007; Coombe et al. 2014; Sharma and Kulhari 2015; Miklós 2017; Coombe and Malik 2018; FAO and EBRD 2018; Doğanlı 2020). GIs have the potential for value creation in various ways like increasing revenue, employment, preserving traditional production/processing systems and biodiversity, enhancing economic, social and environmental sustainability, etc. (Belletti et al. 2007, 2015 Benavente 2010; Miklós 2017). The realization of one or more of the potential benefits of geographical indication registration can be considered a "success". Good management of GIs by all stakeholders in the value chain is of critical importance to reveal its benefits. In order to create value for the GI product, all stakeholders related to this product must be involved in all processes, starting from the preparation of the GI to the marketing. Stakeholders have the right to set the rules. A common vision and a participatory approach are needed for successful implementations (Benavente 2010; FAO and SINERGI 2010; Belletti et al. 2017).

Factors that may affect the success of GI implementations can be listed under some main categories as shown in Table 11.2.

In addition to the factors given in Table 11.2, legal regulations and relevant organizations in Türkiye have an impact on successful practices. However, this section focuses on rights holders in terms of implementation.

Some GI registrants in Türkiye make successful implementations and results from their GI strategy that reveal specific benefits, and actually confirm the key factors related to the value chain for the creation of benefits through GIs: well-established organisations, good management, collaborations with relevant institutions, well-designed registration process, creation of a sound traceable system, continuously control, provide training for each stakeholder in the system, and efficient marketing and promotional activities. Two Turkish GI experiences are presented below to illustrate this success and related key factors.

Table 11.2 Factors affected to successful implementation of GIs

Content of the registry	Registrant	Right owners	Product
Product specifi-cations Production method Inspection method and traceability	Knowledge level Managerial capacity Human resources Financial capacity Relationship with producers Collaboration with stake-holders and related organizations	Knowledge level Interest level Financial capacity Marketing capacity Position in the value chain Number of pro-ducers and stakeholders	Type of the product Production area Shelf-life Specificity Reputation and demand Existence of similar products in the market

Source: Dokuzlu and Söyler (2023)

Fig. 11.4 *Gemlik Zeytini*

11.6.1   Gemlik Zeytini (Black Table Olive of Gemlik)

Gemlik Zeytini is a black table olive (Fig. 11.4) that is very popular in Türkiye due to its high oil content and rich taste. It originates from Gemlik where is a district of Bursa province in the Marmara Region. Gemlik olive was Türkiye's first GI registered olive by Gemlik Commodity Exchange[7] (GTB) in 2003. *Gemlik Zeytini* has facing unfair competition in the market for many years from olives that come from different regions until its GI implemented.

Good Management and Allocation of Sources
The management board of the registrant changed in 2013 and the new board of directors decided the activation of GI as one of its priorities. Management allocated budget to develop GI implementation. A small-scale laboratory was established by the registrant in order to better control of the GI product quality and decrease control cost. GTB has employed a permanent staff to handle GI-related work, including laboratory analyses (Dokuzlu and Söyler 2023). With the establishment of the laboratory, analysis costs decreased 5 times and the number of products analysed increased 10 times compared to before.

---

[7]Commodity exchanges are public legal entities established for engaging in purchase and sale of goods that fall under a commodity exchange and as well as determination, registration, and announcement of the prices of such goods occurring in the commodity exchange in accordance with the principles stated in the legal regulations. Companies that trade the relevant product are members of Commodity Exchanges (TOBB 2024).

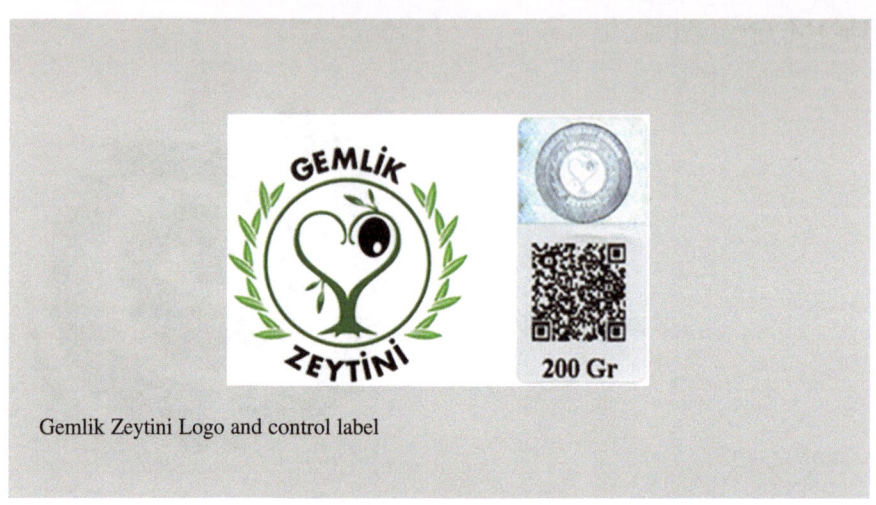

Gemlik Zeytini Logo and control label

Collaboration with Related Organizations
In 2014, GTB initiated a project with Bursa Uludag University to establish a traceable GI control system for Gemlik Zeytini, financed by its own resources. A Gemlik Zeytini promotional project was carried out using grant funds from the BEBKA (Bursa, Eskişehir, Bilecik Development Agency). Due to these two projects, control effectiveness was increased by establishing a traceable system, and approximately 1500 consumers and more than 30 retailers in Türkiye's 5 largest cities were informed through promotional activities.

Content of the Registry
At the time of Gemlik Zeytini registration, there was not enough experience with GIs in Türkiye. Therefore, the specifications, production method and geographical area were not well determined and there was no GI label use or inspection activities until 2014. The content of the registration file was corrected by conducting new research. In this way, criteria that reflect the real specifications of the product have begun to be applied and product quality has been preserved.

Establishment of a Traceable System
The fact that there were approximately 15 000 olive producers and 300 traders in the region made inspection difficult, and the exact amount of product to be inspected was unknown. Therefore, firstly, an inventory of all olive producers and traders/packers within the GI geographical area was determined. A computer software that works with inventory control logic was prepared and producer/trader/packer information was uploaded to this software. A special *Gemlik Zeytini* logo was developed and printed on labels with holograms and Quick Response (QR) code (Dokuzlu 2016). Anyone who wants to sell "*Gemlik Zeytini*" that was procured and packaged within the geographical area can use the labels (e.g. traders, packers, and retailers). If a trader/retailer, or any third party, wants to use a PDO label on its product's package, he/she must sign an agreement with the GTB that commits to the use of the label only for the appropriate quality standards of the PDO. The agreement

contains several penal provisions such as warnings, prohibition, and compensation (Dokuzlu and Söyler 2023). With the establishment of a traceable system, market inspections more rigorous and many brands that unfairly used the registered '*Gemlik Zeytini*' GI were identified and were removed from market shelves.

How Does the Labelling System Work?
The following steps need to be carried out after signing the agreement; the trader comes to GTB with a producer's receipt, and subsequently gaining permission to use the logo; the operator enters the producer's information into the system; the system automatically deducts the amount of product quantity from the producer's total production; when the producer's receipts declared by the trader exceed the total production amount of the producer, the system gives an excess-production-amount warning, and labels for the overage are not given to the trader; and if everything is correct, labels are printed and delivered requiring a signature for acknowledgment. The labels can then be used on a products' package (e-Ambrosia 2023a).

Capacity Building
The management of GTB contacted Universities and Turkpatent to increase its own knowledge about GI. Afterwards, during the years when GTB started to develop GI, multiple trainings, seminars, etc. were held every year, attended by all stakeholders. These studies not only increased the level of knowledge about GI throughout the sector, but also increased the level of participation of stakeholders in GI practices.

While no olive companies used GI before these efforts, within a year 12% of GTB member companies launched their products with GI. Although there is no precise figures, according to interviews with companies using GI, an increase in the number of orders and a decrease in order intervals were observed after the use of GI (Dokuzlu 2016; GTB 2021).

In the long term (after 5 years) there was an increase in the demand of the labelled products and product quality were improved. *Gemlik Zeytini* was also registered as PDO in the EU in 2023 (e-Ambrosia 2023a).

11.6.2   *Bursa Siyah İnciri/Bursa Siyahı / Siyah Bursa İnciri (Bursa Black Fig)*

Bursa Siyah İnciri is a large, dark coloured, sweet fig with an intense aroma, produced in Bursa province of Marmara Region and enjoys recognition on national and foreign markets. The GI process of *Bursa Siyah İnciri* started at 2016 with a project carried out in cooperation with Food and Agriculture Organization of the United Nations (FAO), European Bank for Reconstruction and Development (EBRD) and Bursa Uludag University (FAO 2018). Before the project proposal, some large-scale packaging companies that export *Bursa Siyah İnciri* willing for GI registration, but there was no organization that took action to obtain this registration. Turkish Industrial Property Law does not allow private companies to apply for registration (except if they prove that they are the single producer). While companies demand GI for *Bursa Siyah İnciri*, producers were not. Although approximately 45% of fig producers are members of a producer organization, 85% of the producers sell their products to intermediaries. Since most of the producers did not have access directly to the markets and their awareness of GI was weak, they did not have any demands regarding GI. Therefore, the *Bursa Siyah İnciri* GI initiative started with a top-down approach.

Engagement and Capacity Building
During the GI preparation process, the project team worked together with the management and prominent producers from the S.S.[8] Bursa Bölgesi Köy Kalkınma ve Diğer Tarımsal Amaçlı Kooperatif Birliği (Bursa Region Village Development and Other Agricultural Purpose Cooperative Union) as the core group (hereinafter referred to as Producer Organization – PO). The PO is the cooperative union that includes 88 small-scale agricultural cooperatives.

The processes of collecting samples and conducting analyses in order to determine product specifications carried out together with the PO. At the same time, face-to-face meetings were held with many fig producers to determine the production method. These interactions served not only to gather information from producers but also to educate them about the GI concept and its potential advantages. During the project, the specifications and documents for application to GI registration were prepared with the joint participation of all stakeholders in the *Bursa Siyah İnciri* value chain. Stakeholders' knowledge and awareness of GI practices were increased through continuous training and both national and international field visits.

Efficient Marketing and Promotional Activities
After the completion of the GI application file, a week-long marketing tests began. Specially designed logos and packaging (Fig. 11.5) were developed for GI *Bursa Siyah İnciri* together with PO and project team. PO had only worked with intermediaries and had no marketing experience from its establishment in 1998 to 2018. Thus, project team acted as mediators for PO to make agreements with supermarket chains. As a result of the negotiations, an agreement was reached to work with Metro market that is an international chain, and Özdilek that is a national chain market. Based on the success of the marketing tests, the period was extended and it continues today. Consumers' interest in packaging materials, presentation, product taste, geographical indication and being a cooperative product ensured the continuity of sales (Dokuzlu et al. 2019a). Supermarkets have started to carry out their own promotional activities for GI *Bursa Black Fig*. PO also entered new market chains without the help of any mediators. One of the most important problems of PO in the

[8]S.S: Limited Liability.

Fig. 11.5 Bursa Siyah İnciri. (*Source:* Dokuzlu et al. 2019a)

marketing phase was procurement. The fact that PO has not commercial experience before has created hesitancy among producers about delivering their products to PO. To solve this problem, PO requested only a portion of the producers' product. Consequently, some producers allocated 5–10% of their products to the PO for collective marketing, while continuing to sell the remainder through traditional intermediaries. The fact that PO obtained a better price in the market and paid a higher price to the producer than the trader increased the amount of products sold through PO in the following years.

After registration of the *Bursa Siyah İnciri* as GI, product sales price increased 50%, the price received by the producer increased between 25% and 30%, trade volume of the PO increase 15%, It was a real win-win situation. The international market chain, where *Bursa Siyah İnciri* promotional activities were carried out, increased the number of consumers by 11%, 5% sales volume and 49% turnover. At the same time, the number of customers in the retailer's hotel, restaurant and catering (HORECA) group increased (Uluşan 2018). PO applied for EU GI registration of *Bursa Siyah İnciri* in 2022. After EU GI registration complete, PO is planning carry out promotional activities in international markets.

Effective Communication

There are many small and large-scale producers, producer organizations, traders and exporters in the *Bursa Siyah İnciri* value chain. The registrant PO established and maintained communication with stakeholders at different scales during the GI preparation process. The PO worked in harmony with all other stakeholders and ensured their use of GI by auditing all parties making notifications.

Technical support, collective action, awareness of GI, and common marketing efforts are among the key elements of success in this case study. Establishing and maintaining correct communication with all stakeholders in the value chain ensured the sustainability of success.

11.7   Conclusion

Türkiye has a wealth of origin-linked products, and as a result, the number of GI registered products and applications have developed rapidly, especially since 2017, when the new Industrial Property Law came into force. The interest of institutions and organizations related to origin-related products has also been a factor that accelerated this development. Despite the increase in the number of registrations, registrants and relevant public institutions have realized that GIs are not being utilized sufficiently. Thereupon, some registrants started to work on using GIs effectively, and relevant public institutions supported these efforts in different ways.

GI is a collective right and can create collective benefits, but asymmetries in the value chain regarding scale, information, market access, and bargaining power make management and collective action challenging. GIs not only offer collective benefits but also carry collective risks, particularly in cases of poor management. In Türkiye, the difference between small-scale and large-scale producers/processors, intermediaries, and sellers create unbalance in management. Stakeholders often have difficulties to understand that GI is a collective right. Weak link between registrant and actors in the value chain are important barriers to implement GIs successfully. According to Turkish experience, strengthening social and economic linkages between local actors is not easy, because most of the actors on the buyer and seller sides of the value chain perceive that they have opposite interests. This problem was solved by some registrants by holding continuous information meetings and involving different actors in the value chain in common marketing efforts. In fact, achieving collective action is one of the keys to success, but it is not limited to that.

Turkish experience, especially through the two success stories presented, has shown that GIs can serve as powerful tools for differentiating products and securing competitive advantages in the market. However, realizing these benefits requires concerted efforts to address challenges related to management, collective action, traceability, and control. Despite the challenges, the proactive steps taken by registrants and support from public institutions have shown that effective GI utilization can lead to substantial collective benefits. This underscores the need for continued efforts to strengthen the GI system, emphasizing producer involvement and consumer awareness to further elevate the status of Türkiye's GI products on both domestic and international platforms. Another highly emphasized issue in recent years has been traceability and control. Because a poor quality product with the GI logo will damage the reputation of all products bearing this logo. Some GI registrants in Türkiye have realized that the reputation of their GI products is damaged by products that are of poor quality or that come from other regions and do not reflect the characteristics of the GI product. Registrants who were capable of effectively monitoring their GI products have gained a competitive advantage in the market.

GIs have immense potential in diverse nations like Türkiye. Türkiye's GI protection system, while comprehensive, could benefit from integrative strategies, prioritizing producer involvement, and amplifying consumer awareness. This holistic approach holds the promise of elevating Türkiye's GI products both domestically and internationally. Some suggestions to the challenges of the current GI protection system in Türkiye are as follows:

- To make the GI process more inclusive: rather than becoming registrant for GIs, public bodies should strain to empower producer groups to effectively manage GIs.
- Local public bodies should start working with small volunteer groups that include leaders of each value chain actor and enlarging the group step by step. A consortium can then be established by using the most inclusive legal status. Once the collective movement is built on solid foundations, GI management will work easier and can focus on marketing efforts, creating awareness among consumers, participating in larger national and international markets.
- For NGOs such as Chambers of Commerce/Industry and Commodity Exchanges who hold the majority of GI registrations in Türkiye, it is suggested using impartial public institutions such as institutes, and/or universities as mediators to increase the inclusiveness of the GI. Developing measures to provide a fair distribution of costs and benefits according to the production scales of the producers and effectively using legal instruments against infringements of GI rights would ensure the trust and the confidence of both producers and consumers.
- To address the challenge of non-distinctive products being registered as GIs stricter evaluation criteria must be implemented for assessing product uniqueness and verifying the market presence of registered products, together with ex-officio provisions for cancelling GIs may be beneficial to maintain the integrity of the GI system and ensure it protects products with genuine significance.
- Consumers' awareness about GIs and their recognition level of GI emblems should be increased. Local governments, registrants, GI product related institutions, retailers and public institutions should play a role in promotional activities. These activities should be carried out through multiple media tools such as social, written and visual media. In this context, the work of Türkpatent and MAF, which have been carrying out activities to promote and raise awareness from past to present, should continue.

References

Belletti G et al (2007) The impact of geographical indications (PDO and PGI) on the internationalisation process of agro-food products. In: 05th EAAE Seminar "International Marketing and International Trade of Quality Food Products", Bologna, pp 8–10

Belletti G, Marescotti A, Sanz-Cañada J, Vakoufaris H (2015) Linking protection of geographical indications to the environment: evidence from the European Union olive-oil sector. Land Use Policy 48:94–106

Belletti G, Marescotti A, Touzard JM (2017) Geographical indications, public goods, and sustainable development: the roles of actors' strategies and public policies. World Dev 98:45–57

Benavente D (2010) The economics of geographical indications: GIs modeled as club assets. Graduate Institute of International and Development Studies Working Paper (No. 10/2010), Geneva

Çakaloğlu M, Çağatay S (2017) Coğrafi işaretler ve marka değerine sahip ürünlere yönelik tüketici algısı: Finike portakalı ve Antalya tavşan yüreği zeytini örnekleri. Tarım Ekonomisi Araştırmaları Dergisi 3(1):52–65

Candan B (2017) Intellectual property legislation in the ottoman era and its effects on knowledge production. Athens J Mediterr Stud 3:3–14

Coombe RJ, Malik SA (2018) Transforming the work of geographical indications to decolonize racialized labor and support agroecology. UC Irvine L Rev 8:363–412

Coombe RJ, Ives S, Huizenga D (2014) Geographical indications: the promise, perils and politics of protecting place-based products. In: Coombe RJ, Ives S, Huizenga D (eds) Sage handbook on intellectual property. Sage, Thousand Oaks, pp 207–223

Çukur T, Kızılaslan N, Çukur F, Kızılaslan H (2020) Tüketicilerin Coğrafi İşaretli Ürünler İçin Ödeme İstekliliğine Etki Eden Faktörler: Niksar Cevizi Örneği. Turk J Agric-Food Sci Technol 8(11):2476–2481

Denk E, Bilici Sanalan N (2023) Coğrafi İşaretler Hakkında Yerel Yönetici Görüşlerinin Analizi (Analysis of local executive opinions about geographical ındications). J Gastr Hosp Travel 6(1): 319–334

Doğanlı B (2020) Coğrafi işaret, markalaşma ve kırsal turizm ilişkileri. J Hum Social Sci 3(2): 525–541

Dokuzlu S (2016) Geographical indications, implementation and traceability: Gemlik table olives. Br Food J 118(9):2074–2085

Dokuzlu S, Söyler İ (2023) Coğrafi İşaretlerde Denetim Etkinliğini Artırma Yöntemleri. Tarım Ekonomisi Araştırmaları Dergisi 9(EKS 1):15–25

Dokuzlu S et al (2019a) Tüketicilerin yöresel ürün satın alma davranışları: DAP Bölgesi ürünleri. Tarım Ekonomisi Dergisi 25(1):97–108

Dokuzlu S et al (2019b) TRA2 Bölgesi Yöresel Ürün Pazarlama Stratejileri, 1st edn. Serhat Kalkınma Ajansı (SERKA), Kars

e-Ambrosia (2023a) Gemlik Zeytini Single Document [Online]. Available at: https://ec.europa.eu/agriculture/eambrosia/geographical-indications-register/details/EUGI00000017468

e-Ambrosia (2023b) European Commission, EU geographical indications register [Online]. Available at: https://ec.europa.eu/agriculture/eambrosia/geographical-indications-register/

FAO (2018) Protecting Turkey's Bursa black figs and Bursa peaches [Online]. Available at: https://www.fao.org/support-to-investment/news/detail/en/c/1127716/

FAO (2021) FAOSTAT-Crops and livestock products [Online]. Available at: https://www.fao.org/faostat/

FAO and SINERGI (2010) Linking people, places and products, a guide for promoting quality linked to geographical origin and sustainable geographical ındications, 2nd edn. FAO, Rome

FAO—EBRD (European Bank for Reconstruction and Development) (2018) Strengthening sustainable food system through geographical indications. An analysis of economic impacts. FAO, Rome. Available at: www.fao.org/3/I8737EN/i8737en.pdf

GTB (2021) Dünyada Sofralık Zeytin Üretimi ve Coğrafi İşaret Raporu. Gemlik Ticaret Borsası, Bursa

Kadanalı E, Dağdemir V (2016) Tüketicilerin yöresel gıda ürünleri satın alma istekliliği. J Agric Fac Gaziosmanpaşa Univ (JAFAG) 33(1):9–16

Kalekahyası S, Göktaş B (2022) Coğrafi İşaret Almış Yöresel Ürünlerin Bilinirlik Düzeyi ve Tüketici Tutumlarına Etkisi: Bayburt İli Örneği. Stratejik ve Sosyal Araştırmalar Dergisi 6(3): 673–702

Kalkınma Ajansları (2023) Kalkınma Ajansları Portalı [Online]. Available at: https://www.ka.gov.tr/

Kan M, Kan A, Kütükoğlu Ş (2021) Kastamonu ili Merkez ilçesinde gıda ürünleri tercihinde coğrafi işaretlerin etkisi. Tarım Ekonomisi Araştırmaları Dergisi 7(1):40–51

Meral H (2023) Tüketicilerin Coğrafi İşaretlere İlişkin Bilgi Düzeyinin Belirlenmesi: Adana İli. Çanakkale Onsekiz Mart Üniversitesi, Çanakkale, pp 393–402

Meral Y, Şahin A (2013) Tüketicilerin coğrafi işaretli ürün algısı: Gemlik zeytini örneği. KSÜ Doğa Bilimleri Dergisi 16(4):16–24

Mevzuat (2015) Perakende Ticaretin Düzenlenmesi Hakkında Kanun [Online]. Available at: www.mevzuat.gov.tr/MevzuatMetin/1.5.6585.pdf

Mevzuat (2016) Intellectual Property Right Law. s.n, s.l

Mevzuat (2017) Coğrafi İşaret ve Geleneksel Ürün Adi Amblem Yönetmeliği. s.n, s.l

Miklós I (2017) The apricot story: patterns in a local circular food chain in North Hungary. Soc Econ 39(4):549–571

Nizam D, Tatari MF (2022) Rural revitalization through territorial distinctiveness: the use of geographical indications in Turkey. J Rural Stud 93:144–154

Official Gazette (1920) 7.10.1920/11951. Republic of Türkiye, Ankara

Official Gazette (1930) 19.06.1930/1524. Republic of Türkiye, Ankara

Official Gazette (1995) 27.06.1995/22326. Republic of Türkiye, Ankara

Official Gazette (2009) 30.05.2009/27243. Republic of Türkiye, Ankara

Official Gazette (2017) 10.01.2017/29944. Republic of Türkiye, Ankara

SBB (2023) Türkiye Cumhuriyeti Cumhurbaşkanlığı Strateji ve Bütçe Başkanlığı [Online]. Available at: www.sbb.gov.tr/wp-content/uploads/2022/07/On_Birinci_Kalkinma_Plani-2019-2023.pdf

Sharma RW, Kulhari S (2015) Marketing of GI products: unlocking their commercial potential. Centre for WTO Studies IIFT, New Delhi

T.C Kültür ve Turizm Bakanlığı (2023) Türkiye Kültür Portalı [Online]. Available at: https://sgb.ktb.gov.tr/TR-158707/turkiye-kultur-portali.html

Ticaret Bakanlığı (2018) Türkiye Cumhuriyeti Ticaret Bakanlığı Kurumsal Haberler [Online]. Available at: https://ticaret.gov.tr/kurumsal-haberler/ticaret-bakanligi-ile-turkpatent-arasinda-cografi-isaret-ve-geleneksel-urun-adlar

TOBB (2024) The Union of the Chambers and Commodity Exchanges of Türkiye [Online]. Available at: https://www.tobb.org.tr/Sayfalar/Eng/AnaSayfa.php

Toklu İT (2016) Tüketiciler coğrafi işaret için daha fazla ödemek ister mi? Artvin balı üzerine bir araştırma. Karadeniz Araştırmaları 52:171–190

Türkpatent (2023a) Coğrafi İşaretler Portalı [Online]. Available at: https://ci.turkpatent.gov.tr/Statistics/RegistrationAndApplication

Türkpatent (2023b). Türk Patent ve Marka Kurumu [Online]. Available at: https://www.turkpatent.gov.tr/plan-program-raporlar

Uluşan B (2018) Metro Market ve Coğrafi İşaretler. Unpublished Seminar Presentation 04 December 2018, Bursa

Yörex (2023) Yöresel Ürünler Fuarı—Yörex [Online]. Available at: https://yorex.com.tr/

The opinions expressed in this chapter are those of the author(s) and do not necessarily reflect the views of the [NameOfOrganization], its Board of Directors, or the countries they represent.

Chapter 12
Promotion and Protection of Products of Origin in Chile. The Role of the State

Paola Guerrero Andreu

Acronyms

INAPI	National Institute of Industrial Property, Chile
SAG	Agricultural and Livestock Service of the Ministry of Agriculture
AO	Apellation of Origin
SUBDERE	Undersecretariat of Regional and Administrative Development
AFC	Family Farming
DFL	Decree with Force of Law
GORE	Regional Government
CASEN	National Socioeconomic Characterization Survey

12.1 Introduction

Chile is a tricontinental country with a length of 4270 km and a coastline of over 6000 km. This heterogeneity of territories has contributed to an unparalleled wealth and variety of products. However, the country is internationally recognised—from the perspective of its products of origin—mainly for its wines and Pisco,[1] whose regulations are very specific, sectoral and are contained in special laws.

---

[1]Brandy produced and bottled, in consumer units, in Regions III and IV of the country, made by distillation of genuine drinkable wine, coming from the varieties of vines determined in this regulation, planted in these regions. Decree 521, Sets Regulation of the Apellation of Origin Pisco, Ministry of Agriculture, 30 December 1999. https://www.bcn.cl/leychile/navegar?idNorma=169561

P. Guerrero Andreu (✉)
National Institute of Industrial Property, Chile, Santiago, USA

© The Author(s) 2025   171
E. Vandecandelaere et al. (eds.), *Worldwide Perspectives on Geographical Indications*, https://doi.org/10.1007/978-3-031-71641-6_12

In developing countries like Chile, with this great diversity of traditional products both of a food nature and crafts, promotion and protection programmes for their products of origin are ideal instruments to boost territorial development, enhance and highlight the country's traditional and unique products.

However, the development of public policies involving protected local products is posed as a challenge, that these policies remain over time and that they are part of comprehensive local development designs.

12.2 Systems of Protection of Products of Origin in Chile

The system of protection of products of origin in Chile, we could call it a triple helix. The general system available for products in general, is contained in the Industrial Property Law N° 19.039[2] with the competent authority being the National Institute of Industrial Property, INAPI and recognises protected products of origin through geographical indications and apellations of origin. On the other hand, there are special laws for particular products, such as Pisco, Pajarete (genuine generous wine) and sun-dried wine (genuine generous wine) whose regulation is contained in the Alcohol Law.[3] There is also a particular regulation for wines. In all these cases the competent authority is the Ministry of Agriculture through a specialised agency which is the Agricultural and Livestock Service, SAG. The last helix, is composed of the international treaties signed by Chile, which have considered the protection of products of origin through a mechanism known as *"exchange of lists"*.[4]

The Chilean system in terms of geographical indications and apellations of origin, is characterised because it is a *sui generis*,[5] system that considers as a relevant factor the link of the product with a specific territory, from which unique qualities, characteristics and reputation attributable to that geographical origin derive. They are collective rights so that—in theory—they benefit all producers, artisans, groups or associations of them that carry out their activity directly with the protected products. It is a system open to all kinds of products, without restrictions as to their nature (crafts, agricultural products, food, wines, spirits). Finally, it is a system that also opens up to foreign products that want to obtain their protection in Chile, obviously complying with the specific requirements of the legislation.

---

[2]Decree4, Sets consolidated, coordinated and systematised text of Law No. 19.039, on Industrial Property, Ministry of Economy, Development and Tourism; Undersecretariat for Economy and Small Businesses, 6th August 2022 https://www.bcn.cl/leychile/navegar?idNorma=1179684

[3]Law No. 18.455, Sets rules on production, processing and marketing of ethyl alcohols, alcoholic beverages and vinegars, and repeals Book I of Law No. 17.105, Ministry of Agriculture, 31st October 1985 https://www.bcn.cl/leychile/navegar?idNorma=29859

[4]Intellectual Property, Undersecretariat for International Economic Relations https://www.subrei.gob.cl/ejes-de-trabajo/propiedad_intelectual

[5]Considers geographical indications, appellations of origin, collective and certification marks

12.3   Seal of Origin Programme

In 2012, the Ministry of Economy, Development and Tourism and the National Institute of Industrial Property (INAPI) launched the Origin Seal programme,[6] aimed at protecting, promoting and valuing traditional Chilean products with a high local connection, through the appropriate use of industrial property rights. The Origin Seal itself is a certification mark. To be part of the programme and use the mark, it is necessary: (i) that it is a local product; (ii) that it has a collective basis; (iii) that it complies with all the requirements and conditions established in the Industrial Property Law, to be protected through a geographical indication, apellation of origin; it also extends to collective and certification marks, as long as they are linked to products of specific territorial origin.

The reasons that were taken into account for the design of the programme were varied. On the one hand, it was found that there was little use by producers and artisans of geographical indications or apellations of origin, despite being available in national legislation since 2005 after a modification to the Industrial Property Law. The aim was also to promote entrepreneurship, the productive development of local communities and the safeguarding of traditions that could be linked to the products. Likewise, it was considered that the Seal of Origin as a protected trademark, provided effective protection against its misuse and would facilitate the positioning of the products among consumers.

In short, it could be said that the products of the programme have a kind of *plus protection* because, on the one hand, they have the individual legal status of the right that allowed them to be part of the programme (GI, AO, collective or certification mark) and on the other hand, the protection status of a brand like the Origin Seal *"created to group a wide range of traditional Chilean products and make them more easily recognisable in the market"* (Olivos and Carrasco 2016) whose ownership corresponds to a public entity, which is, the Undersecretariat of Economy and Small Businesses,[7] which is part of the administrative structure of the Ministry of Economy, Development and Tourism of Chile.[8]

It should be noted that in the design stage of the programme, INAPI and the Undersecretariat of Economy and Small Businesses, worked coordinatedly with the Undersecretariat of Regional Development and Administrative (SUBDERE, 2013) entity that considered that programmes like the Seal of Origin were effective in issues *"identity and heritage as essential and active elements of economic development and culture"* linked to decentralisation and local development[9] taking into consideration elements such as: (i) associativity; (ii) regional legitimacy; (iii) local productive practices; (iv) economic-productive projection (SUBDERE, 2013).

---

[6] National Institute of Industrial Property, INAPI https://www.inapi.cl/sello-de-origen

[7] https://www.economia.gob.cl/subsecretaria-de-economia-y-empresas-de-menor-tamano

[8] Records No. 1232103, 1232102, 1232101, 1232100, 1232099, 1262276, 1262275, 1227747, 1262273, 1262272.

[9] https://www.subdere.gov.cl/sites/default/files/documentos/publicacion_programa_identidad_final_27-12-13.pdf

More than a decade after the launch of the programme, more than forty products are part of the Seal of Origin catalogue, these being of various nature (crafts, fresh or processed foods, sea products) and in which the obtaining of the right that enabled them to be part of the programme,[10] involved years of work with the communities and local authorities (Venegas, 2011).

It should be noted that, a significant percentage of the products of origin protected through geographical indications, apellations of origin, collective and certification marks and that are part of the catalogue of the Seal of Origin, are linked to rural areas, with low economic income and in areas of isolation.[11] Likewise, it is possible to find the presence of traditional and ancestral knowledge linked to indigenous people. Regarding agri-food products, there are products strongly linked to family farming (AFC)[12] and subsistence. The COVID 19 pandemic and other natural disasters such as fires or floods, severely affected the economic activity of artisans and producers, causing a decrease and/or suspension of activities as well as the closure of marketing centres. All of the above, coupled with a low associative density on the part of the producers and artisans, which translates into difficulty in exercising ownership and management of enabling rights[13] creates difficult gaps to address. In this sense, the effective support role of the competent state agencies is fundamental.

During the years 2021 and 2022, an evaluation of the programme was carried out, which included the application of different instruments to measure the real impact of the programme (surveys, comparative bibliographic consultations, interviews with relevant actors, among others). Among the most relevant results, it was possible to rescue that the Seal of Origin and the IP rights that allow access, was perceived as a prize or award and not, as Industrial Property assets that require or demand positioning strategies. On the other hand, local actors or stakeholders—including producers and artisans—as well as local authorities showed a low or scarce knowledge of the impact or utility of a geographical indication, apellation of origin, collective or certification mark; as well as the value of collective action. Finally, a cross-cutting element was the need for an active and permanent support role from the linked public agencies, that is, to develop long-term public policies, which consider the local products protected through IP rights within local development strategies.

---

[10] Geographical indications, appellations of origin, collective and certification marks.

[11] Seven of the products in the Origin Seal catalogue are located in areas known as lagging, that is, those territories that present difficulties in accessibility and physical connectivity, low population density, dispersion in the territorial distribution of its inhabitants and low presence or coverage of basic and public services. Convergence Territories Unit of the Undersecretariat for Regional and Administrative Development, SUBDERE https://www.subdere.gov.cl/

[12] Family Farming (AFC) is a way of organising agricultural and forestry production, as well as fishing, grazing, and aquaculture, which is managed and directed by a family and largely depends on family labour, both women and men. The family and the farm are linked, co-evolve and combine economic, environmental, reproductive, social and cultural functions. AIAF-2014 civil society campaign, United Nations Food and Agriculture Organization.

[13] Geographical indications, appellations of origin, collective and certification marks.

A not insignificant element to consider is the confusion that exists between the Seal of Origin trademark that functions as a kind of *umbrella,* with the individual rights that protect the products that are part of the programme. It is not clear to the producers or artisans and to the consumers that Seal of Origin is the name of the programme and that what is intended to highlight is the geographical indication, apellation of origin, collective or certification mark, which is linked to a product of specific origin. In this part, it could be affirmed that the *cobranding* has not worked.

12.3.1   Territorial Distribution

In accordance with Chapter XIV, Article 110 and following of the Political Constitution of the Republic of Chile,[14] the territory is divided into regions and these into provinces.

Currently, there are sixteen regional units, whose administration according to the provisions in Chapter I, Article 13 and following of the DFL 1-19175 that "Sets the consolidated, coordinated, systematised and updated text of the constitutional organic law on government and internal administration" corresponds to a Regional Government (GORE).

The above is of interest because currently, of the sixteen regional units, not all have local products that are part of the catalogue of protected local products. The reasons can be diverse, however, what is curious is that some of them have more than one product and yet, there are no clear regional support policies associated (Fig. 12.1).

12.3.2   Some Learnings

The registration of a geographical indication, apellation of origin, collective or certification mark, is not an end in itself, it requires the design of value creation strategies that allow the products to be distinguishable.

Products of origin protected through Industrial Property rights—in general—do not compete on price, but on value (Samper 2021). This makes it essential to develop identity strategies for the product and for the linked territory.

Protected origin products require associated services integrated experiences. Hence the importance of sectors such as tourism or gastronomy (Villela 2021).

Protected origin products require the design of business models adjusted to their purpose and value proposition.

---

[14] https://www.bcn.cl/leychile/navegar?idNorma=242302&idVersion=2024-01-19&idParte=8563601

Origin label

Some figures

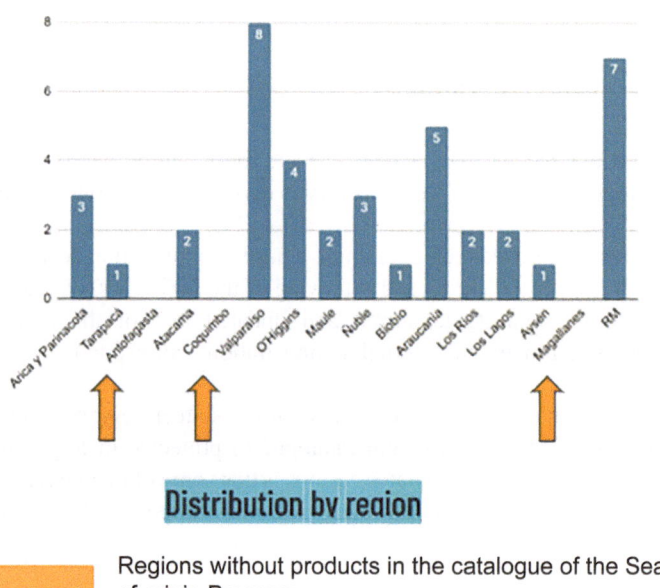

Distribution by region

Regions without products in the catalogue of the Seal of origin Program

Fig. 12.1 Distribution by regional unit of products from the Seal of Origin program

Products of origin recognised through Industrial Property rights can be a hub of local economies, with the consequent recognition of the people and communities involved.

On the other hand, it is possible to indicate that not all local products are mature enough to start a process of geographical indication, apellation of origin and even collective or certification mark. The above, given that being collective rights they demand elements that are not necessarily present in the structures linked to producers or artisans. The associativity, a transversal element in all the rights mentioned, is a basic requirement that is not always present and without which, the possibilities of real impact are scarce. The same reflection can be made about the product itself, as there is the mistaken idea that any product or manifestation can be protected through a geographical indication or apellations of origin, confusing the statutes of these rights for example with that of intangible cultural heritage.[15]

---

[15] (...) uses, representations, expressions, knowledge and techniques—along with the instruments, objects, artefacts and cultural spaces inherent to them – that communities, groups and in some cases individuals recognise as an integral part of their cultural heritage (...). UNESCO Convention for the Safeguarding of the Intangible Cultural Heritage, 2003.

For all the above, the question or reflection is valid if countries like Chile, are prepared for systems like the geographical indications and apellations of origin. Given the evidence to date, the answer would be that a reinforcement of the system is necessary, rethinking the objectives and understanding what the minimum requirements are, in order to avoid the so common *"paper GIs"*, that is, those that are nominal but in practice never work.

12.3.3 Necessary Actions from Public Policy

In coherence with what was stated in the previous paragraph and according to the data available to date, programmes like Sello de Origin, can undoubtedly be very useful, but a reformulation of objectives is urgently required, not only of the programme but from the public policy and the role that the State has to play.

Among the future actions that can be suggested, it is possible to mention some as an example:

Coordination and articulation between the different government agencies linked to agriculture, economy, tourism, culture, among others. This aspect is crucial both for the continuity of the programme and also of the system of geographical indications, apellations of origin, collective and certification marks, that are linked to products of origin.

Greater rigour is indispensable when selecting products *potentially eligible* this task must be joint and coordinated between the different government agencies that are linked to groups of producers and artisans. It is also important to involve local authorities and actors in this task, in order to avoid empty promises of sure successes, without delving into the demands and challenges that are associated (Belletti and Marescotti, 2021).

Continuous support in education and training actions aimed at both producers and artisans, as well as relevant local actors decision makers and authorities of the territory linked to the products. The above so that they understand the value of the Industrial Property right that recognises the product. This can be to gradually implement through certification marks as a preliminary step to GI and AO (Venegas, 2011).

Creation of alliances between artisan and producer groups, rights holders, with academic and research institutions. Innovation for the improvement of production processes is an area to incorporate considering that, for example, climate change is one of the main threats to protected local products.

In the aforementioned alliances, international cooperation and the exchange of good practices among producers and artisans should not be overlooked. In this regard and only as an example, it is possible to mention the continuous support of the World Intellectual Property Organization WIPO, in these matters.

12.4 Final Ideas

The design of programmes such as Seal of Origin from public policy and its permanence over time in developing countries like Chile, can have multidimensional effects. From a social and cultural perspective, they allow traditions or expressions inherited from our ancestors, to be transmitted to future generations. Similarly, it is possible to promote social capital through collective actions.

From an economic perspective, public programmes or initiatives that include protected origin products, considering both those that are currently part of the Seal of Origin programme—protected through geographical indications, apellations of origin, collective or certification marks—as well as those that have special statutes, (e.g. wines) can mobilise local economies and be part of local economic development strategies.[16] The above can represent an opportunity for improvement for regional public policies in areas such as, productive promotion,[17] job creation and decentralisation.[18]

Finally, successful protected origin products, can contribute to overcoming poverty in countries like Chile, where according to the results of the National Socioeconomic Characterisation Survey CASEN, 2022, 16.9% of the population lives in conditions of multidimensional poverty, that is, considering the areas of education, health, work and social security, housing and environment, networks and social cohesion.[19]

---

[16]Process of growth and structural change that, through the use of the existing development potential in the territory, leads to the improvement of the well-being of the population of a locality or a region. When the local community is able to lead the process of structural change, the form of development can be agreed to be called endogenous local development (Vásquez Barquero 2000).

[17]It is a process that reactivates the economy and dynamises the local society, which, by efficiently using the endogenous (internal) resources existing in a certain area, is able to stimulate its economic growth, create employment and improve the quality of life of the local community (Alburquerque 2004).

[18]A good example at this point is the region of Arica and Parinacota. This region, the first in the north of Chile, has three geographical indications: Oregano from the foothills of Putre, Maize from Lluta and Olives from Azapa. The Regional Government (GORE) in its "Regional Innovation Strategy 2022–2030" expressly incorporates these products by stating "(. . .) 67% of the business fabric of the Arica and Parinacota Region is made up of micro and small businesses. In addition, the oregano from the foothills of Putre, the olives from Azapa and the maize from Lluta have something in common: all three have a Seal of Origin (. . .)", National Agency for Research and Development, Government of Arica and Parinacota, p. 4, https://gorearicayparinacota.cl/images/Estrategia%20 Regional/ERI_Arica_y_Parinacota.pdf

[19]A household is considered to be in a situation of multidimensional poverty if it presents 22.5% or more of deprivations, which is equivalent to a traditional dimension. Ministry of Social Development and Family, Social Observatory https://observatorio.ministeriodesarrollosocial.gob.cl/ encuesta-casen-2022

References

Alburquerque F (2004). The Local Economic Development approach. Retrieved from http://biblioteca.municipios.unq.edu.ar/modules/mislibros/archivos/Des.Econ.Alburquerque.pdf

Belletti G, Marescotti A (2021) Evaluating geographical indications. Guide to tailor evaluations for the development and improvement of geographical indications. Retrieved from https://www.fao.org/documents/card/en/c/cb6511en

Olivos C, Carrasco F (2016) Adding value to traditional Chilean products with the Seal of Origin. Retrieved from https://www.wipo.int/wipo_magazine/es/2016/03/article_0003.html

Samper L (2021) INAPI+Origin, online training programme organised by wipo and inapi

Undersecretariat of Regional and Administrative Development, SUBDERE Programme for Strengthening Regional Identity Department of Studies and Evaluation Division of Policies and Studies (2013) Retrieved from https://www.subdere.gov.cl/sites/default/files/documentos/publication_program_identity_final_27-12-13.pdf

Vásquez Barquero, Antonio (2000) Local Economic Development and decentralisation. Towards a conceptual framework. Retrieved from https://repositorio.cepal.org/server/api/core/bitstreams/96269953-87d5-4cff-8984-d5d944981ad2/content

Venegas C (2011) Culture, traditional knowledge and agricultural heritage in Chiloé. Keys to a DTR-IC strategy. Retrieved from https://www.rimisp.org/wp-content/files_mf/1367849281ChiloeCVenegas1junio2011.pdf

Villela A (2021) INAPI+Origin, online training programme organised by Wipo and Inapi

Chapter 13
Two Decades of Dedication: The Story of Registering Karoo Lamb as a GI in South Africa

Johann Kirsten

Abbreviations

DALRRD   Department of Agriculture Land Reform and Rural Development, South Africa
DURAS   Promoting Sustainable Development in Agricultural Research Systems
SADC   Southern African Development Community
EPA   Economic partnership agreement
RSA   Republic of South Africa

13.1 Introduction

The intellectual work on Geographical Indications in South Africa was initiated in 2005 when a Promoting Sustainable Development in Agricultural Research Systems (DURAS)[1] funded project was implemented. While South Africa has a rich diversity of agricultural products with characteristics attributable to their geographical origin, there has been no public legal system for protecting GIs (Bramley et al. 2013). The DURAS project unpacked the characteristic of at least 20 agricultural commodities in South Africa to establish whether they fit the definition and specifications of Geographical Indications (See Bienabé et al. 2008). In the end, after a long period of investigations and stakeholder meetings only six commodities made the final list of commodities with a distinct geographical characteristic and through its name, signal

---

[1] An initiative supported by the French Ministry of Foreign and European Affairs through its Priority Solidarity Fund.

J. Kirsten (✉)
Bureau for Economic Research, Stellenbosch, South Africa
e-mail: jkirsten@sun.ac.za

© The Author(s) 2025
E. Vandecandelaere et al. (eds.), *Worldwide Perspectives on Geographical Indications*, https://doi.org/10.1007/978-3-031-71641-6_13

its connection with a specific region. Eventually only three products *Rooibos*, *Karoo Lamb* and *Honey bush tea* were taken forward for full GI protection. These three products have encountered different issues in their journey to GI registration.

This Chapter provides a detailed account of the legal, regulatory, and consultative processes to get Karoo Lamb registered as a GI in South Africa. The Chapter first describes the initial process post the DURAS project which led to the implementation of a certification scheme for Karoo lamb as a collective initiative and 'placeholder' for the Karoo Lamb GI. Then we provide an historic record of the politics linked to the GI legal frameworks in South Africa and then described the various attempts to get official registration of the Karoo Lamb GI; first under the Merchandise Marks Act and then two attempts under the new GI regulations of 2019 gazetted by the Department of Agriculture. Revised and amended GI regulations published in February 2023 provided another opportunity to register Karoo Lamb as a GI. The new application went out for public comment in September 2023 and with no objections received the Department of Agriculture Land Reform and Rural Development (DALRRD) finally registered *Karoo Lamb* as a Geographical Indication on 27 October 2023.

This is a real-life story of commitment and perseverance which eventually led to the registration of the GI—virtually two decades since the initial work was initiated.

13.2   Karoo Lamb as a Product of Origin

Karoo Lamb has been part of South African culture for more than a hundred years. It is part of South African cuisine and many businesses and towns in the Karoo market themselves as 'the home of Karoo Lamb'.

This lamb reared on natural indigenous Karoo veldt in the central arid part of South Africa is believed to produce meat with a unique flavour (Erasmus 2017). The unique identity of, and the geographical value attached to, Karoo Lamb makes it possible to sell Karoo Lamb at a premium price above an ordinary lamb product.

This unique identity makes the product exceptionally vulnerable to opportunistic behaviour by stakeholders who do not comply with the strict Karoo Lamb production protocols. The misuse of the name means that the geographic advantage of farmers raising lamb, according to the protocols is lost, not only to the farmers but also to the Karoo community. Moreover, this misuse of the name further confuses the consumers, who have no way of authenticating the Karoo Lamb's origin and free-range credence attributes.

13.3 The Early Stages of Protection of Karoo Lamb in South Africa

Following the work of the DURAS project, the Karoo Development Foundation was established, and the first business was to register a certification mark for Karoo Lamb. To roll out this process and to support the implementation of the certification scheme the foundation was supported by the Northern Cape Government and Western Cape government. The launch meeting of the certification scheme and the establishment of Meat of Origin Karoo (MOOK), as the operating company for the certification scheme, took place in October 2010. MOOK is a registered not for profit company to manage the membership and the rules of the certification scheme. In essence its mandate was the same as a typical collective management organisation. The company had to pay the auditors and ensure that membership lists are current and paid-up.

Following this consultative process, the certification mark (or quality indication) "Certified Karoo Meat of Origin" and its production protocol was registered in March 2011 with the Department of Agriculture in terms of the "*Regulations regarding the classification and marking of meat intended for sale in the Republic of South Africa*" (No. R. 863 of 1 Sept 2006 and later revised to R. 55 of 30 January 2015). This registration as a "quality indication" was the recommended route because at that time South Africa had no other legal framework in place to register a GI. This was therefore the first formal public protection of Karoo Lamb. Before this process there were several private entities that registered trademarks including the words "Karoo Lamb". These are all private trademarks with private production protocols. By the time the certification mark was registered all these trademarks were dormant and not in use.

Under the requirements of Regulation R.55 of 2015, the Department of Agriculture appoints an 'assignee' to do independent audits to verify the origin claim of Karoo. The audits of the farms, abattoirs and retailers registered under the certification scheme, cost the organisation a considerable amount of money.

By the end of 2015 and through the assistance and funding provided by the two provincial departments of agriculture, there were 6 abattoirs enrolled under the certification scheme and 209 farmers with 417 farms covering 2,05 million hectares in the core region of the Karoo (estimated area of 7 million hectares). The registration and audit costs were covered by the Government grants with no costs incurred by abattoirs, retailers and farmers. At its peak in 2018 a total of 20 000 Karoo Lamb carcasses were sold as 'certified' Karoo Lamb.

13.4   Karoo Lamb and the Negotiations on the Economic Partnership Agreement, 2013–2016

One of the pre-conditions during the negotiations on the Economic Partnership Agreement (EPA) between the Southern African Development Community (SADC) and the EU was that South Africa should provide evidence that it has the legislative means and processes in place to protect geographical indications. The South African government offered to the EU, as an interim and immediate measure, the Merchandise Marks Act (Act 17 of 1941) as the legal instrument by which it will protect European GI names in South Africa. To understand how that will work the EU negotiators requested that South African GIs should first be protected under this Act. At that stage there were no GIs identified or registered and the government was desperately looking for products to illustrate the workings of the Act. Word got the Minister of Trade and Industry of the DURAS project, and he then requested the research team to provide the 'rules of use' for the three GI products identified in the project: Karoo Lamb, Rooibos and Honey Bush.

The notice announcing the proposed prohibition of the use of the words 'Karoo Lamb' was published in the Government Gazette in terms of the Merchandise Marks Act on 1 November 2013 (Notice 1074 of 2013). The notice contains the 'rules of use' for Karoo Lamb which clearly defined the attributes of the product and required processes to guarantee the authenticity of the project. The notice was published for public comment and whereas the notices for Rooibos and Honey Bush received no or minor comments, the Karoo Lamb proposed prohibition was met with some stiff comments. One large retail chain (Woolworths), although supportive of the idea, provided a list of concerns and conditions. One former politician connected to the Minister of Trade and Industry used his political connections to vehemently oppose this 'proposed prohibition' on the grounds that it will give 'unfair private rights to a third party' and that the boundaries of the region were not inclusive enough. All these comments illustrated the lack of understanding of the role of a GI and at the same time it illustrated that this specific Act was not conducive to support collective rights and the protection of heritage.

In the end the Registrar of Trademarks (who oversees all applications and matters related to the Merchandise Marks Act) rejected the application for the protection of the Karoo Lamb name on the grounds of the various contestations with private rights and the political sensitive nature of the case. This created a major problem for the European negotiators as South Africa still could not provide evidence of its ability to protect a GI in the meat or cheese industries.

In a final test of SA's legal system the EU requested the author and his team to supply a full case file of evidence to test the "Karoo Meat of Origin" mark and protocol against the EU's definition of a Geographical Indication. This process was successfully completed, and the EU confirmed the GI status of the quality indication mark "Certified Karoo Meat of Origin". So now the EU was finally satisfied that South Africa's legal system provide sufficient protection for GIs even though the Karoo Lamb registration under the Merchandise Marks Act was unsuccessful. The

EU commission was satisfied that the 'rules of use', the registration under the meat regulations and the trademark protection of the certification mark (Certified Karoo Meat of Origin), provided sufficient protection of the name. What they however did not know was that the Department of Agriculture has since 2013 allowed four new registrations of 'quality indications' containing the words Karoo Lamb. Again, creating market confusion and eroding the collective reputation of the name and thus acting against the intentions of the proposed GI dispensation.

Under Protocol 3 to the SADC-EU EPA, South Africa now protects 251 EU GIs covering food (e.g., meats, cheeses, fruit, vegetables, olive oil), wines and spirits. In return, the EU protects 105 GI names from South Africa. These include 102 wine GIs that are closely aligned to the registered Wines of Origin in South Africa plus three additional non-wine agricultural products—*Karoo Meat of Origin*, *Rooibos* and *Honeybush*. The South Africa GIs protected by the EU and the EU GIs protected by South Africa are all listed in Annex I to the Protocol 3.

13.5   The Introduction of GI Regulations in South Africa

When the SADC-EU EPA was finally signed in 2016 the task was now for the Department of Agriculture to finalise specific regulations for the registration and protection of Geographical Indications so that all GI names[2] can move from the Merchandise Marks Act to these new regulations. The initial drafts of the regulation were prepared towards the end of 2015 and the first full draft published for public comments in early 2016. Comments were received, a second draft published in September 2016 and more comments received. The Regulations were finalised in November 2016, but it took more than 2 years for the Minister of Agriculture to sign off on the regulations and R.447 of 2019 was only gazetted in March 2019.

The first test case of the regulations was the application for the Karoo Lamb GI. As we explain below the Karoo Lamb application highlighted several legal and technical problems with the regulations which again scuttled the process to register the Karoo Lamb GI. Consequently, there had to be amendments to the regulations. The amendments address the constraining rules of 'representativity of the group', the role of auditors, and the introduction of the Republic of South Africa (RSA)-GI logo.

---

[2] By the end of 2016 a total of 112 European GIs were listed and protected under the Merchandise Marks Act.

13.6   Karoo Lamb Applications for GI Registration

Following the publication of the GI regulations in March 2019 the consortium, representing the interests of the Karoo Lamb supply chain members, submitted the application for the official registration of the *Karoo Lamb* GI in June 2019. After several exchanges with the officials in the Department of Agriculture the notice of the 'intention to register Karoo lamb as a GI' was published in the government gazette of 2 August 2019 for public comment. The comments received were rather disappointing and once more illustrated the lack of understanding of South Africans about the meaning and role of a GI. The lack of collective action, the stubbornness of individual farmers and the mere size of the Karoo region (larger than most countries in Europe) clearly contributed to a view that the GI application is likely to infringe on their private rights. However, none of the objections have complied with the GI regulations in as far as these objections are based on conjecture and did not reveal any scientific proof or supporting facts as is required. The objections typically came from the following groups that may experience a limitation in their current practises when the GI Rules for Karoo Lamb are gazetted:

> Lamb that are produced outside the Karoo region but is claimed to be otherwise.
> Lamb that are produced in the Karoo region but are raised on pastures or in feed lots.
> Lamb that are produced and sold with neither traceability nor a verification in place.
> Lamb that are produced, slaughtered, or processed on farms, abattoirs or processing facilities that do not comply with prescribed rearing conditions, minimum government hygiene regulations or acceptable slaughtering practices.

Despite a well-argued rebuttal and counterstatement by the applicants, the Department of Agriculture still thought it wise to reject the application in early January 2020 on the grounds that insufficient buy-in was obtained from all stakeholders and that an inclusive stakeholder meeting should be called to obtain final agreement on the rules for the GI. This stakeholder meeting took place on 28 February 2020 where the rules were agreed upon, a more representative board of directors elected, and the name of the representative organisation changed to the "Karoo Lamb Consortium" (from "Meat of Origin Karoo" previously). The board of directors now had representatives from abattoirs, retailers/processors as well as representatives from communities previously excluded from commercial agriculture in South Africa. A minority of representatives questioned whether the Consortium should submit the application. Nevertheless, the resolution that the Karoo Lamb Consortium should apply for the GI registration was passed with more than 90% of delegates voting in favour.

The 2nd application for the GI registration was submitted in July 2020 and finally (due to COVID-19 delays) published for public comment on 9 October 2020. After a long process of engagement with lawyers on the interpretation of the regulations, the Department of Agriculture once again rejected the application on 30 April 2021 on the grounds that the applicant group fails to comply with the regulation that "at least 50% of the production volume of the agricultural product concerned". We maintain that this is an incorrect interpretation as the product in question is those lamb carcasses sold and marked as Karoo Lamb. The State lawyers interpreted it as meat produced in the Karoo geographical area that have the <u>potential to be sold</u> as Karoo lamb.

The other reasons provided for the rejection related to the resignation of directors and the internal auditor after the official publication of the notice. These latter reasons are rather strange and not related to the regulations but taken together illustrates the government's unwillingness to protect any GI in South Africa.

The Karoo Lamb Consortium expressed their dismay with the outcome and stated *"...we wish to record that the application to register Karoo Lamb as a GI is an effort to protect a product from the Karoo region from being exploited by individuals, instead of inhabitants and from produce, instead of region."*

This situation obviously created a larger dilemma for the South African government and necessitated a change to the current regulations governing the registration of GIs. This was needed to remove the prohibitive and unnecessary dimensions of the regulations. Draft amendments of the regulations were published for comment in August 2021 and following comments received a 2nd draft was published on 22nd March 2022.

This version was an improvement, showing that the Department of Agriculture has now finally realised the public good and enabling nature of GI rules. They made the following important pronouncement in response to the previous comments and the new proposed regulations:

> *This office strongly supports the principle that a registered geographical indication name shall benefit all producers (and not only certain groups or individuals) of a specific agricultural product produced in that geographical region in South Africa. However, this shall always be subject to the provision that (a) a minimum product specification is compiled and agreed upon by the affected producers in the geographical area, and (b) the minimum product specification is published as an Annexure to the South African geographical indication regulations.*

The 2nd draft amendment regulations therefore entail the revision of the entire regulation R.447 dated 22 March 2019 to accommodate this principle and to ensure that the Department of Agriculture, Land Reform and Rural Development (DALRRD) takes ownership of geographical indications registered in South Africa in terms of the Agricultural Product Standards Act (No. 119 of 1990) ('APS Act')."

The final workshop (3rd) between the Department of Agriculture and stakeholders to discuss the draft government regulations for GI registration in South Africa took place on 9 June 2022.

The meeting discussed the comments and proposals made on the 2nd draft of the amended regulations circulated to stakeholders on 18 May 2022. Most attendees agreed with the Department's responses and then discussed a few remaining issues. These were all amicably resolved, and the meeting agreed to let the Department finalise the regulations. The new regulations excluded the various requirements on representation and only focussed on the evidence related to the regional attributes of the product. The changes to the draft regulations that were agreed upon in the meeting were again circulated amongst stakeholders in July 2022. With no comments received the regulations were approved and forwarded to the Minister of

Agriculture for her approval and signature. This took some time, and it was only towards the end of November 2022 when Ministerial approval was obtained. The new regulations for the registration of GIs in South Africa were finally published in the government gazette of 10 February 2023. It is now known as R. 3023 of 10 February 2023: *Regulations relating to the Protection of Geographical Indications and Designations of Origin used on Agricultural Products intended for sale in the Republic of South Africa*. The publication of the regulations implied that South Africa now for the first time had a workable sui generis system of rules that can govern the registration and protection of GIs in South Africa.

Following the publication of the regulations in February 2023 the Karoo Lamb Consortium spend some time to integrate the definition and rules of Karoo Lamb as was agreed upon in the meeting of March 2022, into the application document and reorganise all the evidence and proofs for registration. The final application for the Karoo Lamb GI was submitted on 5 June 2023. The Department requested some changes to ensure clarity in the various rules of use. They worked together with the members of the Karoo Lamb Consortium to ensure that all empirical evidence are in place. A final stakeholder meeting was also held on 21 July 2023 where last refinements to the rules where discussed. A few outstanding issues were resolved with government officials and the revised final application document was submitted on 4 August 2023.

The process here illustrates how important it is to take all stakeholders and government officials along in the process of developing specific regulations for GIs. It takes some time for individuals that are used to systems of private intellectual property rights and their role as regulators, to understand their new roles in the context of GIs.

On 30 August 2023 the Registrar of Agricultural Products Standards (DALRRD) published notice No. 3828: Intention to Register Karoo Lamb/Karoo Lam as a South African Geographical Indication (GI): Invitation for Objections. The closing date for objections and comments was 30 September 2023. The 30-day period expired without any objection or comment received. We found this rather interesting given that there were no real fundamental changes to the rules of Karoo Lamb from the previous applications. The simple recipe of a lamb produced under free range conditions on Karoo veld and slaughtered in the Karoo remained the core definition. Having received no objections, the Department officially registered *Karoo Lamb* as a GI in the Government Gazette of 27 October 2023—almost two decades since the initial intellectual work was undertaken in South Africa.

Notice 3992 of 27 October 2023 as shown below, confirmed the GI status of *Karoo Lamb* and is the first GI to be registered in terms of the South African GI regulations (Fig. 13.1).

The name "*Karoo Lamb*" is now protected by law and may only be used in accordance with the GI and Designation of Origin regulations. The South African Meat Industry Company (SAMIC) as the designated assignee responsible for the enforcement of the Red Meat Regulations is subsequently responsible for the inspections of producers and groups of producers using the registered SA GI

DEPARTMENT OF AGRICULTURE, LAND REFORM AND RURAL DEVELOPMENT

NO. 3992   27 October 2023

AGRICULTURAL PRODUCT STANDARDS ACT, 1990 (ACT No. 119 OF 1990)

REGISTRATION OF *KAROO LAMB/ KAROO LAM* AS A SOUTH AFRICAN GEOGRAPHICAL
INDICATION (GI)

It is hereby made known that the Executive Officer: Agricultural Product Standards has registered the *Karoo Lamb/ Karoo Lam* as a South African Geographical Indication ('GI') in terms of the "Regulations relating to the protection of Geographical Indications and Designations of Origin used on Agricultural Products intended for sale in the Republic of South Africa" ('The GI and Designation of Origin Regulations') as published under no. R 3023 under *Government Gazette* no. 48015 on 10 February 2023.

Further made known is that the Executive Officer: Agricultural Product Standards in terms of the GI and Designation of Origin Regulations hereby inserts after regulation 24, the following Annexure in respect of the "Applicable Minimum Product Specifications for the registered South African GI *Karoo Lamb/ Karoo Lam*".

ANNEXURE A

THE APPLICABLE MINIMUM PRODUCT SPECIFICATIONS FOR 'KAROO LAMB'/ 'KAROO LAM'

1. Type of Agricultural Product

Karoo Lamb/ Karoo Lam is a (primary or processed) lamb meat product regulated in terms of sections 3(1) and 15 of the Agricultural Product Standards Act, 1990 (Act No. 119 of 1990) and that is intended for sale in the Republic of South Africa.

2. Descriptions of the characteristics of KAROO LAMB/ KAROO LAM

2.1 Minimum Requirements for the claim
The name *KAROO LAMB/ KAROO LAM* applies to all meat cuts from lamb produced under free range conditions and slaughtered in the geographical area as defined herein. Only lamb born, raised, and slaughtered in the geographical area as defined qualifies as the product concerned. The product must comply with the following specifications prescribed by the "Regulations regarding the Classification and Marking of Meat intended for sale in the Republic of South Africa" as published under no. R. 55 under *Government Gazette* no. 38431 on 30 January 2015.

Breed		Preferably meat breed types with good bone: muscle ratio with an even fat distribution.
Carcass mass and age class	A	>12 but < 25kg
Classification	Age classes	A
	Fat classes	1 to 6
	Conformation	3, 4 and 5
Damage		Only F1 damage allowed

Karoo Lamb meat has specific aromatic and sensory attributes, which can be directly attributed to the geographical area. Various scientific studies have shown that the grazing plants from the Karoo region as defined herein impart herbal and musty flavour attributes to meat from sheep breeds of this region.

Fig. 13.1 An extract from the government gazette of 27 October 2023

Karoo Lamb to establish and ensure compliance with the critical elements of control specified under the Annexure A of the Notice No. 3992 mentioned above.

No implementation period of the *South Africa Karoo Lamb* GI upon its registration was prescribed in the GI and Designation of Origin Regulation, therefore the requirements of the GI and Designation of Origin Regulation are effective immediately.

The core aspect of the GI rules for *Karoo Lamb* is situated in the specific production practices that are summarised in the Box 13.1 below.

Box 13.1 Summary of the *Karoo Lamb* Specifications

The traditional production practices for meat to be covered by the proposed GI imply free range grazing on indigenous veld vegetation in the Karoo as defined. Vegetation which is typically associated with the Karoo include the following shrubs:

- Plinthus karrooicus ("Silverkaroo")
- Penzia spinescens ("Skaapbossie")
- Pentzia sphaerocephale ('Berggansie")
- Pentzia lananta ("Blesbokkaroo")
- Pentzia globose ("Vaalkaroo")
- Pentzia calcarea ("Meerkatkaroo")
- Pentzia incana ("Ankerkaroo")
- Eriocephalus ericoides ("Kapokbossie")
- Eriocephalus spinescens ("Doringkapokbossie")
- Salsola glabrescens ("Rivierganna")
- Pieronia glauca / Rosenia humilis ("Perdebos")

The rules for the production for Karoo Lamb are therefore derived from the established fact that these and similar Karoo shrubs are unequivocally responsible for the physical, chemical, and organoleptic characteristics of the product.

The production practices that would sustain this intrinsic attribute of the product are:

- Free range grazing on indigenous Karoo vegetation
- The occasional use of supplementary feeds (such as licks) for animals grazing in the veldt that may contain cereals, silage, or any other natural plant matter with the understanding that the dominant feed intake will still be the natural grazing.
- No animal by-products are allowed in supplementary feeds.
- In the case of long dry spells on specific farms or sub regions of the Karoo where grazing conditions have deteriorated so much that free range grazing is impossible will unfortunately render the production of Karoo Lamb unavailable and farms/regions will not be able to produce Karoo Lamb during that period as there is no vegetation.
- Animals originating from feed lots (as opposed to free range grazing) in the Karoo do not qualify for use of the proposed GI.
- Animals reared on cultivated or planted pastures, do not qualify for use of the proposed GI.
- The use of routine antibiotics and the application of any hormones/growth promoters/stimulants are strictly prohibited.

(continued)

Box 13.1 (continued)

Slaughtering shall take place at an abattoir in the Karoo geographical area that has been registered for classification in terms of the "Regulations regarding the classification and marking of meat intended for sale in the Republic of South Africa" (Government Notice No.R.55 of 30 January 2015). Further processing (cutting and packing) may take place outside the Karoo area, if traceability is maintained.

Source: author creation.

13.7 Conclusion

This Chapter documented the Karoo Lamb producers' quest to get GI protection for *Karoo Lamb* and illustrated the dedication and endurance to finally get the GI registered despite the initial deliberate attempts by individual producers and retailers to obstruct the process of GI registration. In the end the Karoo Lamb team persevered and ensured that history was made by ensuring the registration of Karoo Lamb as the first GI in South Africa in terms of the newly drafted GI regulations. The lessons learned during the Karoo Lamb process will be valuable for any agricultural product in South Africa with true origin characteristics to protect their regional identity by means of a GI registration.

The initial opposition to the GI process was clearly embedded in the existing system of intellectual property rights protection in South Africa, which vests all rights in the hands of individuals or individual companies. As a result, there was limited appreciation for the value of collective brands, certification marks and GIs per se. In addition, there was considerable mistrust and confusion amongst producers and abattoirs about the role of GI organisations and their role in helping individual entrepreneurs expand their business to protect the reputational value of a GI product. It is hoped that the GI registration and the growth in the market for *Karoo Lamb* will overcome these wrong perceptions.

References

Biénabe E, Bramley C, Kirsten J, Troskie D (2008) Linking farmers to markets through valorisation of local resources: the case for intellectual property rights of indigenous resources. IPR Duras Project Final report, Montpelier/Pretoria

Bramley C, Bienabe E, Kirsten JF (eds) (2013) Developing geographical indications in the south: the Southern African experience. Springer, Dordrecht

Erasmus S (2017) The authentication of regionally unique South African lamb. PhD thesis, University of Stellenbosch, Stellenbosch

The opinions expressed in this chapter are those of the author(s) and do not necessarily reflect the views of the [NameOfOrganization], its Board of Directors, or the countries they represent.

Chapter 14
Better Interinstitutional Coordination for the Efficient Operation of the Delegated Entities of the PDOs in Colombia

Claire Philippoteaux and Yudy Paola Pineda Suarez

Acronyms

AIN	Regulatory Impact Analysis
PDO	Protected Designation of Origin
EUIPO	European Union Intellectual Property Office
GI	Geographical Indication
INVIMA	National Institute for Food and Drug Surveillance
IP	Industrial Property
SIC	Superintendency of Industry and Commerce
EU	European Union

14.1 Introduction

In Colombia, protection through Protected Designations of Origin (PDO) is declared by the Superintendency of Industry and Commerce (SIC), which is also the entity in charge of delegating the power to authorize the use of this sign to a third party that represents the beneficiaries of the PDO and can demonstrate that it complies with the legal requirements and suitability to carry out promotion, surveillance, and control tasks.

To date, only 18 of the 29 declared PDOs have a delegated entity, and among these, 12 are delegated entities of agro-industrial PDOs. Few comply with all the

C. Philippoteaux (✉)
Colombian-Swiss Intellectual Property Project – COLIPRI, Bogotá, Colombia

Y. P. Pineda Suarez
Company La Selección SAS, Moniquirá, Colombia

© The Author(s) 2025
E. Vandecandelaere et al. (eds.), *Worldwide Perspectives on Geographical Indications*, https://doi.org/10.1007/978-3-031-71641-6_14

requirements established by the SIC, such as implementing a control and surveillance system linked to their usage regulations. And they have not been able to operate effectively due to two main reasons: the lack of interinstitutional coordination and the little support from the State to the delegated entities.

In this text, we propose to illustrate the topic through examples. We will first address the case of health standards: in the country, apart from flowers, agro-industrial PDOs are food products, which must comply with the regulations issued by the Ministry of Health and Social Protection and respond to the health control authority, that is, the National Institute for Food and Drug Surveillance, INVIMA.

In this study, it can be seen that health standards prevent strict compliance with the PDO usage regulations and, therefore, limit their visibility, the control role of the delegated entities, and, consequently, the good economic development of the PDOs.

We will then look at the international aspect of Colombian PDOs, through their recognition by the European Union. There are several ways for Colombian PDOs to have the PDO or the Geographical Indication in the European Union. One way is direct application and the other is inclusion in the list of recognized PDOs through the Free Trade Agreement between the EU and Colombia. Again, the country's PDOs are not backed enough to be able to use and enforce their collective distinctive sign in the EU.

This study offers a qualitative exploratory approach, as the subject of study has been little investigated. It is worth highlighting the importance of this in the research to be developed, as the objective is to obtain information that allows understanding the behavior of the actors, considering the need to know the operational processes carried out in the delegated entities and their relationships with the institutions, as well as the relationships between these institutions regarding the PDOs.

The method used was the case study for both examples: taking as references the health standards and the coordination between the SIC, INVIMA, and the delegated entity for the first example; and the communications between the Ministry of Industry and Commerce, the SIC, the European Union, and the delegated entity of the *Bocadillo Veleño* PDO for the second.

14.2 The Case of the Conflict Between Health Standards and the Control of the Paipa Cheese and Bocadillo Veleño's PDOs

14.2.1 Methodology

The sample selection was made for convenience, considering the ease and access to information and the availability of time of the people. In this way, an organization that already has the power to authorize the use of the PDO by the SIC was selected, and another organization that is in the process of obtaining it. The first is

Fedeveleños, the delegated entity for the management from the PDO *Bocadillo Veleño*, and the second is based on the experience of the PDO of *Queso Paipa*.

The following information gathering techniques were used:

- Observation: the authors have interacted with different delegated entities and have been able to identify areas where there is a lack of interinstitutional coordination.
- Interviews: Semi-structured interviews were conducted with members of the two selected organizations within the sample.

14.2.2   Results

In Colombia, every processed food producer must be registered in INVIMA, including the manufacturer, product name, and product brand, among other data. In the product name, some manufacturers register a name corresponding to a PDO, regardless of the business location or the relationship with the PDO. To date, INVIMA accepts such name in the records, arguing that the name that the manufacturer assigns to its product does not affect its safety, and therefore, does not pose sanitary risks for INVIMA.

The conflict arises in the product labeling when containers or labels are checked for producers who do not have the authorization to use the designation of origin, either by the delegated entity, or directly by the SIC. INVIMA requires that the labeling be consistent with the information registration they have made before them, while the delegated entity for the authorization of use and administration of the PDO has not allowed them to use such name. For the producer, it is more important to comply with the sanitary entity, which has the power to sanction them at the time of inspection, while the delegated entity has little power at the time of their visit and little capacity for action, either due to ignorance or lack of technical, human and financial capacity.

Conflicts were also found between what is stipulated in the sanitary norms and the usage regulations: reading the two documents in parallel, it is impossible to comply with both. This shows that producers navigate between two guidelines, not knowing which one they should follow more strictly.

It was identified that the health standards that create conflict with the PDOs have been established before the recognition of the PDO, which allows us to think that there really is a lack of coordination and communication between the industrial property office and the health control authority.

On the one hand, it seems that the SIC alone is the one that carries out the study of the conditions of a PDO without carrying out a regulatory impact study of it, in which it is verified whether this is or is not in conflict with other standards of other institutions. On the other hand, during the process, when the specifications are published to present oppositions, there is no response from other government entities, in this case the INVIMA.

The conflicts mentioned are only illustrative of other possible ones, as well as of inconveniences that the delegated entities may have when exercising their control actions and follow-ups to the authorized producers.

14.3 The Case of the Use of the EU GI Seal for Colombian PDOs

14.3.1 Methodology

For this case, the protection process in the European Union of the PDO of *Bocadillo Veleño* was taken as a central example, as it is a case close to the authors, and the last Colombian PDO protected in the EU to date.

The following information collecting and analysis techniques were used for the case:

- Observation: The stages between the declaration of the PDO of *Bocadillo Veleño* in Colombia in 2017, and its protection in the EU at the end of 2022 were noted.
- Interviews: Calls were made and information was sought through officials from Colombian public entities and the European Union, in order to understand the processes and show the lack of information and communication about the proper use of the European PDO seal by the Colombian PDOs protected in the EU.

14.3.2 Results

Since September 2012 and the publication of chapter 7 of the circular unique of the Superintendence of Industry and Commerce, in Colombia, when a PDO is declared, an entity that represents the beneficiaries of this is delegated at the same time with the power to authorize the use of the PDO. This delegation entails many responsibilities for the delegated entity, having to develop an authorization system, control and observance of its PDO. The delegated entity becomes the institution responsible for promoting the PDO, but also for ensuring the proper use of its name and make sure it is respected.

In the case of the PDO of *Bocadillo Veleño*, the same day that the PDO declaration was given, in June 2017, the power to authorize its use was delegated to Fedeveleños, the Federation of the productive chain of Bocadillo Veleño. Since then, Fedeveleños has developed initiatives to publicize the PDO, has authorized five companies for its use and initiated several actions against PDO infringers. All these activities have taken place in the territory of Colombia, since PDO, like all IP rights, is territorial.

In parallel to these activities, and on its own, the Ministry of Commerce, Industry and Tourism of Colombia initiated the process before the European Union so that

Bocadillo Veleño could be included in the European register, by virtue of the trade agreement between the European bloc and Colombia. This process took almost 5 years, and in November 2022, the appendix 1 of annex 13 of the Trade Agreement was modified, in order to include Bocadillo Veleño as Colombian product number 14, being protected in the EU. Before and during this long process, Fedeveleños was never consulted, nor notified once the protection was achieved.

The problem arises at the moment of using this protection, which for Colombian PDOs like *Bocadillo Veleño* involves the use of the European Geographical Indication seal, as Colombian PDOs recognized through multilateral agreements are geographical indications in Europe. In good faith, the companies authorized for the use of the PDO of *Bocadillo Veleño* in Colombia, like the authorized companies of several other products with Colombian PDOs registered in the EU, have used both logos on their packaging, both the national and the European one, in order to give greater visibility to the collective sign, to open possible international markets and to indicate to the consumer that they are indeed authorized companies.

But in the interviews and discussions held with officials from the European Commission and the EUIPO, the use of the European GI seal is not automatically granted when a PDO is included in the multilateral trade agreement. According to the same conversations, the process is done by official request to the European Commission. On the side of the Colombian authorities, the process is not clear and the IP office, the SIC, among others consulted, recommend to delegated entities like Fedeveleños, to consult directly with the European Commission.

An incongruity is then identified between a country that for years carries out procedures to protect its PDOs at an international level via trade agreements and the lack of accompaniment during and after the process with the entities to which the same country delegated the management, promotion and defense of the PDOs. One may question the introduction of this PDO of *Bocadillo Veleño*, like several Colombian PDOs, in the trade agreement with the European Union, instead of leaving the option to the delegated entity, in this case Fedeveleños, to carry out the process directly, if they wish, and thus having the possibility to use the European GI seal. For PDOs as small as that of *Bocadillo Veleño*, the process to become a GI in the EU would not start so soon directly due to the high cost it entails. In addition, the European market is not a significant market for many Colombian PDOs, and cases of usurpation of the PDO name are rare. Therefore, international registration is not a priority for the delegated entities of the PDOs, but it is for the national authorities, who see in it the possibility of demonstrating management and making visible the country's intellectual property system.

As in the case of the conflict between health regulations and the control of the PDOs of Queso Paipa and Bocadillo Veleño, the lack of interinstitutional coordination and clear communication between the entities of the IP system, those of international trade and the beneficiaries of the PDOs, is noticeable. As a result, there may be PDOs that are not certain of their protection at an international level, do not know how to take advantage of this protection and even commit infringements.

14.4   Conclusions

The initial research conducted in just 2 PDOs out of the 29 that exist in Colombia illustrates the need to better understand the sign as a mark of collective vocation more broadly. That is, not only should and can the delegated entity be responsible for the good use, authorization, control, and promotion of its PDO, but it should be able to count on technical support from different state entities: the IP office, but also the health authority or the authority of negotiation of free trade agreements, in our examples.

In addition to this, in other PDO cases, there could be a need to link other entities such as the Ministry of Industry and Trade, Ministry of Agriculture, Ministry of Culture, the Ministry of Social Protection, the Ministry of Mines and Energy, among others, not only for support to the delegated entity but from the beginning to participate during the process of evaluation of the recognition of the PDO, in order to achieve regulatory harmony through a process of Regulatory Impact Analysis (RIA) or by agreeing to exempt the application of rules prior to the geographical indication.

PDOs should be able to dialogue with the respective IP office, the SIC in Colombia, to establish a practical work plan, beyond compliance with intellectual property law. Only in this way can we have strong delegated entities and PDOs with a real impact and significant visibility. The size, the administrative organization and reality of the delegated entities in Colombia have not allowed them, for the moment, to have this discussion, and it is time to have it.

In practice, a short-term solution would be to unite the Colombian PDOs into a single group, in order to have the necessary dialogue with public entities, and on the side of the SIC, to carry out permanent training for the delegated entities on the responsibilities and duties that come with the delegation of the power of authorization of use of the PDO.

In addition, it is the Colombian State, through the different competent institutions, which must take ownership of the PDOs and finance dissemination processes so that citizens recognize what a PDO is, its importance, and which products have this distinctive sign.

References

Colombo-Swiss Intellectual Property Project COLIPRI II—[Online]. Available: https://www.ige.ch/de/recht-und-politik/entwicklungszusammenarbeit/laufende-projekte/kolumbien-phase-2. Last accessed 16 Oct 2023

INVIMA. https://www.invima.gov.co/8-pasos-para-obtener-su-registro-sanitario-de-alimentos [Online]. Available: https://www.invima.gov.co/8-pasos-para-obtener-su-registro-sanitario-de-alimentos. Last accessed 16 Oct 2023

Ministry of Commerce, Industry and Tourism [Online]. Available: https://www.mincit.gov.co/prensa/noticias/comercio/ue-indicacion-geografica-para-el-bocadillo-veleno. Last accessed 16 Oct 2023

Superintendency of Industry and Commerce. Denominations of Origin [Online]. Available: https://www.sic.gov.co/marcas/denominaciones-de-origen. Last accessed 16 Oct 2023

Superintendency of Industry and Commerce. SIPI—Intellectual Property System [Online]. Available: http://sipi.sic.gov.co/sipi/Extra/Default.aspx?sid=637738127350207084. Last accessed 16 Oct 2023

Chapter 15
Empirical Investigation of Fraud and Unfair Competition Practices in France and Vietnam: Actors, Types and Drivers

Abbreviations

AOC	Appellation d'Origine Contrôlée
COFRAC	Comité Francais d'Accréditation
DGCCRF	Direction Générale de la concurrence, de la consommation et de la répression des fraudes
DOST	Department of Science and Technology
EU	European Union
EUIPO	European Union Intellectual Property Office
INAO	National Institute of Origin and Quality
GI	Geographical Indication
IP	Intellectual Property
MOST	Ministry of Science and Technology
NOIP	National Office of Intellectual Property
OAPI	Organisation Africaine de Propriété Intellectuelle
ODG	Organisation for the Defence and Management
PGI	Protected Geographical Indication
PDO	Protected Denomination of Origin
STAMEQ	Directorate for Standards, Metrology and Quality
TRIPS	Trade-Related Aspects of Intellectual Property Rights
WTO	World Trade Organization

B. Pick (✉)
Centre International de Recherche Agronomique pour le Développement (CIRAD), Rome, Italy

© The Author(s) 2025
E. Vandecandelaere et al. (eds.), *Worldwide Perspectives on Geographical Indications*, https://doi.org/10.1007/978-3-031-71641-6_15

15.1 Introduction

The protection of geographical indications (GIs) is traditionally justified on an informational efficiency basis (OECD 1996). In the context of experience or credence goods, for which consumers can ascertain quality only after buying and using them, and credence goods, for which consumers cannot ascertain quality even after using them, as is the case for many types of agricultural and food products, consumers usually rely on trust, expertise or reputation of the seller, professionals, third-party certifications or labels to gauge the quality of such products (Nelson 1970). By correcting market failure due to information asymmetries, GIs meet an important social need in enhancing product information to consumers as per the geographical origin of the labelled products, thus saving their search costs in making choices (Landes and Posner 2003). From that perspective, GI protection derives from the imperfect information theory (Stiglitz 1989; Tirole 1988), in a context where connection to place has become increasingly important in marketing and consumer behaviour (Pike 2009). Several surveys, carried out predominantly in Europe, show that consumers are increasingly sensitive to the origin of products. For example, the latest Eurobarometer survey carried out in 2022 shows that the respect of local traditions and know-how is an important factor when buying food products for a great majority of Europeans (from 56% to 97% depending on the country), that consumers are more likely to buy food products that come from a geographical area that they know, and that having a specific label ensuring the quality of the product is important for at least two thirds of respondents (European Commission 2022). GIs have thus become valuable economic assets and useful marketing tools to increase market access and capture price premiums through product differentiation owing to the link between products' specific origin and their unique quality, characteristic or reputation (Bramley 2011). In this context, the producers' goodwill and the brand's reputation constitute the underlying 'valuable intangible that is being protected' (Gangjee 2012: 145) against free-rider competitors. Referred to as the 'institutionalisation of reputation' (Belletti et al. 2000: 239), GIs primarily aim to protect consumers and producers' interests against fraud and unfair competition practices by preventing name usurpation and diversion of income (Bramley et al. 2009). While the opportunity to capture price premiums is 'often one of the first aims of supporting a strategy for an origin-linked product' (European Commission 2011), the potential of GIs to bring about economic and commercial benefits also explains why GI products have been increasingly prone to fraud and unfair competition practices. The more the reputation and specificity of a product are acknowledged by consumers, the more likely its name may be usurped and misappropriated (Belletti et al. 2000). To take a few examples, it is estimated that there are over 100 trademark offences of the name Basmati in over 30 countries Jena and Grote 2010), that the usurpation of the name 'Karoo lamb' from South Africa is commonplace (Biénabe et al. 2011), and that about forty million kilograms of tea are sold worldwide as Darjeeling tea every year, while the production of genuine *Darjeeling tea* is only ten million kilograms (Das 2006). Unfair business practices, which stem from the commercial success of origin names in relation to greater market access and possible price premiums, result in loss of revenue for the genuine producers while misleading consumers in their purchasing decisions (Das 2009). In the European Union, the value of GI infringing products represents € 4.3 billion per year, constituting 9% of the total volume of products sold as GI (EUIPO 2016).

Whereas there is a rich literature on the economic and legal justification and effects of GI protection, including the relationship between the product reputation before the GI legal protection and the development of fraud and unfair competition practices in the marketplace (Belletti et al. 2007), very few authors have studied the non-economic drivers of such practices. Further, no empirical investigation has been conducted on the various types and root causes of such practices nor on the plurality of actors involved therein. Building upon the existing research, the objective of this paper is to understand better fraud and unfair competition practices by unpacking the economic and non-economic factors contributing to their development across two contrasting protecting systems in France and Vietnam. Why and how fraud and unfair competition practices develop in both contexts? Which actors are involved and when? Using a qualitative methodology, this paper is based on a comparative analysis of five case studies in France and Vietnam. Primary data were collected through semi-structured interviews that were conducted in France and Vietnam in 2014 and by phone and email communications in 2017–2024. Secondary data were generated through desk study of legal texts, product specifications, registration regulations, and evaluation reports. This comparative case study approach is critical for understanding whether and how the broader socio-economic, legal and institutional environment in which GI products are embedded in each country impact differently on the development of fraud and unfair competition practices.

15.2   Enforcement of GI Rights in France and Vietnam

At the international level, the TRIPS Agreement sets forth the legal framework for GI protection in its 164 members, including France and Vietnam, through two types of protection that have important consequences in the fight against fraud and unfair competition. Article 22.2 provides the standard protection for all GIs against misleading conduct and unfair competition. A higher-level protection is afforded in Article 23.1 for wines and spirits only. Within this category of products, any use of the indication on goods that do not originate from the indicated place is strictly forbidden, whether or not consumers are deceived, 'even where the true origin of the goods is indicated or the [GI] is used in translation or accompanied by expressions such as 'kind', 'type', 'style', 'imitation' or the like'. Consequently, Article 22.2 requires proving consumer confusion or unfair competition within a specific context, while Article 23.1, in providing protection '*per se* or in absolute terms' (Watal 2001: 268) beyond misconception or unfair competition, treats GIs as 'objects, regardless of their connotations in a specific context' (Gangjee 2012: 238).

The obligation for the WTO Members to implement the TRIPS provisions through the method of their choice [TRIPS Agreement, Article 1.1] has led to very diverse approaches for the enforcement of GI rights (O'Connor 2004), including regulations focusing on business practices, including unfair competition, consumer protection and passing off; collective and certification marks; and *sui generis* systems that acknowledge GIs as a distinct category of right to be protected outside typical legal protection systems. *Sui generis* systems usually impose a formal registration process that goes beyond the minimum standards of protection of the TRIPS Agreement (Correa 2007) and make provisions for an official control of the product specification. Both France and Vietnam have adopted a *sui generis* system for protecting GIs, which will be the focus of this section.

The terminology that applies to the *sui generis* protection of GIs for agricultural products and foodstuffs products in France is particularly complex with different terms in use due to the coexistence of the French and European legislation. Historically, the French *sui generis* protection system of GIs has derived from a long, drawn-out process initiated by local producers which was first designed to fight against the increased levels of fraud and artificial wines following the phylloxera epidemic at the end of the nineteenth century. A few decades later, the Decree-Law of 30 July 1935 on the defence of the wine market and the economic regime of alcohol created the concept of controlled appellations of origin (*appellation d'origine contrôlée*—AOCs). It also established the National Institute for Appellations of Origin (INAO—which became the National Institute of Origin and Quality in 2006), a public institution operating under the authority of the French Ministry of Agriculture and Food that is specifically dedicated to the recognition, control and defence of AOCs which cover all types of agri-food and forestry products since 1990. The French AOC legislation subsequently inspired European law (Allaire et al. 2011) which introduced the concepts of protected denominations of origin (PDOs) and protected geographical indications (PGIs) for agricultural products and foodstuffs through the Council Regulation (EEC) No 2081/92 of 14 July 1992, as repealed by Council Regulation (EU) No 1151/2012 which was itself recently repealed by Regulation (EU) 2024/1143 of 11 April 2024 on GIs for wine, spirit drinks and agricultural products and other quality schemes for agricultural products, now in force. The French and European legislation have also recently provided for the protection of GIs for industrial and artisanal products, however these will not be detailed here considering that industrial and artisanal products are not included in our French case studies.

In Vietnam, the Intellectual Property Law 50/2005/QH11 [hereafter 'IP Law'] was adopted on 29 November 2005 to comply with the TRIPS Agreement. It covers any type of local products, including raw materials, agricultural, food and drink products, industrial and handicraft products. All GIs fall within the competence of the National Office of Intellectual Property (NOIP) under the authority of the Ministry of Science and Technology (MOST).

The present section will set out the relevant legal context in France and Vietnam for the enforcement of GI rights through *sui generis* protection, with a specific focus on identifying the right-holders, the type of infringements, and the pre-market and post-market quality control mechanisms in each country.

15.2.1 Right-Holders in French and Vietnamese Law

In French and European law, only those stakeholders who are located inside the concerned region and who comply with the product specification have the right to use the AOC [Rural Code, Article L.641-5], PDO or PGI logos [Regulation 2024/1143, Article 36], taking into account that the use of an AOC, PDO and PGI logo is also subject to results of the quality controls [Rural Code, Article L.642-3].

In Vietnam, the state is the owner of all GIs, and as such it has the right to manage GIs [IP Law, Article 121.4.], which includes the right to grant licenses to use GIs [IP Law, Article 123.2(a)]. The right to manage can be transferred to the People's Committee(s) of the Province(s) or city(ies) to which the GI pertains. It can also be transferred to a representative organisation of all collectives and individuals conferred with the right to use the GI [IP Law, Article 121.4], as authorized by the People's Committees of the Province or city to which the GI pertains. The right to use GIs can be granted to organisations or individuals located in the relevant area and involved in the production and marketing of the GI product [IP Law, Article 121.4]. Local stakeholders must apply individually to the management organisation to obtain the right to use the GI which is not transferrable [IP Law, Article 139.2]. Other emerging and developing countries have adopted a similar system. For instance, in India, where the GI registration process is led by public authorities, producers join in the GI process only after they have applied for the right to use the GI (Bramley et al. 2013).

15.2.2 Types of Infringements in French and Vietnamese Law

The scope of protection is similar in French and European law, yet European legislation is more detailed. French legislation prohibits the use of any evocative term of the appellation on any similar product, as well as on other establishment, product or service when such use is likely to misappropriate or weaken the reputation of the name [Rural Code, Article L.643-1]. Likewise, European law protects against any direct or indirect commercial use on comparable products; any misuse, imitation or evocation, even if the true origin is indicated or if the protected name is translated or accompanied by an expression such as 'style', 'type', 'method', 'as produced in', 'imitation' or similar, including when those products are used as an ingredient; any other false or misleading indication as to the provenance, origin, nature or essential qualities of the product; and any other practice liable to mislead the consumer as to the true origin of the product [Regulation 2024/1143, Article 26]. The protection of an AOC may be requested by any person authorised to use it, the Organisation for the Defence and Management of the AOC (*Organisme de Défense et de Gestion*, ODG) or INAO.

In Vietnam, the law prohibits the use of the denominations on products from the geographical area that do not have the same characteristics and do not meet the quality criteria of the GI product; on similar products when such use is likely to misappropriate or weaken the reputation of the name, regardless of whether they originate from the geographical area; and on products not originating from the geographical area and misleading consumers as to the true origin of the product, even if such products meet the same quality and production criteria as the GI products [IP Law, Article 129.3]. Alike in French and European legislation, the Vietnamese law does not require establishing the risk of confusion among the relevant public to prohibit the use of a GI on products that do not come from the geographical area or that do not meet the quality criteria of the GI product. What is protected by the law is the name or sign that constitutes the GI itself, i.e. not the combination of the name or sign with the type of product and even less so the associated logo. For instance, only the indication '*Lạng Sơn*' is protected and not 'star anise from Lạng Sơn'. The use of the GI label on similar products originating from the geographical area and that meet the same characteristics and quality criteria of the GI product is also considered an infringement when no right to use the GI has been conferred to producers by the managing organization (IP Law, Article 213.2). Unlike in French and European law, the Vietnamese law does not provide protection when the protected name is translated for goods other than wines and spirits (Marie-Vivien 2018). When comparing signs and products to determine infringements of GIs, several factors and criteria are considered, including visual and phonetical similarity, which involves examining the spelling, pronunciation, and overall appearance of the signs and products; semantic similarity, which includes evaluating whether the sign conveys a similar meaning or message as the GI; likelihood of confusion among consumers regarding the origin, quality, or characteristics of the products; intent and context of use, which requires examining whether the sign was used intentionally to mislead consumers or to unfairly benefit from the reputation of the GI; and evidence of use of the GI, including on products, packaging, promotional materials, or in advertising campaigns (Decree No. 105/2006/ND-CP, Article 37). Actions for infringement can be brought by the state, the managing organisation, as well as the organisations and individuals who have the right to use the GI [IP Law, Article 125.1].

15.2.3 Pre-market Controls in French and Vietnamese Law

Prior to placing GI products on a market, Regulation 2024/1143 requires that, in addition to self-monitoring by the operators on their own products, verification of compliance with the product specification shall be carried out by one or more competent authorities [Article 39.3].

These must offer adequate guarantees of objectivity and impartiality and may delegate the control tasks to one or more control bodies that shall be accredited in accordance with Standards EN ISO/IEC 17065 or 17020 [Article 41.1]. Member States are free to decide whether these controls shall be carried out by public or private certification entities, which was an important reform introduced by Regulation 510/2006. In France, where the establishment of control mechanisms is a prerequisite for the recognition of AOCs, PDOs and PGIs [Rural Code, Article L.641-5], INAO is the national competent authority responsible for official controls before GI products have been placed on the market [Rural Code, Article L.642-5, 4°]. Since 2006, controls must be delegated and carried out by private, independent certification bodies that must be chosen by the ODG [Rural Code, Article R.642-37, 1°], approved by INAO [Rural Code, Article L.642-5, 4°], and accredited according to the relevant national and European technical accreditation standards [Rural Code, Article R.642-53]—i.e. the French Accreditation Committee (COFRAC) in accordance with EN 45011. Control plans must be elaborated by the certification body in cooperation with the ODG [Rural Code, Article L.642-32 and Article R.642-39] and approved by INAO [Rural Code, Article L.642-5, 3°]. They shall provide for three types of controls: (i) self-monitoring carried out by the operators on their own products; (ii) internal controls conducted under the supervision of the ODG; and (iii) external controls performed by the certification bodies [Rural Code, Article R.642-39] under the authority of INAO and whose costs are borne by producers/processors [Rural Code, Article L.642-27]. Controls may include (i) administrative and documentary controls, aimed at assessing the accuracy of the documentation submitted by producers, including GI application, declarations and documentation related to the production processes and quality control measures; (ii) physical inspections on-site aimed to verify compliance of the production processes with the product specifications; and (iii) analytical controls at INAO-approved laboratories including, for instance, chemical and microbiological analysis, authentication of origin and quality parameters including colour, texture, and sensory characteristics. In addition, whenever necessary, blind organoleptic examination of GI products shall be carried out by a commission made up of competent professionals and experts.

By contrast, GI applications in Vietnam must include information on self-control mechanisms only [IP Law, Article 106.2]. There is no legal requirement as per the adoption of control regulations, including internal and external control plans. Yet, as the responsible governmental agency for developing and implementing national standards, technical regulations, metrology, and quality assurance policies across different sectors of the economy, the Directorate for Standards, Metrology and Quality (STAMEQ), which operates under the authority of the MOST, has developed numerous requirements and standards for product quality, safety, and labeling related to GIs [Decision No. 27/2014/QD-TTg of 4 May 2014 on the functions, tasks, powers and organizational structure of STAMEQ, Article 1.1]. These may serve as benchmarks for compliance and guide the inspection activities of market surveillance agencies. STAMEQ also oversees the accreditation of testing laboratories, inspection bodies, and certification agencies involved in verifying compliance with quality standards and regulations, including those pertaining to GIs.

15.2.4 Post-market Controls in French and Vietnamese Law

To verify monitor the use of registered names on the market, Regulation 2024/1143 requires EU Member States to designate the competent authority(ies) responsible for official controls [Article 42]. In France, the Directorate-General for Competition Policy, Consumer Affairs and Fraud Control (*Direction Générale de la concurrence, de la consommation et de la répression des fraudes*—DGCCRF), which operates under the authority of the French Ministry of Economy and Finance, is the central authority responsible for consumer protection and for monitoring the safety and reliability of products and the use of GI names in the marketplace (EUIPO 2016). Controls can be carried out at all stages (import, export, distribution, wholesale, internet, restaurants) and include (i) labeling controls and inspections aimed to verify that products labeled with GI names adhere to labeling requirements, including ensuring accurate indication of the geographical origin, compliance with specific labeling guidelines, and proper use of AOC, PDO, and PGI designations; (ii) product quality and composition checks aimed to assess the quality and composition of products bearing GI names through analytical testing and sensory evaluation to verify compliance with the product specifications; (iii) traceability audits that involve examining the documentation and records maintained by producers and distributors to trace the origin and journey of GI products from production to the point of sale; and (iv) market surveillance activities to detect and investigate instances of fraud, mislabeling, or unauthorized use of GI names. The DGCCRF collaborates with INAO, regional authorities, and other relevant bodies responsible for the certification and control of GI products.

In Vietnam, the Ministry of Industry and Trade is the central authority responsible for managing market surveillance and inspection agencies (Marie-Vivien 2018). These include (i) Vietnam Directorate of Market Surveillance, under the authority of the Ministry of Industry and Trade, who conducts regular inspections at retail outlets, wholesale markets, and distribution centers to verify the accuracy of product labeling, including GI designations, and check for counterfeit or substandard products; (ii) Vietnam Food Administration, including local authorities, who conducts inspections of food products, including those bearing GIs, to assess compliance with food safety standards and regulations, through the examination of production facilities, storage conditions, and hygiene practices; (iii) local Departments of Intellectual Property who carries out labeling inspections to detect infringements, unauthorized use of GI names, or misleading claims regarding geographical origin; and (iv) local Departments of Industry and Trade who perform local inspections at markets, fairs, and trading venues to monitor compliance with trading standards, including regulations governing the sale and promotion of GI products and address issues such as misrepresentation, deceptive practices, and unfair competition; and (v) the Anti-Smuggling Investigation Department, under the authority of the Ministry of Finance, who monitors imports and exports of goods, including products bearing GIs, at border checkpoints and ports of entry.

15.3 Fraud and Unfair Competition Practices in Practice: Case Studies in France and Vietnam

This section explores fraud and unfair competition practices in five case studies in France and Vietnam which were selected randomly and cover a variety of products as follows:

- France: *Pélardon cheese* (AOC in 2000; PDO since 2001); *Bouchot mussels from Mont-St-Michel Bay* (AOC in 2006; PDO since 2011);
- Vietnam: *Star anise from Lạng Sơn* (GI since 2007); *Conical hat from Huế* (GI since 2010); *Fried calamari from Hạ Long* (GI since 2013).

The objective is to understand better the different types of fraud and unfair competition practices and the reasons underpinning their development. Fraud and unfair competition practices were found in all our case studies across the two countries both before and after the GI registration.

15.3.1 Fraud and Unfair Competition Practices Before the GI Registration

Fraud and unfair competition practices were found before the GI registration process in three of our case studies, namely *fried calamari from Hạ Long*, *Bouchot mussels from Mont-St-Michel Bay*, and *Pélardon cheese*. Mussel producers in the Mont-St-Michel Bay started noticing in the 1980s that 20,000–30,000 tonnes of mussels were sold every year under the name 'Mont Saint-Michel' while only 10,000–12,000 tonnes were—and still are—grown and harvested in the area (Frangoudes 1999). For *Pélardon cheese* that is produced in the Cévennes area in French Languedoc-Roussillon Region, it was found in the 1980s that the name 'Pélardon' was increasingly being used on goat cheeses produced in Spain and other parts of France where production costs were lower, which resulted in cheeses sold as 'Pélardon' with rock-bottom prices (Benkahla et al. 2004). Similarly, the widespread use of the name "fried calamari from Hạ Long" on products that do not come from Hạ Long City, particularly in the neighbouring Quang Yên District, was a serious concern for the legitimate producers before they started the GI protection process (Van Thinh 2017).

A common characteristic of these three products is their long-standing reputation among consumers, and/or the reputation of the place they come from, prompting actors outside the area of origin to use their names since well before the GI registration. In the case of *fried calamari from Hạ Long* and *Bouchot mussels from Mont-St-Michel Bay*, both products originate from regions that have been protected as UNESCO World Heritage Sites since before the GI protection process (Hạ Long Bay and Mont-St-Michel Bay, respectively) and which attract millions of tourists every year. The choice of the name to be protected as a GI for these two products was particularly strategic as it took advantage of the renown linked to a place. This has contributed to increasing the commercial success of the products on the market because the UNESCO recognition has been attractive to distant domestic or foreign consumers who associate it with quality products and/or because the area of origin attracts a high number of tourists who add to the 'local' demand for these products. In these instances, free-riders take advantage of the reputation of the products or the origin area or both (Pick 2022).

Although *Pélardon cheese*, one of the oldest goat cheeses in Europe, is also produced on a UNESCO World Heritage Site, the breathtaking Cévennes area in French Languedoc-Roussillon Region, its long-lasting reputation does not derive from the cachet of its production area. In fact, the Cévennes area was included in the UNESCO prestigious list in 2011, i.e. 10 years *after* the GI registration. Rather, as most of the production of the cheese has traditionally been sold on local markets through traditional distribution channels, either on-farm or in local markets, the reputation of the cheese has long been built upon geographic proximity between consumers and producers. Spatial closeness has contributed to increasing consumer knowledge and awareness while building the reputation of the product and individual producers through trust and personal interaction within the area of origin (Benkahla et al. 2004).

For all three products that have a long-standing reputation among consumers, and/or that come a famous and prestigious area, it is the perception of a common risk derived from competitors free-riding on the established reputation of the name in the marketplace that led local actors to seek a GI protection. Unsurprisingly, the primary objective attached to the GI legal protection for these products was to protect producers against the misuse and misappropriation of the name. In such instances, the GI protection is used as a legal tool to protect the collective reputation attached to the product (Bramley et al. 2009).

15.3.2 Fraud and Unfair Competition Practices After the GI Registration

After the GI registration, fraud and unfair competition practices were observed for *Fried calamari from Hạ Long, Star anise from Lạng Sơn, Conical hat from Huế,* and *Pélardon cheese*. By contrast to the sole involvement of external actors in such practices in a pre-registration context, these case studies show the involvement of actors from both outside and inside the GI system, including certified and non-certified producers, retailers and local authorities involved in the management of the certified products.

15.3.2.1 Involvement of Actors from Inside the GI System

Various causes give rise to the development of fraud and unfair competition practices by actors from inside the GI system, including production constraints, lack of producer understanding, and disagreement among GI stakeholders.

In the initiative of *Fried calamari from Hạ Long*, the costs of ingredients and the availability of raw material were found to be important production constraints for some producers, leading them to adopt unfair competition practices. Indeed, some producers use more flour and less calamari than prescribed by the product specifications with a view to decreasing their production costs without subsequent adequate controls from the authorities [phone interview interviews with the President of the producer association. September 2020. Interviews in person with producers. Hạ Long, Vietnam. March 2014]. Besides, according to the product specification of *Fried calamari from Hạ Long*, at least 70% of calamari should be fished in the Gulf of Tonkin [Regulations on the management and use of the GI for fried *calamari from Hạ Long*, Article 3.3.1]. However, in recent years, increased consumer demand has led to over-exploitation of calamari in the Gulf of Tonkin, resulting in a significant reduction in the number of calamari in the area. Because their supply is quickly decreasing, producers have increasingly sourced calamari from Central Vietnam, China, Indonesia, and Malaysia [phone interview interviews with the President of the producer association. September 2020. Interviews in person with producers. Hạ Long, Vietnam. March 2014]. Not only these practices show how increased consumer demand for GI products can lead to excessive pressure on the biological resources needed for their production, which may ultimately result in their over-exploitation and erosion (Pick 2022). They also mislead the consumer as to the true origin of the product. Interestingly, although the quality and origin of the calamari are specified in the product specification [Regulations on the management and use of the *GI for fried calamari from Hạ Long*, Article 9], fishermen were excluded from the initiative, as was also the case for *Phu Quoc fish sauce* (protected as an appellation of origin since 2002). As a matter of fact, providers of raw materials do not usually participate in the construction phase of the Vietnamese GI initiatives, including in the elaboration of the product specification, which may lead to quality control issues.

If fraudulent practices primarily aim to reduce production costs, they may also stem from producers' misunderstanding as per the meaning and function of a GI, as with *star anise from Lạng Sơn*. In this initiative, one company who has the right to use the GI was found to affix the logo not only on the packaging of star anise products, but also on that of cinnamon products sold in its retail shops. This practice was justified by the sale manager of the company on the basis that 'the GI logo enables [their] consumers to identify that the products are sold by [his] company' [phone interview with the Sale Manager, Vietnam Star Anise Processing and

Exporting Co. Lạng Sơn, Vietnam. October 2020]. In this instance, the GI logo is used as a marketing (rather than legal) tool to identify the *commercial* origin of the products, whether they be cinnamon or star anise products, both being spices, and build the reputation of the company as if the GI were its private trade or commercial name. This practice not only amounts to a direct commercial misuse of the GI label on comparable products (aromatic spices), but also shows the company's lack of misunderstanding of the very meaning and function of a GI and the lack of checks and inspections [interview in person with the Director of STAMEC, Lạng Sơn, Vietnam. May 2014].

Finally, disagreement over the requirements of the product specification among the very actors responsible for protecting, managing and promoting the GI, can also lead to fraudulent practices, as in the case of *conical hats from Huế*. Conical hats have enjoyed a high international visibility among Western tourists who identify conical hats as symbolic and representative of Vietnam in general, particularly its farming culture, and not of Huế (Thirumaran et al. 2014). This exposure to foreign tourists has greatly contributed to turning conical hats into souvenir products, keeping in mind that Huế Imperial City has been listed as a World Heritage Site since 1993 and as such attracts hundreds of thousands of visitors every year. As tourists' interest in cultural representations of the visited country often translates into the purchase of handicraft souvenirs (Dung 2016), foreign tourists make up the large majority of customers. While three categories of conical hat are produced in Thừa Thiên-Huế Province, the GI-qualified hat involves the most expensive and time-consuming production process. In particular, the product specification specifies that only those hats made of coconut-palm leaf with three stitches per 1 cm knitted down from the top brim to the 15th brim can be sold with the GI label [Document describing the characteristics of conical hats from Huế, 8], thereby excluding the other categories of hats made of a lower quality leaf and with more space between stitches that are produced in the same area. However, foreign tourists are generally not connoisseurs and as such are not aware of the characteristics and methods of production of the genuine *conical hat from Huế*. This is especially true considering that, as acknowledged in the GI application document, conical hats from Huế are very similar to other conical hats produced across the country, despite having its own characteristics in terms of colour, design, size, weight and durability which only connoisseurs are aware of. As a result, most producers have taken advantage of tourists' ignorance and switched to the production of the cheapest categories of hats to save time and money [phone interview with the President of the Association, President of the Provincial Women's Union. September 2020. Interviews in person with producers. Huế, Vietnam. April 2014]. This practice spread quickly, especially because the large majority of producers are small households whose production is very little and hence who combine hat production with other jobs to increase their incomes. Such practices would not qualify as fraud and unfair competition practices if there is no use of the GI label nor other acts involving deceptive advertising or misleading representation. However, at the association level, the producer association, which is the GI managing organisation, was found to sell all types of conical hats with the GI logo because it actively supports its use on all categories of hats as a way to support the local economy [phone interview with the President of the Association, President of the Provincial Women's Union. September 2020]. These practices allegedly led to tensions with the local Department of Science and Technology [interview in person with the Head of the IP Division, Department of Science and Technology (DOST). Huế, Vietnam. April 2014], who registered the GI and drafted the product specifications, showing a conflict between the very actors involved in the GI registration and protection (DOST) and those responsible for its management and promotion (the producers' organisation).

15.3.2.2 Involvement of Actors from Outside the GI System

In the case of *Pélardon cheese*, unfair competition or trading practices are more diffuse and difficult to grasp, mainly because they are found in oral communication involving deceptive advertising or misleading representation at the farm level by external actors in a context where the largest part of the production has traditionally been sold inside the production area, either on-farm or in local markets. In this initiative, most producers have limited material, human and organisational capacities, which generally contributes to explaining the adoption of local marketing channels (Aubert and Enjolras 2013). One important consequence is that the local clientele has a personal knowledge and experience of both goat cheese and their producers and therefore does not rely on labels. Indeed, at the farm gate, trust in the producer, which is not mediated by brands and labels but through personal interaction, is the most important marketing argument (Pick 2022). Market success at the local level is thus largely based on spatial proximity which, by contributing to develop consumers' knowledge and skills, diminishes the communicative function of the origin label inside the area of production [phone interview with the Coordinator of the ODG. July 2020].

Adding to the weak communicative value of the PDO label, inside the area of origin, PDO Pélardon producers, who account for about one-third of all 'Pélardon type' cheese producers in the Cévennes area [phone interview with the Coordinator of the ODG. July 2014], find themselves increasingly competing with non-PDO goat cheese local producers who either never joined the initiative, or who left it because they find the compliance rules included in the product specification too prescriptive. For instance, the prohibition of using mechanical molds and concentrated or powdered milk, as well as the practice of taking goats out on the territory for a minimum number of days, in consistency with traditional pastoral and grazing practices in the area, are mandatory requirements in the product specification. To ensure that producers observe agro-pastoral practices, the quantity of feed concentrate to be distributed to goats is limited to 400 grams per day per liter of goat's milk produced [Technical rules for the implementation the product specification for 'Pélardon', Article 1]. However, the obligation to take goats out on the territory to feed on local herbs for a minimum number of days is viewed as too burdensome by some producers (Napoléone and Boutonnet 2004). The prohibition of sale of the cheese before 11 days of maturation after curdling is also being challenged by a few producers who find this rule too difficult to comply with. As a result, many producers have decided to leave the ODG who has 66 members in 2024 down from 120 in 2001 [Coordinator of the ODG, E-mail received on 9 February 2024]. While, in such circumstances, it is particularly challenging for the PDO initiative to generate interest in the origin label among local producers and thus to keep and attract new members in the association, those who are not part of the initiative still benefit from the reputation of, and collective communication done on, the PDO *Pélardon*. In effect, many non-PDO producers claim the same origin and similar values of authenticity and tradition when selling their cheese without having to support the costs related to the GI initiative [phone interview with the Coordinator of the ODG. July 2014]. As explained by the coordinator of the ODG: '[i]n short distribution channels, information is diluted and gets lost. Consumers get confused and ultimately buy cheese from a specific producer because they like the producer and the product, whether it is labelled or not. It happened that producers had joined the PDO initiative, developed their clientele and subsequently left the PDO initiative to avoid its obligations but kept their clientele all the same. In some cases, they even keep promoting their cheese as a PDO *Pélardon cheese* even though they do not comply with the PDO requirements' [phone interview with the Coordinator of the ODG. July 2020]. Another farmer stated, 'when people here want to buy some goat cheese, they say "I want a Pélardon", not "I want a goat cheese". For them, goat cheese is a Pélardon, even if it does not have the PDO label. Pélardon has become a name commonly used by local people' [interview in person with producers. Cévennes, France. June 2014]. Consequently, because of the direct competition between PDO and non-PDO cheese producers at the farm gate, where sales are based on a relationship of trust rather than on labels, the price difference between the PDO and non-PDO cheese is only marginal. In this context, not only misleading oral communication by some non-PDO local producers advertising goat cheese which does not comply with the PDO requirements as "*Pélardon*" constitutes unfair competition and deceptive commercial practices. It also contributes to further weakening the value of the origin label inside the PDO area.

15.4   Discussion

While confirming that fraud and unfair competition practices exist, or have existed, in all five GI initiatives across both France and Vietnam at different times (pre- and/or post-registration), our findings show the variety of types, root causes and drivers of such practices, as well as the plurality of actors and interests involved. This section will discuss three main issues arising out of the fraud and unfair competition practices observed after the GI registration: (i) the inefficiency of quality controls; (ii) the lack of producer understanding and engagement; and (iii) consumer ignorance.

15.4.1 Inefficiency of Quality Controls

When occurring after the GI registration, fraud and unfair competition practices usually point to the inefficiency or inexistence of quality controls, as is the case for *Fried calamari from Hạ Long*, *Star anise from Lạng Sơn* and *Conical hat from Huế*. In this respect, a study found that about 85% of the authorities in charge of the external controls and 75% of the collective organisations in charge of internal controls in Vietnam do not perform well (Đức Huấn and others 2017). The lack of efficient quality controls has also been reported in many other emerging and developing countries, for instance in India and in OAPI countries (Marie-Vivien and Biénabe 2017).

The comparison between France and Vietnam allows to appreciate the differences between the quality control systems across the two countries. While the establishment of quality control mechanisms based upon the product specifications is a prerequisite for the recognition of GIs in France (Rural Code, Article L.641-5), the elaboration of internal and external control plans is not legally required in Vietnam. Consequently, there is no control plan and hence no technical standards and quality criteria upon which state authorities in charge of the external quality controls, or associations and managing organisations responsible for internal controls, can base their controls. One notable exception is found for the *Phú Quốc fish sauce* that is subject to a detailed control plan. For other GI initiatives, this significant legal loophole leads to inconsistent, inefficient, non-transparent or even non-existent external quality controls (Đức Huấn and others 2017), which challenges the process of building consumer trust.

Besides, the market surveillance and inspection agencies responsible for market inspection and controls in Vietnam have limited staff, technical expertise, funding and equipment to perform their role efficiently (Marie-Vivien 2018). For instance, Vietnam Directorate of Market Surveillance lacks technical knowledge in intellectual property, whereas the local Departments of Intellectual Property lack agricultural expertise. Adding to this, the high number of state agencies involved in market inspection and controls with unclear, undefined or redundant functions poses problems of transparency, efficiency and coordination (Pick 2022). For instance, some producers of *fried calamari from Hạ Long* reported being inspected more than ten times a year by state delegations, which further creates opportunities for non-transparent practices [interview in person with producers. Hạ Long, Vietnam. March 2014].

In France, with regard the unfair competition practices with *Pélardon cheese*, effective quality control systems face significant challenges when addressing oral communication involving deceptive advertising or misleading representation at the farm level. The complexities lie in verifying the accuracy of verbal assertions and marketing claims made verbally about the product and the production processes involved.

15.4.2   Lack of Producer Understanding and Engagement

In the initiative of *Star anise from Lạng Sơn*, fraudulent practices are driven by producers' lack of understanding of the GI concept. This is a serious concern that derives from the broader issue of producers' lack of participation in and engagement with the GI initiatives in Vietnam. In effect, by contrast to the French bottom-up approach to GIs, Vietnamese law does not contain specific requirements aimed at ensuring producers' representativeness or promoting their collaboration for the elaboration of the product specifications and/or the design of the GI logo, and external experts have emerged as key players in such processes. While the product specifications and the GI logo are elaborated and designed by outside actors follow-ing Vietnam's top-down approach to GIs local stakeholders are usually consulted either directly or through their representatives as part of the surveys, data collection and meetings that the technical expert must organise in the pre-application phase under the supervision of the local authorities (NOIP 2009). In most cases, the participation of local stakeholders is limited to approving the pre-drafted production rules, the GI logo and the statutes of the association (Durand 2016). However, in the initiative of *Star anise from Lạng Sơn*, the association was established in 2008, i.e. one year after the registration of the GI, and the elaboration and design process of the product specification and the GI logo did not involve consulting local stake-holders. As a result, some stakeholders reported not being aware of its content or of the GI logo itself, and producers' lack of misunderstanding of the very meaning and function of a GI has led to fraudulent practices [interview in person with producers. Lạng Sơn, Vietnam. May 2014]. This case study illustrates how Vietnam's top-down approach to GIs makes it difficult for local producers to understand the GI concept and the rules contained in the GI specifications, take ownership and participate willingly in the initiatives, resulting in their low awareness of the functioning of the initiative which itself threatens its sustainability (Pick 2022). Top-down approaches, which are still common in many emerging and developing countries, often fail to mobilise local actors because of their lack of empowerment and little knowledge of both the characteristics of their own product and the meaning of the GI protection (Sautier et al. 2011). In effect, a large majority of origin-labelling initia-tives in Vietnam show little understanding, interest and commitment of local actors. For instance, in analysing eight case studies in Vietnam, a project funded by the French Agency for Development also found that almost 90% of producers were not even aware that a GI had been registered for their product (Đức Huấn and others 2017: 56). Generally speaking, the lack of awareness among intellectual property rights (IPR) holders, including those associated with GIs, remains a significant challenge in Vietnam (Marie-Vivien 2018). Many IPR holders, particularly small-scale farmers and producers, often lack awareness of their rights and responsibilities due to limited access to information, inadequate legal education, and cultural barriers, leaving them vulnerable to infringement, misappropriation, and unfair competition in the marketplace. Lack of producer understanding of the GI concept was further exacerbated by the lack of a national GI logo until end of 2022.

15.4.3 Consumer Ignorance

In the case study of *Conical hats from Huế*, the producer association's practice of displaying the GI label on all categories of hats for economic and market purposes is facilitated by tourists' ignorance of the quality and characteristics of the 'genuine' GI-protected conical hats. Similarly, in the Pélardon initiative, consumers' confusion and ignorance of the very existence of the GI label when buying goat cheese at the farm gate also provides opportunities for unfair competition practices to develop through oral misleading communication. These two case studies point to the need to educate consumers as per the characteristics of the GI product and build both its reputation and that of the GI label through investments in advertising and promotion. However, consumers' education is particularly challenging in both initiatives. On the one hand, the broader status of conical hats as a national symbol has overshadowed their specific territorial origin. This status has resulted in a weak territorial identity, especially considering that most customers are foreign tourists who buy conical hats as iconic souvenirs from *Vietnam* rather than from *Huế*. In this context, the concept of origin is closer to that of country of provenance, as it is the case for famous 'country-GIs' such as *Café de Colombia* that is 'based on the general image of the country instead of on real links of the coffee production process to local specific resources' (Marescotti and Belletti 2016). As such, Vietnamese conical hats, including but not limited to those from Huế, arguably lack geographical distinctiveness at least among foreign tourists, which generally explains why local producers have turned away from the traditional local product and the origin label. If there is no consumer awareness about, and demand for, the GI-labelled product, producers see no reason to produce it especially since the manufacturing process of the GI-qualified product is more costly and time-consuming (Pick 2022). Consequently, the production of the genuine conical hat from Huế has been decreasing rapidly, which not only puts at risk the preservation of traditional methods of production, but also underpins fraud practices with the active involvement of the producer association.

While this case study highlights the role of ignorance among foreign tourists, there is generally very limited consumer awareness about GIs which poses a significant challenge to the recognition and protection of GI products in the country (Marie-Vivien 2018). Factors contributing to this limited awareness include insufficient education and information dissemination about GIs which was further exacerbated by the lack of a national GI logo until end of 2022, language barriers, and the predominance of price-based purchasing decisions. Adding to this, the prevalence of counterfeit goods, often priced significantly lower than authentic products, appeals to budget-conscious consumers seeking affordability over authenticity. This reflects broader challenges related to consumer trust in enforcement agencies. Indeed, many Vietnamese consumers perceive enforcement agencies as ineffective, leading to a lack of confidence in the authenticity and quality of goods in the marketplace (Marie-Vivien 2018), which further perpetuates a cycle of counterfeit consumption and undermines the GI protection system.

On the other hand, in the case of *Pélardon cheese*, the importance of physical proximity between traders and consumers has resulted in consumer confusion as per the differences between PDO and non-PDO goat cheese and the dilution of the meaning of the PDO label inside the area of origin, which is compounded by the low consumer awareness and ignorance of the PDO and PGI signs in France and other European countries in general. Indeed, although 78% of French consumers believe that having a specific label ensuring the quality of the product is important or very important, only 27% and 17% of French consumers recognize the PGI and PDO logos, respectively (European Commission 2022).

These case studies show the importance of consumer education and awareness about GIs and unfair competition practices, thereby making informed purchasing decisions and promoting fair competition on the market.

15.5   Conclusion

By providing insights into the legal context in France and Vietnam for the enforcement of GI rights and a variety of empirical case studies, this paper builds a more nuanced understanding of the actors, types and drivers of fraud and unfair competition practices in the context of GIs. It has evidenced that fraud and unfair competition practices may derive in both countries from both inside and outside the GI system for a range of reasons. Before the GI registration, outside actors may free-ride on the reputation of the products and the commercial success of the origin names. In such instances, the GI protection is used as a legal tool to protect the collective reputation attached to the product and fight against unfair competition practices. After the GI registration, these practices may be conducted by both inside and outside actors in contexts characterized by ineffective control systems, production constraints, lack of producer understanding, disagreement among GI stakeholders, and consumer confusion and ignorance. The lack of engagement and awareness of legitimate stakeholders is a particularly serious issue considering that the motivation, understanding and commitment of local actors have been identified in the GI literature as a main factor of successful GI mobilisations (Barjolle and Sylvander 2002).

Addressing these concerns requires not only strengthening legal frameworks, including quality control mechanisms before and after the marketing of GI products, increased enforcement capabilities through capacity building, and greater collaboration among enforcement agencies. It may also be necessary to resort to mediation and extension services in certain contexts to facilitate the negotiations and adoption of common standards of production thereby ensuring the adhesion and commitment of producers. Finally, it also entails enhancing producer and consumer education and awareness about GI protection and the consequences of fraud and unfair competition through targeted awareness-raising campaigns, public outreach initiatives, capacity-building initiatives, and tailored support services.

References

Allaire G, Casabianca F, Thevenod-Mottet E (2011) The geographical origin, a complex feature for agro-food products. In: Barham E, Sylvander B (eds) Labels of origin for food: local development, global recognition. CAB International

Aubert M, Enjolras G (2013) Déterminants de la commercialisation en circuits courts : quels exploitants, sur quelles exploitations? Paper presented at the 7th Journées de Recherches en Sciences Sociales SFER-INRA, Angers, France, 12–13 December 2013

Barjolle D, Sylvander B (2002) Some factors of success for 'origin labelled products' in agri-food supply chains in Europe: market, internal resources and institutions. Economies et sociétés 36(9–10):1441

Belletti G (2000) Origin labelled products, reputation, and heterogeneity of firms. In: Sylvander B, Barjolle D, Arfini F (eds) The socio-economics of origin labelled products in agrifood supply chains: spatial, institutional and co-ordination aspects, vol 17(1). INRA, Actes et Communications, p 239

Belletti G, Burgassi T, Manco E, Marescotti A, Pacciani A, Scaramuzzi S (2007) The roles of geographical indications (PDO/PGI) on the internationalisation process of agro-food products. In: 105th EAAE seminar on international marketing and international trade of quality food products. Bologna, Italy

Benkahla A, Boutonnet JP, Napoléone M (2004) Proximités et signalisation de la qualité: approaches croisées pour l'étude d'une AOC. Le cas du Pélardon. 4th Congress on Proximity Economics: Proximity, Networks and Co-ordination, Marseille, 17–18 juin 2004

Biénabe E, Bramley C, Kirsten J, Troskie D (2011) Linking farmers to markets through valorisation of local resources: the case for intellectual property rights of indigenous resources. IPR DURAS Project Scientific Report, April 2011

Bramley C (2011) A review of the socio-economic impact of geographical indications: considerations for the developing world. WIPO Worldwide Symposium on Geographical Indications, Lima Peru

Bramley C, Biénabe E, Kirsten J (2009) The economics of geographical indications: towards a conceptual framework for geographical indication research in developing countries. In: The economics of intellectual property: suggestions for further research in developing countries and countries with economies in transition. WIPO, p 109

Bramley C, Marie-Vivien D, Biénabe E (2013) Considerations in designing an appropriate legal framework for GIs in southern countries. In: Bramley C, Biénabe E, Kirsten JF (eds) Developing geographical indications in the South: The Southern African experience. Springer, pp 15–47

Correa C (2007) Trade related aspects of intellectual property rights. A commentary on the TRIPS Agreement. Oxford University Press

Das K (2006) International protection of India's geographical indications with special reference to 'Darjeeling' tea. JWIP 9(5):459–480. https://doi.org/10.1111/j.1422-2213.2006.00300.x

Das K (2009) Socio-economic implications of protecting geographical indications in India. Centre for WTO Studies 3. https://papers.ssrn.com/sol3/papers.cfm?abstract_id=1587352. Accessed 8 Feb 2024

Đức Huấn Đ, Đức Chiến Đ, Phúc Giang Đ, Sỹ Đạt N, Thế Bảo P, Văn Ba N., Văn Tuấn T., Kim Đồng B., Thị Minh Huyền H., Marie-Vivien D. and Sautier D., 'Rapport—Etude des modèles de gestion des indications géographiques du Vietnam' (NOIP/AFD, 2017)

Dung TT (2016) Représentation culturelle et souvenir artisanal, Expérience des touristes au centre du Viet Nam. Master's thesis, Université du Québec à Montréal

Durand C (2016) 'L'émergence des indications géographiques dans les processus de qualification territoriale des produits agroalimentaires—Une analyse comparée entre l'Indonésie et le Vietnam', PhD thesis, Institut national d'études supérieures agronomiques de Montpellier—SupAgro Montpellier

EUIPO (2016) Infringement of protected geographical indications for wine, spirits, agricultural products and foodstuffs in the European Union, EUIPO Study, April 2016. Available at https://

euipo.europa.eu/tunnel-web/secure/webdav/guest/document_library/observatory/documents/ Geographical_indications_report/geographical_indications_report_en.pdf.   Accessed 8 Feb 2024

European Commission (2011) Workshops on geographical indications—development and use of specific instruments to market origin-based agricultural products, making particular use of GIs in African ACP countries. European Commission

European Commission, 'Europeans, Agriculture and the CAP' (2022) Special Eurobarometer 520. Available at https://europa.eu/eurobarometer/surveys/detail/2665. Accessed 8 Feb 2024

Frangoudes K (1999) L'occupation du domaine public maritime par des cultures marines, le cas de la Baie du Mont-Saint-Michel. Coastman Working Paper n° 11. http://coastman.free.fr/wp/ Coastman_WP11.pdf. Accessed 8 Feb 2024

Gangjee D (2012) Relocating the law of geographical indications. CUP

Jena P, Grote U (2010) Changing institutions to protect regional heritage: a case for geographical indications in the Indian agrifood sector. Dev Pol Rev 2(28):217. https://doi.org/10.1111/j. 1467-7679.2010.00482.x

Landes W, Posner R (2003) The economic structure of intellectual property law. The Belknap Press of Harvard University Press

Marescotti A, Belletti G (2016) Differentiation strategies in coffee global value chains through. Reference to territorial origin in Latin American countries. Cult Hist Digit J 5(1):1. https://doi. org/10.3989/chdj.2016.007

Marie-Vivien D (2018) Assessment of geographical indications enforcement in Vietnam, study for EUIPO. European Union, Montpellier

Marie-Vivien D, Biénabe E (2017) The multifaceted role of the state in the protection of geographical indications: a worldwide review. World Development 98(1)

Napoléone M, Boutonnet JP (2004) AOC Pélardon: du compromis vers l'émergence d'actions collectives. Dynamiques de systèmes de production et des stratégies de commercialisation', Séminaire SFER, Les systèmes de production agricoles: performances, évolutions, perspective, Lille, France, 18–19/11/2004

Nelson P (1970) Information and consumer behaviour. J Polit Econ 78:311

NOIP (2009) Guide à la construction du project "Gestion et Developpement des IGs". NOIP

O'Connor B (2004) The law of geographical indications. Cameron

OECD (1996) Appellations of origin and geographical indications in OECD member countries: economic and legal implications. (COM/AGR/APM/TD/WP(2000)15/FINAL)

Pick B (2022) Intellectual property and development: geographical indications in practice. Routledge, Abingdon

Pike A (2009) Brand and branding geographies. Geography Compass 3, 190(1)

Sautier G, Biénabe E, Cerdan C (2011) Geographical indications in developing countries. In: Barham E, Sylvander B (eds) Labels of origin for food: local development, global recognition. CAB International, p 138

Stiglitz JE (1989) Imperfect information in the product market. In: Schmalensee R, Willig RD (eds) Handbook of industrial organization. Elsevier Science Publishers

Thirumaran K, Dam MX, Thirumaran CM (2014) Integrating souvenirs with tourism development: Vietnam's challenges. Tourism Plan Dev 11(1):57–59. https://doi.org/10.1080/21568316.2013. 839471

Tirole J (1988) The theory of industrial organization. MIT Press

Van Thinh N (2017) Analysis of factors affecting the income of households making GI products: a case study of Hạ Long grilled squid—Quang Ninh Province. Proceedings of the International Conference for Young Researchers in Economics and Business, Da Nang, Vietnam,

Watal J (2001) Intellectual property rights in the WTO and developing countries. Kluwer Law International

The opinions expressed in this chapter are those of the author(s) and do not necessarily reflect the views of the [NameOfOrganization], its Board of Directors, or the countries they represent.

Chapter 16
A Model of Geographical Indication's Product Specification for ASEAN Countries

Miranda Risang Ayu Palar

Abbreviations

AEC	ASEAN Economic Community
AO	Appellation of Origin
ASEAN	Southeast Asian Nations
FAO	Food and Agriculture Organization
GI	Geographical Indication
GIP	Geographical Indication Potential
IP Code	Intellectual Property Code
IP	Intellectual Property
Lao PDR	Lao People's Democratic Republic
OriGin	Organization for an International Geographical Indications Network
TM	Trademarks
TRIPs Agreement	Agreement on the Trade-Related Aspects of Intellectual Property Rights

16.1 Introduction

On August 8, 1967, the Association of Southeast Asian Nations (ASEAN) was established in Bangkok, Thailand. Indonesia, Malaysia, Philippines, Singapore, and Thailand signed the ASEAN Declaration which then known as the Bangkok

M. Risang Ayu Palar (✉)
Intellectual Property Centre on Regulation and Application Studies, Faculty of Law, Universitas Padjadjaran, Bandung, West Java, Indonesia
e-mail: miranda.risang.ayu@unpad.ac.id

© The Author(s) 2025   223
E. Vandecandelaere et al. (eds.), *Worldwide Perspectives on Geographical Indications*, https://doi.org/10.1007/978-3-031-71641-6_16

Declaration (ASEAN Declaration 1967). Later, Brunei Darussalam, Cambodia, Lao PDR, Myanmar, and Vietnam also joined ASEAN and forming the 10 member countries of ASEAN today. On December 15, 1995, Brunei Darussalam, Indonesia, Malaysia, Philippines, Singapore, Thailand, and Vietnam signed the ASEAN Framework Agreement on Intellectual Property (IP) Cooperation to enhance closer cooperations between ASEAN members in the area of IP (ASEAN Framework Agreement on IP Cooperation 1995), including geographical indications (GI). In 2016, the ASEAN Intellectual Property Rights Action Plan 2016–2025 was introduced. ASEAN member countries viewed that the ASEAN Economic Community (AEC) was in prospect and IP, including GI, should play a strategic role in it (ASEAN IP Register to officially launch, 2024).

16.2 Legal Means to Protect GI in ASEAN Countries

Geographical Indication (GI) is a subject matter of conventional Intellectual Property (IP), or the IP that is based on international conventions. GI is mainly regulated by the World Trade Organization's Agreement on the Trade-Related Aspects of Intellectual Property Rights 1994 (TRIPs Agreement, 1995) and the World Intellectual Property Organization's Geneva Act of the Lisbon Agreement on Appellations of Origin and Geographical Indications 2015 (Geneva Act of the Lisbon Agreement, 2015). The two international legal instruments have slightly different definitions of GI that lead to the different requirements to obtain a GI protection.

In regard with GI protection, ASEAN countries have their own peculiarities and issues (Delphine Marie-Vivien 2020). Several of them combine the definitions and/or requirements from TRIPS Agreement and Geneva Act of the Lisbon Agreement, regardless of whether they have become members of both conventions or the members of the TRIPS Agreement only.

TRIPS Agreement provides one way to protect a sign of geographical origin, that is, protecting it as a GI. In this regard, article 22 (1) of TRIPS Agreement defines a GI as,

> … indication which identify a good as originating in the territory of a member, or a region or locality in that territory, where a given quality, reputation, or other characteristic of the good is essentially attributable to its geographical origin.

Differently, Geneva Act of the Lisbon Agreement provides two alternatives of protection: protecting the sign as a Geographical Indication (GI) or as an Appellation of Origin (AO). In this regard, article 2 (1) (ii) of Geneva Act of the Lisbon Agreement defines a GI as:

> … any indication protected in the Contracting Party of Origin consisting of or containing the name of a geographical area, or another indication known as referring to such area, which identifies a good as originating in that geographical area, where a given quality, reputation, or other characteristic of the good is essentially attributable to its geographical origin.

Article 2 (1) (i) of Geneva Act of the Lisbon Agreement defines an AO as:

> ... any denomination protected in the Contracting Party of Origin consisting of or containing the name of a geographical area, or another denomination known as referring to such area, which serves to designate a good as originating in that geographical area, where the quality or characteristics of the good are due exclusively or essentially to the geographical environment, including natural and human factors, and which has given the good its reputation.

The requirements to obtain protections from the definitions of GI in the international legal instruments further determine the elements of a GI product specification. In this regard, although there is no explicit requirement about providing a GI product specification in TRIPS Agreement as well as in the Geneva Act of the Lisbon Agreement, the usage of the elements of a GI product specification has been becoming practices in the majority of ASEAN member countries to proceed applications for GI registrations.

16.3 The Importance of a GI Product Specification

A GI Product Specification, or a code of practice, is "a document that describes specific attributes of the GI product in relation to its geographical origin. It describes the product and its production process, and lays down requirements regarding production and/or processing method, packaging, labelling, etc." (FAO and OriGin 2024).

GI product specification (Denis 2019) is important because of several reasons. *First*, in many countries, it is a substantial requirement for GI registration. *Second*, it forms a legal basis to grant an exclusive right to use a GI. *Third*, it also establishes the geographical originality of the product: the causal link between the GI product and its geographical origin. *Fourth*, it forms a legally binding contract that guides the production process and quality control on the products. *Fifth*, it provides the consumers with a solid guarantee regarding the GI product's reputation, quality and/or characteristic, and if applicable, the sustainability of the product's geographical environment.

GI Product Specification is neither literally mentioned in the legal instruments of the Lisbon system, nor in the TRIPs Agreement. However, GI product specification has also been widely used because of its usefulness in describing the geographical origin's attribution to the protected product.

In the context of enhancing the ASEAN Economic Community, the more similar GI product specifications used amongst ASEAN member countries, the easier the transborder GI registrations between ASEAN member countries will be. It will also speed up the occurrence of joint holders if a GI covers a locality that is of a transborder origin. The example of GIs with a transborder origin are Adan Krayan Rice from Indonesia (Mubarok et al. 2021) and Bario Rice from Malaysia (Macdalyna et al. 2022) those are originated from an area on the borderline of the two countries in Kalimantan (Borneo) Island.

Adan rice is a traditional variety grown by indigenous Dayak sub-tribe, especially Dayak Lundayeh, whose place of origin covers Krayan District, Nunukan Regency, Indonesia, to the Bario and Kelabit regions in Sarawak, Malaysia.

Until 2015, Adan rice from Indonesia was produced and sold under the name Adan rice, while in Malaysia, this rice was known as *Bario or Kelabit rice*. In that time, Adan rice from Krayan Indonesia was actually more widely marketed by the Krayan local farmers for Malaysian market but labelled by Malaysian traders as the Malaysian origin. To avoid potential tensions between the two countries, Adan Rice has been protected as GI of *Bario Rice* by Malaysian government and *Adan Krayan Rice* by Indonesian government. Due to different environmental factors, the two rices are stated to have their own unique characteristics, even though both are still produced by the same indigenous people (Antara KBI 2015).

Until the end of 2023, ten member countries of ASEAN still have different legal means to protect GI in their national levels. They all have been members of TRIPS Agreement. But so far, only two of them have ratified Geneva Act of the Lisbon Agreement: the Kingdom of Cambodia and Lao People's Democratic Republic (WIPO, Lisbon, 2023).

In general, various legal means to protect GI can be divided into three versions.

The first version is protecting GI as a part of the general Trademarks (TM) protection system. Substantially, GI is considered a descriptive mark with a low capability to distinguish. Yet, registration of a TM substantially requires a high quality of distinctive element. Therefore, if a GI is to be registered as a TM, especially a Collective TM, the GI should display a secondary significance that makes it able to be registered as a TM, that is, a reputation. So, certain degree of reputation of a GI in TM registration is the additional requirement. Besides that, a GI can also be protected as a Certification Mark.

The second version is establishing an independent or standalone regime specialized for GI, separately from the TM regime. The legal basis of the regime can be in the same or separate laws or regulations with the TM, but the administrative and substantial requirements are completely different. In particular, GI protection does not require a distinctive element as TM, but the strong link with its geographical origin. Furthermore, a GI product specification is substantial in the administration procedure of obtaining the protection, whereas in TM, no such document is required.

The third version combines the protection of GI and TM together. It means that GI has a special law, regulation, or provisions differ from TM, but both GI and TM are regarded as signs that are used in trade. So, several aspects of them are the same, such as: requirement of distinctiveness, term of protection, the right holder/s, and/or the eligibility of licensing.

In the light of the abovementioned versions of legal means to protect the GI, ASEAN member countries can be divided as follows (see Table 16.1):

Brunei Darussalam, until early 2024, protected GI as a part of general TM protection system, especially as collective mark and/or certification mark. Differently, Cambodia, Indonesia, Thailand, and Lao PDR protected GI by a standalone GI system. Malaysia, Singapore, and Vietnam protected GI separately from TM, but they also combined several aspects of the GI and TM protection systems to be

Table 16.1 Variation of GI protection system in ASEAN member countries

Trademarks	Geographical Indication	Trademarks & Geographical Indication
Brunei Darussalam	Cambodia	Malaysia
	Indonesia	Singapore
	Thailand	Vietnam
	Lao	
Philippines		
Myanmar		

applicable for both, for example: 'the distinctiveness' requirement in TM that was called 'the distinctive character' requirement in GI, the nature of the right holder that could be an individual or collective, the term of protection that was 10 years renewable, and/or the licensing system. Differently, Philippines and Myanmar protected GI by establishing a sub system of GI under the TM system.

16.4   GI Product Specifications in ASEAN Member Countries

GI product specifications in ASEAN members countries are varied. It can be in a form of one document, one document with several supporting documents, or a bundle of elemental documents. It has different names too: a book of GI requirements, a book of GI specification, a GI technical specification, a GI description document, manual of GI specifications, or a bundle of documents for GI registration.

In the countries where a GI protection is an integral part of the general trademarks protection systems, a GI potential cannot be straightforwardly protected as a collective mark because the nature of a GI name or logo is descriptive. A descriptive mark is a weak TM that lacks strong distinctiveness. So, to enable a GI potential to be registered as a collective mark, it should have a certain degree of reputation as a secondary meaning or secondary significance to strengthen its capability to distinguish through use (Lynne 2007). In this regard, GI product specification can be used as an additional document/s to ascertain the reputation of the GI.

The national GI protection systems in ASEAN member countries with the legal bases for GI product specification are briefly explained below.

16.4.1 Brunei Darussalam

Until the start of 2024, Brunei Darussalam has no specific form of GI product specification. It is because GI in Brunei is protected under the general trademarks protection system, especially collective marks, and the protection against unfair competition in business practices, especially certification marks.

Legal basis for the protection of collective mark and certification mark is the Brunei Darussalam's Trademarks Act (Cap 98). Section 50 of the Act defines a collective mark as "… a mark distinguishing the goods or services of members of an association which is the proprietor of that mark from those of other undertakings".

Section 52 of the Act defines a certification mark as "… a mark indicating that the goods or services in connection with which it is used are certified by the proprietor of that mark in respect of origin, material, mode of manufacture of goods or performance of services, quality, accuracy, or other characteristics."

GI protection in Brunei includes agricultural and non-agricultural products. However, GI protection does not applicable for wines and spirits, as the alcoholic beverages are subjects to special domestic public policy of Brunei government.

Recently, the government of Brunei Darussalam is in the process to endorse a sui generis legal system to protect GI (Jennifer 2019).

16.4.2 Philippines

In the Philippines, GI used to be protected solely under the general trademark protection system, especially collective mark and certification mark. It was based on the Intellectual Property Code of the Philippines (IP Code) [Republic Act No. 8293, as amended]. Then, on 5 October 2022, the Philippines successfully endorsed the Intellectual Property Office of the Philippines' Memorandum Circular No. 2022-022 about Rules and Regulations on Geographical Indications, that is furtherly cited as GI Regulations. So, the legal means to protect GI in the Philippines become more varied: as a collective mark, as a certification mark, or as a GI. Based on the IP Code provisions, the administrations of them are conducted by the Bureau of Trademarks of the IP Office of the Philippines.

Rule 2.h. of the GI Regulation defines a GI as, "… any indication which identifies a good as originating in a territory, region or locality, where a given quality, reputation, or other characteristic of the good is essentially attributable to its geographical origin and/or human factors." Further, Rule 2.p. of the regulation defines Manual of GI Specifications as, "… a document contains the name to be protected as geographical indication description of the goods, the delimited geographical area where the goods are produced, and an explanation of the link between the said area and its quality, reputation or characteristics description of its production processes, quality control process and standards and labelling rules, among others."

Elements of the manual of GI specifications can be derived from the details of requirements of GI applications. They are substantiated in the Rule 10 of the Philippines' GI Regulation (IPOPHL Memorandum Circular No. 2022-022 2022).

16.4.3 Cambodia

GI in Cambodia is protected by standalone legislations, those are: the Law on Geographical Indications 2014, the Ministerial Regulation (Prakas) on Procedures for Registration and Protection of GI 2016, TRIPS Agreement, and the Geneva Act of the Lisbon Agreement.

According to articles in the Law and the Ministerial Regulation, GI refers to a distinctive name, symbol and/or any other sign which is a name or represents a geographical origin and identifies the goods as originating in such geographical area where a given quality, reputation or other characteristic of the goods is essentially attributable to its geographical origin.

GI product specification in Cambodia takes forms of a GI Book of Specification and the summary of the GI Book of Specification, whilst the holder is regarded as a GI association. The elements of the Cambodia's GI Book of Specification are regulated in Article 10 of the Cambodia's Prakas of GI (Ministerial Regulation on the Procedure for the Registration and Protection of GI 2016).

16.4.4 Indonesia

GI in Indonesia is protected by Law on TM and GI Number 20 Year 2016, Government Regulation Number 51 Year 2007 about GI, the Minister of Justice and Human Rights' Regulation Number 12 Year 2019 about GI, and the Minister of Justice and Human Rights' Regulation Number 10 Year 2022 amending the Minister of Justice and Human Rights Regulation Number 12 Year 2019 about GI (Miranda et al. 2021).

Article 1.6 of the Indonesian Law on Trademarks and GI 2016 defines a GI as," ... any indication which identifies a good and/or a product as originating from a particular region, of which its geographical environment factors including natural factor, human factor, or combination of both are attributable to a given reputation, quality, and characteristics of the produced good and/or product."

Elements of the Indonesia's GI Product Specification, that is called Description Document of GI, is regulated in the Republic of Indonesia's Trademarks and GI Law Number 20 Year 2016 Article 1.11 about the definition of the Description Document, and the Regulation of the Ministry of Justice and Human Rights of the Republic of Indonesia Number 12 Year 2019 Article 3 (6) about the contents of the Description Document of GI (Law of the Republic of Indonesia Number 20 2016).

In the late 2022, Indonesian government endorsed the Government Regulation Number 56 Year 2022 about Communal Intellectual Property. Communal IP systems include Geographical Indication Potential (GIP) (Miranda et al. 2023). GIP enjoys a defensive protection mechanism by uploading the GIP in the official database of Communal IP managed by the Directorate General of Intellectual Property of the Ministry of Justice and Human Rights, Republic of Indonesia. The objects of the GIP are similar with the objects of registered GI, although it requires a reputation attributable to the product's geographical origin only.

16.4.5 Thailand

GI in Thailand is protected by GI Protection Act B.E. 2546 (2003), Ministerial Regulation 2004, Ministerial Notification 2004, and the Department of Intellectual Property Notification 2004.

Section 3 of the Thailand's Geographical Indications Protection Act of 2003 defines a GI as,

> … a name, symbol or any other thing used for calling or representing a geographical origin and can identify the goods originating in such geographical origin where the quality, reputation or other characteristic of the goods is attribute to the geographical origin.

Elements of GI Product Specification in Thailand are regulated in the Thailand's Geographical Indications Protection Act B.E. 2546 (2003) Chapter II about Application for Registration of Geographical Indications Clause 9 and 10. The elements are incorporated in a form of application for registration that is accompanied by several supporting documents and evidence (FAO 2024).

16.4.6 Lao PDR

GI in Lao PDR is protected by the amended Law on Intellectual Property (Decree of the President of the Lao People's Democratic Republic No. 322/P, 25 December 2017), Decision of the Minister of Science and Technology on the Implementation of GI under the Law on IP No. 1119 (25 October 2016), Agreement of the Minister of Science and Technology on GI (2019), Paris Convention, TRIPS Agreement, and Geneva Act of the Lisbon Agreement 2015.

Article 18 of Lao PDR's Law on IP defines a GI as, "… a sign used to indicate a good as originating in the territory of a country or region or locality in that territory, where a given quality and reputation or other characteristic of the good is essentially attributable to its geographical origin." (OpenDevelopmentMekong 2020).

Elements of a Book of GI Specifications in Lao PDR are regulated in the Law of Intellectual Property Article 18 (revised) about the eligibility requirements for GI Certificate and Article 35 about the application for registration of GI. The elements are submitted as a bundle of documents.

16.4.7   *Myanmar*

GI in Myanmar is protected by the Pyidaungsu Hluttaw Law (Union Law) No. 3, 2019, the 10th Waning Day of Pyatho, 1380 M.E. (30 January 2019) on Trademark and GI Law, especially Chapter 16 Sections 53 to 60 specialized for GI (Spruson and Ferguson 2024).

Chapter 1 Section 2 Sub Section (o) of the Myanmar's Trademark and GI Law defines a GI as, "... any indication which identifies goods as originating in the territory of a country or a region or a locality in that territory where a given quality, reputation or other characteristics of the goods is essentially attributable to its geographical origin." The scope of GI legal protection in Myanmar includes agricultural products, handicrafts, and industrial products.

Elements of GI Specification in Myanmar are regulated in the Myanmar's TM Law and GI Law 2019 Article 53 and 54 about the applicants and the points that shall be specified in the GI application for registration.

16.4.8   *Malaysia*

GI in Malaysia is protected by Malaysia's Act 836—the Geographical Indications Act 2022 (Manique et al. 2023). Part I.2 of the Malaysia's GI Act 2022 defines a GI as, "... an indication which may contain one or more words which identifies any goods as originating in a country or territory, or a region or locality in that country or territory, where a given quality, reputation or other characteristic of the goods is essentially attributable to its geographical origin."

It is interesting to note that according to the Malaysia's GI Act 2022, the right to use a GI can be obtained by a person, a competent authority, or a trade organization or association. Furthermore, the term of protection is similar with a trademark, that is 10 years with indefinite renewability.

Elements of GI Product Specification in Malaysia are based on the Malaysia's Geographical Indication Act 2022 (Act 836) Chapter 6 Section 17 about Registration of GI, especially according to the guideline of Technical Specifications of GI from the Malaysia Intellectual Property Office. In Malaysia, GI Product Specification takes a form of a Technical Specification of GI (Law of Malaysia Number 836 about GI Act, 2022).

16.4.9   *Singapore*

GI in Singapore is protected by GI Act Number 19 Year 2014, GI Rules 2019 and Trademarks Act (Cap. 332).

Section 2 of the Singapore's Geographical Indications Act 2014 defines a GI as, "… any indication used in trade to identify goods as originating from a place, provided that: a. the place is a qualifying country or a region or locality in the qualifying country; and b. a given quality, reputation or other characteristic of the goods is essentially attributable to that place."

In Singapore GI protection system, the registration of a GI is not mandatory, but advisable. Similar with the system in Malaysia, a registered GI in Singapore enjoys the initial period of 10 years, and it is renewable indefinitely (Foo 2021). This term of GI protection is the same as the term of protection of TM.

Elements of GI Specification in Singapore are regulated in the Singapore Act 2014 Article 39 about the application for registration of GI.

16.4.10   Vietnam

GI in Vietnam is protected by Civil Code 2005, Criminal Code 2015, IP Law 2005 (amended and supplemented in 2009), Decrees, Circulars, and IP international legal instruments, notably: Paris Convention, TRIPS Agreement, international regional and bilateral agreements, and memorandums of cooperation related to GI.

Vietnam's IP legal system defines a GI as, "… a sign to indicate products originated from regions, localities, territories or countries, of which the quality or reputation is attributable to the geographical environment or natural characteristics and/or combination of human factor."

Elements of GI Specification in Vietnam are regulated in the Socialist Republic of Vietnam's Ministry of Science and Technology Circular No. 01/2007/TT-BKHCN about Guiding the Implementation of the Government's Decree No. 103/2006/ND-CP (22 September 2006) Detailing and Guiding the Implementation of a Number of Articles of the Law on Intellectual Property and Industrial Property, Section 6.43 about the requirements for applications for registration of GI (Intellectual Property Office of Vietnam, 2024).

16.5   Discussing a Model of GI Product Specification That May Be applicable for ASEAN Countries

Table 16.2 below describe the elements of GI product specification in ASEAN Member countries.

Tables 16.3 and 16.4 below compare 9 (nine) elements of GI Product Specification in ASEAN countries. The elements are described in two tables to make them easier to read.

Table 16.2 Comparison of the Element of GI Right Holders in ASEAN Countries

ASEAN countries		Legal bases	Right holders of GI
1	Brunei Darussalam	Part IX of Trademark Rules (Trademarks Act [-Chapter 98]) R1 1984 Ed. S 27/2000	applicant/proprietor of a registered collective mark/certification mark (collective right holder)
2	Philippines	Art IV Rule 7 of GI Regulations	Producers/producers' organization/association Government agencies/local government units Organizations/associations/indigenous cultural communities/indigenous peoples
3	Cambodia	Article 1o of Prakas of GI	Association, producers, or operators
4	Indonesia	Article 53 (3)a & b of Law on TM & GI	Institution or organization representing community in the geographical origin, such as: Association of producers, cooperation, GI protection society. Province, regency or city government (local government)
5	Thailand	Chapter II Section 7 of GI Act, B.E. 2546 (2003)	Government agency, state agency, state enterprise, local government organization, other body ascribed the status of a juristic person responsible of the geographical origin of the goods Natural person, a group of juristic persons engaged in trade and has a residence in the geographical area of the goods a group or organization of consumers
6	Lao	Article 35, 60 of Decree on the Promulgation of the Law on IP	Owner of registered GI
7	Myanmar	Article 53 Pyidaungsu Hluttaw Law Number 3 Year 2019	Any legal entity representing persons of the locality in which the relevant goods are produced: Persons producing goods by exploiting natural products/resources Producers of agricultural products Manufacturers of handicrafts/industrial products Competent authority from government department organization on behalf of persons
8	Malaysia	Article 2 of the GI Act Number 836 Year 2022	Interested person in relation to goods identified by a GI: a producer of the goods a trader of the goods
9	Singapore	Article 38 of the GI Act Number 19 Year 2014	a person carrying activity as a producer in the geographical origin Association Competent authority

(continued)

Table 16.2 (continued)

ASEAN countries		Legal bases	Right holders of GI
10	Vietnam	Section 1 Point 7 of the Circular on the Implementation of a number of articles of the Law IP and Industrial Property (14 February 2007)	State, or organization granted the right to use by the state to represent the interests of all organizations or individuals

Table 16.3 below compares 6 (six) elements GI Product Specification, those are: GI name, GI objects, link, reputation, and quality and/or characteristic, in ASEAN countries according to their national legal system for GI:

Table 16.4 below depicts the comparison of the 3 (three) elements of GI Product Specification: method, map, and controlling mechanism, in ASEAN countries.

Based on similarities and differences of GI Product Specifications in ASEAN countries, the shared main elements can be used as the bases of a model of GI Product Specification for ASEAN member countries. The shared elements on the GI Product Specification in 7 (seven) ASEAN member countries are comprised of:

- Identity, includes: identity of GI, identity of the GI object, and identity of the right holders.
- Reputation, quality, and/or characteristic/s caused by the GI geographical environment.
- Link between the GI product (reputation, quality and/or characteristic of the product) and its geographical origin (geographical environment, including natural and/or human factor/s).
- Geographical origin, includes: GI map depicting the geographical area where the GI product is produced, and an optional explanation of the origin.

It is interesting to note that in Brunei, where GI is protected as a Collective Mark and/or Certification Mark only, the existence elements are only the identities and reputation. In addition, Thailand and Singapore do not literally require the method or process of production as the compulsory element of GI product specification. Singapore even does not mention about the element of controlling mechanism. Yet, in practice, it is safe to advice that these elements should be accommodated as best practices.

16.6 Concluding Remarks

In the prospect of ASEAN Economic Community, creating one model of GI product specification is suggested. However, the model should be more focused on the completeness of the elements of a GI product specification rather than the unification of the form of the document. The elements are: GI name, GI objects, link, reputation, quality and/or characteristic, method, map, and controlling mechanism.

Table 16.3 Comparison of 6 elements of GI product specification in ASEAN countries

Country	Element		Link				
	GI name	GI object	Natural factor	Human factor	Reputation	Quality	Characteristic
Brunei	Logo	Goods			Reputation	Quality	Characteristic
Philippines	Indication/ name	Goods	Alternative; Geographical origin	Human factor	Specific reputation	Specific quality	Specific characteristic
Cambodia	Name + labelling rule	Products/ goods	Alternative; Natural factor	Human factor	Reputation	Quality	Main character/feature
Indonesia	Indication	Goods/ products	Accumulative / Alternative; Natural factor	Human factor	Reputation	Quality	Characteristic
Thailand	Name/ symbol/ calling	Description + picture of goods	Connection/ description of geographical environment		Alternative; Reputation	Quality	Property/ characteristic
Lao PDR	Indication/ name + GI image	Goods/ condition of products	Accumulative; Natural factor	Human factor	Alternative; Reputation	Quality	Characteristic
Myanmar	Name	Goods	Link between reputation, quality, characteristic, geographical origin, & production method		Alternative; Specific reputation	Specific quality	Specific characteristic
Malaysia	Name	Description + class of goods	Causal link between geographical area and specific product's quality/reputation/ characteristics		Alternative; Reputation	Quality	Characteristic
Singapore	Name (words, image)	Goods + category of goods	How the reputation, quality or characteristic is essentially attributable to the place of origin		Alternative; Reputation	Quality	Particular characteristic differs from other good in the same category
Vietnam	Name	Products	Geographical condition / Accumulative; Natural factor	Human factor	Accumulative; Reputation	Quality	Character/nature

Table 16.4 Comparison of 3 elements of GI product specification in ASEAN countries

Country	Element		
	Method of production	Map	Controlling mechanism
Brunei			
Philippines	Description of production processes	Description of geographical area	Quality control process and standards and labelling rules
Cambodia	Description of method to obtain the product	Geographical area + evidence of the origin of the product	Method & process of quality control + controlling body
Indonesia	Production process	Map of geographical origin	Method of quality control
Thailand		Details of location of geographical origin	Details of GI use (+ GI control system)
Lao PDR	Traditional production methods	Statement of geographical region	Methods of control
Myanmar	Method of production	Geographical area/ region of production	Official control by competent authorities or delegated third party
Malaysia	Specific steps in production	Geographical area/ map + proof of origin	Facultative inspection body
Singapore		Geographical area	
Vietnam	Skills of producers, traditional local production process	Map of geographical area + description	Mechanism of self-control on the characteristic and/ or quality

Regarding ASEAN country/ies whose GI objects are protected by collective marks and/or certification marks, GI product specification can be used as an additional document in the application or implementation of the related marks. It should mainly substantiate the reputation of the product as a secondary significance, so the GI name and/or logo that is regarded as a descriptive mark can get a strong capability to distinguish through use.

References

Antara Kantor Berita Indonesia (2015) Beras Krayan Diklaim Produk Malaysia. antaranews.com/berita/523520/beras-krayan-diklaim-produk-Malaysia

ASEAN Framework Agreement on Intellectual Property Cooperation (1995) https://asean.org/asean-framework-agreement-on-intellectual-property-cooperation-bangkok-thailand-15-december-1995/

Denis S (2019) How to draft geographical indication specifications? Paper presented at the capacity building workshop on geographical indications, promoting intellectual property rights in the ASEAN region. European Union Intellectual Property Office & ARISE Plus ASEAN Intellectual Property Rights. Kuching

Foo WY (2021) The geographical indications system in Singapore. Intellectual Property Office of Singapore. https://ipkey.eu/en/node/1390

Food and Agriculture Organization of the United Nation and Organization for an International Geographical Indications Network (2024) Developing a roadmap towards increased sustainability in geographical indication systems, practical guidelines for producer organizations to identify priorities, assess performance and improve the sustainability of their geographical indication systems, Glossary, p 160

Intellectual Property Office of the Philippines (2022) IPOPHL's GI rules now in effect, signaling strengthened protection and promotion of local products. https://www.ipophil.gov.ph/geographical-indications/

Jennifer EL (2019) Consultation meeting on developing a geographical indications protection system in Brunei Darussalam. IPKey of European Commission. Brunei Darussalam. https://ipkey.eu/sites/default/files/ipkey-docs/2019/Developing-a-Sui-Generis-GI-Protection-System-in-Brunei.pdf

Law of the Republic of Indonesia about Trademarks and Geographical Indications Number 20 Year (2016) https://internationalipcooperation.eu/sites/default/files/arise-docs/2019/Indonesia_Law-on-Marks-and-Geographical-Indications-20-2016.pdf

Lynne B (2007) Geographical indications: the current landscape. Fordham Intellect Prop Med Entertain Law J 17(4)

Macdalyna ER, Ahmad HAA, Nor QIMN, Faridah Y, Hasmadi M (2022) Characterization of Bario rice flour varieties: nutritional composition and physicochemical properties. Innovative food products and processing. J Appl Sci 12(18)

Manique C, Lee JC, Justin JBA (2023) A discourse on the Malaysian Geographical Indications Act. Sriwijaya Law Rev 7(2.2741):368–383. https://doi.org/10.28946/slrev.Vol7.Iss2.2741

Marie-Vivien D (2020) Protection of geographical indications in ASEAN countries: convergences and challenges to awakening sleeping geographical indications. J World Intellect Prop 23(3–4). https://doi.org/10.1111/jwip.12155

Ministerial Regulation (Prakas) of the Kingdom of Cambodia on the Procedure for the Registration and Protection of Geographical Indications (2016) https://wipolex-res.wipo.int/edocs/lexdocs/laws/en/kh/kh042en.pdf

Miranda RAP, Ahmad MR, Dadang ES, Ika CD, Saky S (2021) Geographical indication protection for non-agricultural products in Indonesia. J Intellect Prop Law Pract 16(4–5). https://doi.org/10.1093/jiplp/jpaa214

Miranda RAP, Laina R, Helitha NM (2023) Inclusive rights to protect communal intellectual property: Indonesian perspective on its new government regulation. Cogent Soc Sci 9(2). https://doi.org/10.1080/23311886.2023.2274431

Mubarok AS, Nurlela M, Galih YR (2021) Improving performance of Krayan Rice farming entrepreneurship-based Indonesia-Malaysia border society. IOP Conf Ser Earth Environ Sci. https://doi.org/10.1088/1755-1315/748/1/012020

Spruson & Ferguson (2024) Myanmar's geographical indication law: summary and recommendation, Lexology. https://www.lexology.com/library/detail.aspx?g=6f04c0af-8531-41ef-9109-6b3dc40fcd7f

The ASEAN Declaration (Bangkok Declaration) (1967) https://agreement.asean.org/media/download/20140117154159.pdf

The opinions expressed in this chapter are those of the author(s) and do not necessarily reflect the views of the [NameOfOrganization], its Board of Directors, or the countries they represent.

Part III
Geographical Indications Contributions to Territorial Development

Chapter 17
Contribution of the PDO and PGI of Extremadura (Spain) to the Protection of Biodiversity and the Development of the Green and Circular Economy

Juan J. Ferrero-García and Julia Martín-Cerrato

Abbreviations

AREPO	Association of European Regions for Products of Origin
EAFRD	European Agricultural Fund for Rural Development
EU	European Union
PDO	protected designations of origin
PGI	protected geographical indications
UNESCO	United Nations Educational, Scientific and Cultural Organization

17.1 Introduction

The move towards agroecological transition processes and adaptation to climate change has been promoted in the European Union with the recent publication of the Action Plan for the circular economy (EC 2020). In line with this, the Autonomous Community of Extremadura (southwest of Spain) is implementing its own Green and Circular Economy Strategy, called *Extremadura 2030* (JE 2018). This is a regional action plan that, among many other issues, includes the strengthening of differentiated quality figures, in particular protected designations of origin (PDO) and protected geographical indications (PGI). To achieve this, the regional administration (*Junta de Extremadura*) carries out actions aimed at stimulating the promotion and information actions of these figures, supporting the incorporation of new

J. J. Ferrero-García (✉) · J. Martín-Cerrato
Junta de Extremadura, Merida (Badajoz), Spain
e-mail: juanjose.ferrerog@juntaex.es

© The Author(s) 2025
E. Vandecandelaere et al. (eds.), *Worldwide Perspectives on Geographical Indications*, https://doi.org/10.1007/978-3-031-71641-6_17

producers to the PDOs and PGIs, financially assisting their management entities—producer groups recognised as public law corporations—and, in general, providing them with legal security and an appropriate legal framework. All this taking into account the context of a region where the agri-food sector is its main pillar (CaixaBank Research 2019; Sanguino 2022).

17.2   The PDOs and PGIs and Sustainable Development

The strong commitment to PDOs and PGIs in Extremadura is largely based on the conviction of their significant contribution to the protection of biodiversity and sustainable development. Typically, the specifications of many of these quality figures from the Extremadura region have included requirements that, at least indirectly, protect natural resources or the landscape of their geographical area. Moreover, many of these figures are based on extensive or semi-extensive production methods that require few inputs and generate few emissions and waste. Also, the fact that PDOs and PGIs defend the products of a specific area at a local, regional or regional level, implies a reduction in the dependence on imports of raw materials and natural resources, which translates into a decrease in transport and its associated problems (consumption of fossil fuels, emission of pollutants and greenhouse gases, risk of spills, etc.). In addition, it should be valued that, even those PDOs and PGIs that do not involve particularly optimal environmental practices, at least tend to be located in disadvantaged regions or with some natural limitations (Hinojosa-Rodríguez et al. 2014). In any case, PDOs and PGIs, if well managed, constitute tools that facilitate or encourage sustainable rural development, reducing the most intensive and harmful agricultural practices for the environment (Williams 2007; Vandecandelaere et al. 2010; Sylvander et al. 2012; Belletti et al. 2015). These circumstances, along with the fact that PDOs and PGIs have the ability to fix production to a specific territory and promote the creation of an institutional context in support of the activities of quality figures (Freitas 2016), implies that they usually contribute to preventing depopulation and the deterioration of the rural environment (*e.g.*, Crescenzi et al. 2022). In general, they facilitate the economic growth of the agricultural sector (Li et al. 2023), although existing empirical studies show heterogeneous results (Török et al. 2020). In any case, public authorities must be alert to possible failures and risks of unsustainability due to the disaffection of the producers when certain negative dynamics occur (Belletti et al. 2017; De Rosa et al. 2023).

The legislation regulating quality regimes in the European Union has recently incorporated significant changes to recognise some of these circumstances from a legal point of view. Thus, since the end of 2021, PDOs and PGIs can also refer to agricultural products that, among their characteristics, have attributes that provide added value for their contribution to sustainable development (EP&C 2021), aspects that will be further emphasised from 2024, with the new community regulation on quality: Regulation (EU) 2024/1143 of the European Parliament and of the Council, of 11 April 2024.

17.3   The PDOs and PGIs of Extremadura

The first fully Extremaduran quality figures registered in the Community register of PDOs and PGIs did so in the 1990s of the last century (García-Galán et al. 2006); specifically, these are the PDOs *"Dehesa de Extremadura"* and *"Queso de la Serena"*, registered in 1996. Since then, a total of 12 PDOs and five exclusively Extremaduran PGIs have been registered in this register, especially in the period 2003–2007 (Table 17.1).

Of all these figures, only two are viticultural, while the rest are agri-food. In addition, there are three other PDOs and one PGI whose territorial scope is distributed between Extremadura and other Spanish regions, so they are not the responsibility of any Regional Authority but of the Spanish State (Table 17.2). Together, the PDO and PGI strictly from Extremadura comprise just over a hundred operators who produce protected agricultural food products, and as many who produce protected wines. To these figures, we would have to add several thousand farmers and livestock farmers involved in the production of the raw materials of the respective PDO and PGI, which, in some specific sectors, represent a very significant part. Thus, for example, the total cultivated area of cherry in Extremadura reaches 7500 hectares (Sanguino 2022); of this amount, no less than about 6500 hectares are registered in the PDO *"Cereza del Jerte"* (MAPA 2023).

Many of these PDO and PGI can promote sustainable development and, therefore, a greener and more circular economy. This is because they are often based on

Table 17.1 Protected Designations of Origin (PDO) and Protected Geographical Indications (PGI) present exclusively in Extremadura, by type of protected product, and with indication of their registration date in the European Union Register of PDO and PGI

Differentiated quality figures	Product	EU registration date
PDO *"Torta del Casar"*	Raw sheep's milk cheeses	26.08.2003
PDO *"Queso de La Serena"*	Raw sheep's milk cheeses	21.06.1996
PDO *"Queso Ibores"*	Raw goat's milk cheeses	05.02.2005
PDO *"Queso de Acehúche"*	Raw goat's milk cheeses	05.07.2022
PDO *"Dehesa de Extremadura"*	Iberian pig hams	21.06.1996
PGI *"Cordero de Extremadura"*	Fresh sheep meat	05.10.2011
PGI *"Ternera de Extremadura"*	Fresh beef	12.08.2004
PGI *"Vaca de Extremadura"*	Fresh beef	18.08.2023
PGI *"Cabrito de Extremadura"*	Fresh goat kid meat	10.01.2024
PDO *"Miel Villuercas-Ibores"*	Honeys	20.01.2017
PDO *"Cereza del Jerte"*	Fruits (cherries)	15.12.2007
PDO *"Pimentón de la Vera"*	Spices (paprika)	22.08.2007
PDO *"Gata-Hurdes"*	Extra virgin olive oils	16.02.2007
PDO *"Aceite Monterrubio"*	Extra virgin olive oils	07.03.2007
PDO *"Aceite Villuercas Ibores Jara"*	Extra virgin olive oils	23.10.2023
PDO *"Ribera del Guadiana"*	Wines	14.04.2004
PGI *"Extremadura"*	Wines	16.04.2004

Table 17.2 Protected Designations of Origin (PDO) and Protected Geographical Indications (PGI) present in Extremadura and in other Spanish regions, by type of protected product, and with indication of their date of registration in the European Union Register of PDO and PGI

Differentiated quality figures	Product	EU registration date
PDO "*Jabugo*"	Iberian pig hams	27.01.1998
PDO "*Guijuelo*"	Iberian pig hams	21.06.1996
PDO "*Cava*"[a]	Wines	13.06.1986
PGI "*Carne de Ávila*"	Fresh beef meats	21.06.1996

[a]The geographical scope of the PDO "*Cava*" includes a municipality of Extremadura, Almendralejo, since 1995

the survival of ancestral ecosystems of dehesas and natural pastures, which involves the presence of native livestock breeds raised in extensive or semi-extensive systems. In particular, the dehesa is an agro-livestock landscape exceptionally represented in Extremadura, based on the maintenance of a thinned forest of species of the genus *Quercus* and its pastures, as well as on the sustainable exploitation of its various associated natural resources (Pulido et al. 2007); for this reason, it is given special attention in the Action Plan *Extremadura 2030* (JE 2018). Thus, these agricultural systems are the ones that have made it possible for the cheese protected by the PDO "*Torta del Casar*", "*Queso de la Serena*", "*Queso Ibores*" and "*Queso de Acehúche*" to reach our days. The same can be said for the PDO of Iberian pig hams —fed with acorns from holm oaks and cork oaks— "*Dehesa de Extremadura*", "*Guijuelo*" and "*Jabugo*"; and by the PGI of meat "*Cordero de Extremadura*", "*Ternera de Extremadura*", "*Vaca de Extremadura*", "*Cabrito de Extremadura*" and "*Carne de Ávila*". Other quality figures such as "*Cereza del Jerte*", "*Ribera del Guadiana*", "*Extremadura*", "*Cava*", "*Pimentón de la Vera*", "*Gata-Hurdes*", "*Aceite Monterrubio*" and "*Aceite Villuercas Ibores Jara*" are based on traditional agricultural crops, based on plant varieties perfectly adapted to the climatic and physical characteristics of the Extremadura region and, sometimes, the peculiarities of certain natural environments with specific limitations. Finally, the PDO "*Miel Villuercas-Ibores*" strengthens the beekeeping sector and the survival of bees (*Apis mellifera*) in a unique mountain area that contains more than 50 *geosites*, which is why in 2015 it was declared a World Geopark by UNESCO (one of the 15 that Spain has).

On the other hand, it is no coincidence that many of the geographical areas of the Extremadura PDO and PGI coincide with legally preserved lands due to their high biodiversity value, often designated as special protection areas for birds or sites of Community importance, and their consequent inclusion in the European Union's Natura 2000 Network (Ferrero-García 2016). In general, in these PDO and PGI of Extremadura, there often occur a accumulation of natural and cultural factors (protected areas, hunting resources, monuments, archaeological sites, etc.) with others specific of a gastronomic nature, which enhances their use from a tourist point of view, through routes, tastings, courses, tastings, visits to model farms, to oil mills, ham dryers, wineries and cheese factories, etc. (*e.g.*, Hernández et al. 2016; Tarazona-Valverde et al. 2021). Very positive synergies for tourism, as has already been highlighted in other places on the Iberian Peninsula (Millán and Agudo 2010; Millán et al. 2014; Cava et al. 2019; Dos Reis et al. 2021), and which contribute to mitigating the depopulation of large rural areas. A crucial aspect in the Extremadura region, since 52% of its territory has an average density of about 10 inhabitants per km^2, bordering the limit of what is usually considered a "demographic desert" (CaixaBank Research 2019).

Thus, for example, the PDOs *"Miel Villuercas-Ibores"*, *"Queso Ibores"* and *"Aceite Villuercas Ibores Jara"* share a predominantly mountainous, depopulated landscape characterised by its affiliation to the Natura 2000 network, its membership of the Global Geopark and the inclusion, within this territorial scope, of the Royal Monastery of Santa María de Guadalupe (a place inscribed on the UNESCO World Heritage List). Similarly, in the oak and cork oak forests of the Monfragüe National Park and its surroundings, also affiliated to the Natura 2000 network and also classified as a Biosphere Reserve by UNESCO —they constitute one of the most extensive and best preserved patches of Mediterranean forest in Europe—, part of the geographical areas of, among other quality figures, the PDO *"Dehesa de Extremadura"*, as well as the PGI *"Cordero de Extremadura"*, *"Ternera de Extremadura"*, *"Vaca de Extremadura"* and *"Cabrito de Extremadura"* converge.

17.4 Actions of the Extremadura Administration

From the Autonomous Community of Extremadura, a series of general (Law 4/2010 and Law 6/2015) and specific (Regulations and Statutes of each quality figure) norms have been approved, whose objective is to provide legal coverage to both PDOs and PGIs and their producer groups. Currently, the PDOs and PGIs of Extremadura are legally considered public domain goods, not susceptible to individual appropriation, sale, alienation or encumbrance. As for their producer groups or management entities —called, in Spain, regulatory councils since 1932 (González 2014), they are public law corporations with their own legal personality and economic autonomy. They include all the producers of a PDO or PGI and are governed by democratic principles, there must be parity in the representation of the different interests present and maintain, as a basic principle, their operation without profit motive. Precisely, this participatory approach is crucial to achieve important objectives in the sphere of sustainable development (Vandecandelaere et al. 2021). The oldest regulatory council in the Extremadura region is that of the PDO *"Dehesa de Extremadura"* which, although under another name, was provisionally constituted over 35 years ago (CA&C 1987). Today, this PDO is one of the most important differentiated quality figures in Extremadura and Spain as a whole, with an economic value estimated at over fifteen million euros in 2022, as a result of the marketing during that year of almost 84,000 pieces (hams and shoulder hams) of Iberian pig protected by the PDO (MAPA 2023).

The *Junta de Extremadura* financially supports some of the operating costs of the regulatory councils through "de minimis" aid: Commission Regulation (EU) No

1407/2013 and, currently, Commission Regulation (EU) 2023/2831). For example, it covers the costs associated with its accreditations in the EN/ISO 17065:2012 technical quality standard (product certification), as well as those derived from the verification of compliance with its specifications (in those cases where this task has been delegated by the competent Authority of the Extremadura Regional Government, and under its supervision).

In addition, it encourages the promotional and informational actions of these entities and encourages the incorporation of new farmers, livestock farmers and agri-food industries into the existing quality schemes, through grants framed in the European Agricultural Fund for Rural Development (EAFRD) of the European Union. Specifically, through the aids of the Rural Development Programme of Extremadura, 2014–2020 (Decree 36/2016 and Decree 88/2022) and, recently, through the aids of the National Strategic Plan of the Common Agricultural Policy, 2023–2027 (Decree 75/2023 and Decree 79/2023). In particular, Decree 79/2023 will represent a strong reinforcement of environmental protection and biodiversity, by financially supporting information and promotion expenses that highlight, among other aspects, the rigorous standards of animal welfare and respect for the environment linked to each quality scheme. On the other hand, both in the previous period (2014–2020) and in the current one (2023–2027) part of the registration or maintenance costs of producers in a differentiated quality scheme are subsidised, as well as part of the costs that producers bear for the controls required to verify the corresponding specifications. Undoubtedly, all these measures have contributed to the success of some of the PDOs and PGIs of Extremadura during the last decade. Thus, for example, in 2012 the PDO "*Pimentón de la Vera*" marketed almost one and a half million kilograms of protected product which represented an economic value of just six million euros (MAAMA 2013); however, in 2022 it has marketed more than two million kilograms of protected paprika, which has translated into an economic value of practically double that of ten years ago (MAPA 2023). Both these promotional aids and the previous support for operation are contemplated in the Action Plan *Extremadura 2030* (JE 2018).

The Extremadura Administration also directly organises various promotional and dissemination activities, as well as technical or informational events. Finally, it promotes green tourism through the design of actions based on the joint participation of differentiated quality schemes, shops, leisure and tourism companies, restaurants, hotels and rural accommodations, within the framework of the initiative *Extremadura Gourmet*. This is implemented with various sustainable gastronomic routes that involve several PDOs of Extremadura: the *Iberian pig* and Extremadura pasture route, the virgin olive oils route, the cheeses route and, finally, the wines route. Through these routes, you can get to know the Extremadura region in a different way, through quality foods that invite you to discover the culture, heritage and nature of the places where they are produced (DGT 2022).

On the other hand, in recent years the Extremadura Regional Government has redoubled its efforts to encourage the creation of new PDOs and PGIs, especially those linked to extensive or semi-extensive livestock farming, such as the PDOs "*Queso de Acehúche*" and "*Aceite Villuercas Ibores Jara*", and the PGIs "*Vaca de*

Extremadura" and "*Cabrito de Extremadura*". In all these cases, the majority of producers showed a favourable attitude to this creation from the beginning. Starting from this favourable situation, the Extremadura Administration has provided continuous advice and explanation of the procedures and steps to follow to achieve community registration, which has allowed to maintain the interest of the producers throughout a A process that, given its complexity, tends to be prolonged over time. Finally, these four quality figures have been recognised by the European Union between July 2022 and January 2024, of which three are based on livestock breeds particularly adapted to the environment of southern Iberian Peninsula. One of these new figures, "*Cabrito de Extremadura*", constitutes the first PGI of meat in the goat kid sector in Spain. It is not surprising, therefore, that Cáceres (one of the two provinces that make up the Autonomous Community of Extremadura) is one of the Spanish provinces with the highest implementation of PDO and PGI agri-food products. The result of this renewed commitment by the Extremaduran Administration to differentiated quality figures is the fact that, since October 2021, the presidency of AREPO (Association of European Regions for Products of Origin) has been held by the Region of Extremadura. In fact, the 1st General Assembly of AREPO in 2023 was held in the city of Cáceres, so Extremadura acted as the host region.

17.5   Conclusions

Most of the PDOs and PGIs of Extremadura are agri-food, with a predominance of products based on extensive livestock or traditional crops (cheeses, hams, meats, honeys, oils, fruit trees or spices), and often located in protected areas or with abundant biodiversity (such as the Monfragüe National Park and Biosphere Reserve). The regional administration has actively encouraged the creation of these differentiated quality figures, with the aim of contributing to the sustainable rural development of the Extremaduran region within the framework of a greener and more circular economy. To this end, it supports the PDOs and PGIs through various actions. Thus, the Junta de Extremadura finances some operating, promotion and information expenses of its management entities (regulatory councils), subsidies part of the cost of incorporating new producers and, on occasion, also carries out certain direct promotion actions.

References

Belletti G, Marescotti A, Sanz-Cañada J, Vakoufaris H (2015) Linking protection of geographical indications to the environment: evidence from the European Union olive-oil sector. Land Use Policy 48:94–106

Belletti G, Marescotti A, Touzard JM (2017) Geographical indications, public goods, and sustainable development: the roles of actors' strategies and public policies. World Dev 98:45–57

CA&C [Consejería de Agricultura y Comercio] (1987) Resolución de 10 de septiembre de 1987, de la Dirección General de Comercio e Industrias Agrarias, por la que se designa al Consejo Regulador Provisional de la Denominación de Origen "Montanera de Extremadura". Diario Oficial de Extremadura, 75 [22.9.1987]: 1117

CaixaBank Research (2019) La economía de la Comunidad Autónoma de Extremadura: Diagnóstico estratégico. CaixaBank, Barcelona

Cava JA, Millán MG, Hernández R (2019) Analysis of the tourism demand for Iberian ham routes in Andalusia (Southern Spain): tourist profile. Sustain For 11(16):4278

Crescenzi R, De Filippis F, Giua M, Vaquero-Piñeiro C (2022) Geographical Indications and local development: the strength of territorial embeddedness. Reg Stud 56(3):381–393

De Rosa M, Masi M, Apostolico L, Bartoli L, Francescone M (2023) Geographical indications and risks of unsustainability linked to "disaffection effects" in the dairy sector. Agriculture 13:333

DGT [Dirección General de Turismo] (2022) Extremadura Gourmet. Rutas gastronómicas sostenibles. Dirección General de Turismo de la Junta de Extremadura. https://issuu.com/extremadura_tur/docs/rutasgastronomicas [Consulted: 14.12.2023]

Dos Reis CM, Rengifo JI, Correia JC (2021) La relación de los productos agroalimentarios de calidad diferenciada con el turismo en España y Portugal. Boletín de la Asociación de Geógrafos Españoles 89. https://doi.org/10.21138/bage.3020

EC [European Commission] (2020) Action plan for the circular economy. For a cleaner and more competitive Europe. Office of the European Union Publications, Luxembourg

EP&C [European Parliament & Council] (2021) Regulation (EU) 2021/2117 of the European Parliament and of the Council, of 2 December 2021. Off J Eur Union L, 435 [6.12.2021]: 262–314

Ferrero-García JJ (2016) Aceites de Oliva Virgen con Denominación de Origen Protegida en Extremadura. In: Martín D, López JM (eds) Aceite de Oliva Virgen. Saber y Sabor de Extremadura. Consejería de Medio Ambiente y Rural, Políticas Agrarias y Territorio-Consejería de Economía e Infraestructuras (CICYTEX), Junta de Extremadura, Badajoz, pp 117–137

Freitas S (2016) Las denominaciones de origen como herramienta del desarrollo territorial rural: estudio de casos españoles: Méntrida, Mondéjar y Uclés. Tesis doctoral. Facultad de Geografía e Historia, Universidad Complutense de Madrid

García-Galán M, Chamorro A, Valero V (2006) Las Denominaciones de Origen en Extremadura: Una apuesta por la calidad. Bol Econ de ICE 2889:55–64

González A (2014) El control de las administraciones públicas en el sector vitivinícola: el papel de los consejos reguladores. Tesis doctoral. Facultad de Derecho, Universidad Complutense de Madrid

Hernández JM, Folgado JA, Campón AM (2016) Oleoturismo en la Sierra de Gata y Las Hurdes (Cáceres): análisis de su potencial a través de un test de producto. Int J Sci Manag Tour 2(1): 333–354

Hinojosa-Rodríguez A, Parra-López C, Carmona-Torres C, Sayadi S (2014) Protected designation of origin in the olive growing sector: adoption factors and goodness of practices in Andalusia, Spain. New Medit 13(3):2–12

JE [Junta de Extremadura] (2018) Extremadura 2030. Estrategia de Economía verde y circular. Plan de Acción de la Junta de Extremadura. Junta de Extremadura. https://www.juntaex.es/documents/77055/406857/Action-+Action+Plan+of+the+Junta+de+Extremadura+2030.pdf. Accessed 20.07.2023

Li C, Gao J, Ge L, Hu W, Ban Q (2023) Do geographical indication products promote the growth of the agricultural economy? An empirical study based on meta-analysis. Sustain For 15:14428

MAAMA [Ministerio de Agricultura, Alimentación y Medio Ambiente] (2013) Datos de las Denominaciones de Origen Protegidas (D.O.P.) e Indicaciones Geográficas Protegidas (I.G. P.) de Productos Agroalimentarios. Año 2012. Ministerio de Agricultura, Alimentación y Medio Ambiente, Madrid

MAPA [Ministerio de Agricultura, Pesca y Alimentación] (2023) Datos de las Denominaciones de Origen Protegidas (D.O.P.), Indicaciones Geográficas Protegidas (I.G.P.) y Especialidades Tradicionales Garantizadas (E.T.G.) de Productos Agroalimentarios. Año 2022. Ministerio de Agricultura, Pesca y Alimentación, Madrid

Millán MG, Agudo EM (2010) El turismo gastronómico y las Denominaciones de Origen en el sur de España: Oleoturismo. Un estudio de caso. Pasos 8(1):91–112

Millán MG, Morales E, Pérez LM (2014) Turismo gastronómico, Denominaciones de Origen y Desarrollo Rural en Andalucía: situación actual. Boletín de la Asociación de Geógrafos Españoles 65:113–137

Pulido F, Sanz R, Abel D, Ezquerra J, Gil A, González G, Hernández A, Moreno G, Pérez JJ, Vázquez FM (2007) Los bosques de Extremadura. Evolución, ecología y conservación. Consejería de Industria, Energía y Medio Ambiente, Junta de Extremadura, Mérida

Sanguino R (2022) Informe sobre la Agricultura y la Ganadería Extremeñas. 2021. Fundación CB/Ibercaja, Badajoz

Sylvander B, Wallet F, Isla A (2012) Under what conditions geographical indications protection schemes can be considered as public goods for sustainable development? In: Torre A, Traversac JB (eds) Territorial governance, local development, rural areas and agrofood systems. Springer, Heidelberg, pp 185–202

Tarazona-Valverde BY, Campón-Cerro AM, Di-Clemente E (2021) Análisis de las posibilidades gastronómicas del AOVE como base para el diseño de experiencias de oleoturismo en Extremadura. Rotur, Revista de Ocio y Turismo 15(2):61–82

Török Á, Jantyik L, Maró ZM, Moir HV (2020) Understanding the real-world impact of geographical indications: a critical review of the empirical economic literature. Sustain For 12:9434

Vandecandelaere E, Arfini F, Belletti G, Marescotti A (2010) Linking people, places and products. A guide for promoting quality linked to geographical origin and sustainable indications. FAO & SINER-GI, Rome

Vandecandelaere E, Samper LF, Rey A, Daza A, Mejía P, Tartanac F, Vittori M (2021) The geographical indication pathway to sustainability: a framework to assess and monitor the contributions of geographical indications to sustainability through a participatory process. Sustain For 13:7535

Williams RM (2007) Do geographical indications promote sustainable rural development? Lincoln University, Canterbury

Chapter 18
"Madd de Casamance": The Collective Construction of a GI on a Picked Product to Develop a Sustainable Industry

Stéphane Fournier, Pape Tahirou Kanouté, Maimouna Sambou, and Fanta Sow

Acronyms

AFD	French Development Agency
APPIGMAC	Association for the Protection and Promotion of the Geographical Indication *Madd de Casamance*
ASPIT	Senegalese Agency for Industrial Property and Technological Innovation
CNIG	National Committee for Geographical Indications in Senegal
GIE	Economic Interest Group
ETDS	Economy, Territories and Development Services
OAPI	African Intellectual Property Organisation
WIPO	World Intellectual Property Organisation

S. Fournier (✉)
Institut Agro Montpellier, Montpellier, France
e-mail: stephane.fournier@supagro.fr

P. T. Kanouté
Economy Territories and Development Services, Ziguinchor, Senegal

M. Sambou · F. Sow
Association for the Protection and Promotion of the Geographical Indication "Madd de Casamance" (APPIGMAC), Ziguinchor, Senegal

© The Author(s) 2025
E. Vandecandelaere et al. (eds.), *Worldwide Perspectives on Geographical Indications*, https://doi.org/10.1007/978-3-031-71641-6_18

18.1   Introduction

The madd, or *Saba senegalensis,* is a wild vine growing in the forest and its fruits are consumed fresh or processed into juice or preserves. The registration of a Geographical Indication (GI) *"Madd de Casamance"* (*Casamance madd*) is requested by an association comprising pickers and processors of madd, the APPIGMAC (Association for the Protection and Promotion of the Geographical Indication *Madd de Casamance*).

Senegal, a signatory to the Bangui Agreement of the African Intellectual Property Organisation (OAPI), has wanted to develop a policy of GI protection for several years. The Senegalese government and the Senegalese Agency for Industrial Property and Technological Innovation (ASPIT) have thus worked on the establishment of a national legal framework. On 12 December 2019, a decree created and set the rules of organisation and operation of the National Committee for Geographical Indications in Senegal (CNIG). Following an official installation meeting of the CNIG (on 18 November 2020 in Dakar), and then a training of members (organised on 1 and 2 July 2021), the system is now operational. Senegal's signing in 2023 of the Geneva Act of the Lisbon Agreement of the WIPO reinforces the country's commitment to the protection of GIs.

This chapter seeks to show how the request for registration of The GI *"Madd de Casamance"* was constructed and how the registration of this GI would contribute to the construction of a supply chain positively impacting the pillars of sustainable development.

18.2   What Organisational Device for the Construction of the GI *Casamance madd*?

Various GI projects have been under study in Senegal since the beginning of the 2010s: honey from Casamance, salt from the pink lake, madd from Casamance, etc. A study of the potential as a GI of various Senegalese products was carried out (Bagal et al. 2018).

The project to register a Geographical Indication (GI) *"Casamance madd"* has been the subject of the most significant investments, as a strong demand from local actors emerged in the mid-2010s. An initial study of the feasibility and interest of setting up this GI was then carried out in 2017 by the study office "Economy, Territories and Development Services" (ETDS) and Institut Agro Montpellier, thanks to FAO funding (Bermond 2017; Bermond et al. 2018). It was able to begin to establish the specificity and reputation of both the fresh fruit and locally and nationally processed products, but also raised many questions, particularly related to the delimitation of the geographical area, traceability and the nature of the product(s) to be protected in the GI. The interest of this protection of the GI, however, clearly appeared, in view of the challenges of promoting the Casamance origin of the madd as well as fighting against usurpations, structuring the sector, maintaining added value in the area and preserving the resource (through more eco-responsible picking practices). This study was presented at a regional seminar on GIs organised in Casamance by the FAO, WIPO, OAPI and ASPIT in November 2017, following which the madd processors massively expressed their keen interest in a GI "*Casamance madd* ".

In July 2018, new FAO funding allowed to continue the prospective evaluation of this GI *Casamance madd*, using a tool co-constructed with the University of Florence. This evaluation was conducted in a participatory manner and aimed to analyse the potential effects of different scenarios of structuring the collective organisation, zoning and name of the GI and nature of the GI product (fresh and/or processed madd), as well as to reflect on the possibilities of control and traceability (Teyssier and Kanouté 2018).

In June 2019, a training was organised for the actors of the *Casamance madd* supply chain, as part of the "Training through Action" component of the African Union's continental strategy for GIs, implemented by EUIPO in coordination with OAPI. This training aimed to strengthen the capacities of the actors in the value chain of *Casamance madd* on the issues of product specifications and structuring of collective GI management organisations, while also analysing success factors (Carimentrand 2019).

An action plan to carry the project through to the registration of the GI was then funded by WIPO from 2019 to 2021. It was first necessary to support the structuring and official registration of a carrier association (Fournier 2019a, b). Once the constitutive assembly of the Association for the Protection and Promotion of the Geographical Indication *Madd de Casamance* (APPIGMAC) was held, on 28 and 29 November 2019, the latter was able to to elect from within its ranks, based on the expertise of the elected individuals, a small committee tasked with working on the construction of the specifications and then presenting this work at the General Assembly for validation before sending it to the competent authorities. This group had previously received training and its work sessions were supported (Fournier 2021a, b). A study of the specificity and reputation of the madd from Casamance and its processed products was also carried out during this period (Hernandez et al. 2020) as well as a study of the market for products under the GI *Casamance madd*.

This action plan was slowed down by the pandemic, but it allowed the submission of the GI registration request to the CNIG in mid-2022.

Since 2022, this GI is supported by the "Facilité IG" project (Cirad/French Development Agency—AFD).

18.3 Challenges Overcome for the Registration of the GI *"Casamance madd"*

The draft specifications provide that the GI *Casamance madd* covers the fresh fruit picked in the forest and five processed products (nectar, syrup and three types of preserves: sweet madd preserve, sweet-salty madd preserve and sweet-salty spicy madd preserve). It requires that the picking and processing must take place in the three administrative regions comprising natural Casamance (Ziguinchor, Kolda and Sédhiou regions), with also constraints in terms of distance (200 kms maximum between the place of picking and that of processing) and time (2 days maximum between the picking and the delivery to the processing unit, then 3 days maximum before the start of the processing). It finally sets the required quality criteria and the methods of obtaining for the fresh fruits and the processed products.

To achieve this result, it was necessary for a sponsoring association to structure itself and establish decision-making rules, that the nature of the product(s) protected by the registration be defined and that solutions be found to ensure the traceability of a picked product. This part successively addresses these three questions.

18.3.1 What Structure for an Inclusive and Efficient Sponsoring Association?

A first challenge to overcome for the registration of the GI *Casamance madd* was the establishment of an association bringing together actors scattered over a vast area, Casamance (28,350 km^2), structured into three administrative regions (Regions of Ziguinchor, Kolda and Sédhiou), these actors working at different stages of the supply chain (pickers, processors, collectors). As seen above, discussions took place between 2017 and 2019 to find the best way to overcome this challenge. The issue was to find rules allowing good representation of different categories of actors and geographical areas, while ensuring an efficient mode of operation.

Created at a Constitutive General Assembly on 29 November 2019, the Association for the Protection and Promotion of the Geographical Indication *Madd de Casamance* (APPIGMAC) brings together the economic actors of the madd supply chain and its processed products from the three administrative regions of natural Casamance.

It includes four types of members:

- groups of 5–10 pickers or collectors (these groups join collectively; a group is considered as a member);
- Economic Interest Groupings (GIE) for processing (the GIE also join collectively);
- individual artisans;
- processing companies.

It is structured into two colleges: college 1 "pickers and collectors" and college 2 "individual processors, EIGs and processing companies". Each college holds 50% of the voting rights at the General Assembly (the decision-making body of the Association, which meets at least once a year). This implies reaching a consensus between the two colleges for any decision-making.

The General Assembly elects an Extended Board for 3 years. This includes eight members and meets once a quarter. Three auditors (one per region) are also elected. The mandate of each member is renewable once.

The following rules have been defined to ensure good representation of the different professions and administrative regions:

- The Extended Board must include at least three representatives from college 1 and at least three representatives from college 2;
- Each administrative region must have at least two representatives.

The Extended Board elects an Executive Board of 4 members, which meets every month. It must respect the following rules:

- The Executive Board must include two representatives from the pickers and two representatives from the processors;
- The Executive Board must include at least 1 representative from each region.

The General Assembly, the extended board, as well as the executive board are assisted in their operation and for the implementation of activities related to the management of the GI sector by a salaried coordinator.

The applicant for the registration of the GI *Casamance madd*—the APPIGMAC—is thus an association offering the expected guarantees of a defense and management organisation of an IG, namely:

- Openness to all interested sector actors, ensured by membership costs and annual contributions set at affordable amounts for the different categories of members and by information/awareness campaigns regularly carried out by the members of the APPIGMAC Board and the coordinator, supported by ETDS;
- Representation of all the professions of the local madd supply chain: pickers, collectors, processors (individuals/GIE/companies) and the different geographical areas (the three administrative regions of natural Casamance)
- Democratic management, ensured by a structure in two colleges and rules for balancing the representation of the different professions and different regions in the executive bodies.

As soon as the APPIGMAC was constituted, it was a matter of setting up a legitimate and operational "specifications committee" to define the local IG rules. This was made up of 12 members of the APPIGMAC: four representatives from the association's Board (President, Treasurer, Secretary and Ziguinchor Focal Point), two co-opted people due to their skills, three representatives from the processors (one per area) and three representatives from the pickers (one per area). The committee received training and then worked on all points of the specifications during six workshops, which took place between November 2019 and June 2021. These meetings were spaced far enough apart for the members of this "specifications committee" to discuss with the other members of the association of problematic points.

18.3.2   A GI Concerning Fresh Fruits and/or Processed Products?

The question of the nature of the product or products protected by the GI registration (fresh fruit and/or processed products) initially posed a problem. The legal possibility of registering a GI for both fresh fruit and processed products was not obvious. Moreover, if they seemed established in the eyes of local actors, the specificity and/or reputation of the fresh product and all the processed products would have to be demonstrated.

Different scenarios were then studied: a GI concerning the fresh madd from Casamance only, a GI concerning the processed products only, and a GI concerning both the fresh fruit and the processed products.

The "fresh Casamance madd only" scenario would presumably leave the possibility of transforming this madd throughout Senegal and even beyond. It does not meet the expectations of the Casamance processors who initiated the GI. This scenario would have been imposed, however, if the specificity or reputation of the products could not be demonstrated.

The scenario of a GI concerning only the processed products was also explored. The reputation of the Casamance processed products seemed easy to establish, as well as the legitimacy of the protection of the associated intellectual property rights, on the names "nectar of Casamance madd", "Casamance madd preserve" ... Such a GI could thus have been registered, but without restriction on the origin of the raw material. Indeed, imposing in the specifications of a GI of this type that the raw material comes from the area in which the transformation must be carried out requires establishing also the specificity of this raw material; and if this is established, the third scenario is preferable. A GI on the processed products without restriction on the origin of the raw material would not fulfil the expected role at the level of structuring a local sector bringing together the Casamance processors and gatherers.

The third scenario, that of a GI covering both the fresh fruit and the processed products, therefore seemed preferable. This scenario seems however difficult to construct on the sole basis of reputation: if the fresh Casamance madd is protected by the GI as well as the processed products based on fresh Casamance madd, it is because this one is specific, and not just reputed. There are then also requirements for traceability of the raw material, difficult to establish for a gathered product.

The study of the specificity and reputation of the fresh Casamance madd and the products processed in Casamance, carried out in 2020 (Hernandez et al. 2020), was therefore decisive for the GI *Casamance madd*. Surveys among consumers and a bibliographic study were able to show the historical reputation of the fresh Casamance madd and its processed products.

A sensory analysis was carried out on four types of products: fresh fruits, madd nectars, madd syrups and sweet-salty madd preserve. The fresh fruits were transformed into natural filtrates (to facilitate the tests) and four products were tested: fresh Casamance madd respecting the terms of the future Specifications, madd sold as Casamance madd on the Dakar market, madd from Kédougou (South-East of Senegal, outside the 3 Casamance regions) and madd from Mali. For the nectars, the syrups and the preserves, products made by a Casamance GIE and products made by a private company from the Dakar region, working with "all-coming" madd, were compared. The sensory analysis tests took place in three phases:

- tests to establish the sensory profile of the different products;
- hedonic tests to know the preferences of the panelists;
- and triangular tests to highlight the differences perceived by the panel members for each product.

The criteria were defined with the help of the pickers and processors of the APPIGMAC association to best identify the characteristic criteria of madd and each of the processed products.

The sensory profiles are presented in Table 18.1. They show a real specificity of fresh Casamance madd and processed products. Tested blind during hedonic tests, the Casamance products were preferred over other madds by the panellists. They were also judged to be different (these differences being statistically significant) during the triangular tests.

Based on this study, local actors were able to build a GI request for *Casamance madd* and the five processed products, which is important in relation to the expected impacts, this will be seen in the next section.

This same study also justifies that the transformation must necessarily take place in Casamance, as the processed products that result from it do have a specificity. As seen above, constraints in terms of distance and time between picking and processing have been introduced into the specifications to guarantee the specificity of the processed products. It will still be possible for craftswomen/semi-artisanal processing units outside the area to make products "based on *Casamance madd* GI ".

18.3.3 What Traceability for a Picked Product and What Control?

The question of traceability and the parts of the supply chain on which it had to be established was also problematic. This traceability can be difficult to establish at the extreme upstream of the supply chain for a picked product like madd, whose supply chain is also characterised by mobile groups of pickers.

Table 18.1 Sensory profiles of fresh madds, madd nectars, madd syrup and sweet-salty madd preserves from different origins

Source: Hernandez et al. (2020)

Two solutions appeared. It is first of all possible to establish a "parcel" or "zone" traceability and to seek to establish and control the plots where the madd was picked. This traceability The parcel-based approach was the one chosen for another picked product, the French PGI *Provence Thyme*, registered in 2018. In the case of the *Casamance madd*, this option involves asking pickers to obtain authorisation forms for these plots from the village chiefs, and then to have the quantities picked validated by these same village chiefs. The advantage of this solution is that thanks to an estimation of the potential of these plots, the match between the declared quantities and this potential could be verified, which would strengthen traceability. The disadvantage is the associated administrative burden. Added to this is the difficulty of establishing a parcel-based control on a forest species whose spatial mapping is still very poorly mastered.

Another solution is to aim for a "global" traceability, in other words, to seek to establish only that the product comes indeed from the area defined as a whole. This is the choice that was made for another French GI, the PDO *Jura wood*, registered in 2019.

For the *Casamance madd*, it is this latter solution which has been chosen. It was considered, very realistically, that the risk of madd not from Casamance being present in Casamance is very low. This very low risk then only required a measure to match it in the Control Plan. A Charter of good picking practices has been put in place, including a commitment to only pick the *Casamance madd* in the defined area, and will be signed by the pickers. To this commitment will be added surprise checks of picking groups based on random draws carried out each year by the Extended Bureau of the APPIGMAC, similar to what is planned for the processing units. This also allows for control of picking practices and the quality and maturity of the fruits picked.

The control in Casamance of the madd declared to the APPIGMAC as *Casamance madd* by the pickers is therefore carried out by this self-check and this internal check. However, this traceability must be guaranteed up to the consumption areas, and this is indeed where the main challenge lies. The identified solution relies on the mandatory passage of the *Casamance madd* fresh through conditioning centres (located in Casamance) within which the APPIGMAC can control its quality. Once this check is carried out and the certification obtained, the fresh *Casamance madd* is packed in sealed nets, and it is in this form that it is transported to the consumption areas. Final consumers can then check that the nets are intact and be assured in this case of traceability.

The establishment of traceability, but, beyond that, the control of the entire specifications, thus relies on a self-check by the pickers/collectors and processing units and a strong internal control ensured by the Extended Bureau of the APPIGMAC. An external check of the association can at any time be carried out by the CNIG.

18.4   What Are the Expected Impacts of the GI *Casamance Madd*?

From an environmental and health perspective ("One health" approach), the registration of this GI will allow, thanks to the better valorisation of this non-timber forest resource that is the madd, the preservation of the Casamance forests, as opposition to forest fires lit by hunters and farmers and to illegal logging will be stronger. The specifications have also been designed to impose good picking practices, guaranteeing both health safety and the preservation of the resource: marketing under GI is only possible for fruits picked at maturity, without artificial ripening or post-harvest treatment (the fruits picked green and artificially ripened causing food poisoning) and the cutting of lianas is forbidden.

From an economic and social perspective, the registration of this GI is capable of building a quality product supply chain guaranteeing a good remuneration for local economic actors, a distribution of added value between the different links in the chain that is more fair and a retention of this added value in the territory of origin. Indeed:

The coordination between gatherers and processors will be strengthened thanks to the structuring of the APPIGMAC guaranteeing a representativeness and involving a consensus between the different professions.

The protection of names is requested for fresh madd and for locally processed products, the specificity of these having been demonstrated. This will help to fight against the usurpations of the name "*Casamance madd*" observed and thus to increase the selling prices of products under GI with established traceability.

The certified transformation of the *Casamance madd* is only possible within a radius of 200 kms around the gathering area, which will avoid competition between the Casamance craftswomen and the semi-industrial units in the north of the country. This is necessary to have fresh fruits and not damaged by long transport. The possibility for the transformation outside the area to make products "based on *Casamance madd* GI" will not restrict the market for the gatherers of the association.

The GI *Casamance madd* will certify a product of origin, natural and ethical, picked and processed/consumed at maturity, and the product should thus benefit from increased demand. The monopoly that local economic actors will have on this certified origin product will allow for more balanced relationships with downstream actors of the value chains.

18.5   Conclusion

The registration of the GI *Casamance madd* is capable of building a quality product supply chain guaranteeing a preservation of forests and resources, a good remuneration for local economic actors, a distribution of added value between the different links in the chain that is more fair and a retention of this added value in the territory of origin, as well as good product health safety.

By retracing the history of the construction of this GI, this chapter has sought to show the importance of preliminary studies and of the identification of the challenges that the GI must meet. The importance of the establishment of good governance of the GI must also be highlighted; it is important in the case of the *Casamance madd* as in many other GIs that all categories of actors are represented and have the same decision-making power in the different bodies.

This example also allows us to emphasise the importance of collective mobilisation and the commitment of local actors. The success of this construction of GI comes indeed from the fact that these actors have remained strongly committed to a process that took 5 years between the first studies and the submission of the registration request, due to the importance of the work that had to be carried out beforehand but also because of the pandemic that raged during these years.

Acknowledgements The authors thank WIPO and FAO for the support provided.

References

Bagal M, Kanouté PT, Slaterry S (2018) Rapport sur la classification des indications géographiques potentielles au Sénégal. FAO, 67 p

Bermond L (2017) Evaluation ex ante de la création d'une indication géographique sur le madd (Saba senegalensis) de Casamance. Dissertation written for the attainment of the ISAM specialisation Engineer diploma, Montpellier SupAgro, 66p

Bermond L, Kanouté PT, Fournier S (2018) Etude ex ante de la création d'une IG sur le madd (Saba senegalensis) dans la région naturelle de Casamance au Sénégal. FAO, 10p

Carimentrand A (2019) Specific GI training by action—GI Madd of Casamance. EUIPO, 12 p

Fournier S (2019a) Rapport de la mission effectuée du 14 au 18 octobre 2019. WIPO, 18 p

Fournier S (2019b) Analyse de scénarios possibles pour la construction de l'IG et définition d'un plan d'action. WIPO, 19 p

Fournier S (2021a) Appui à la formation du CNIG, à la structuration de l'APPIGMAC et à la rédaction du cahier des charges pour l'IG "madd de Casamance"—Rapport intermédiaire. WIPO, 32 p

Fournier S (2021b) Appui à la formation du Comité National des Indications Géographiques du Sénégal, à la structuration de l'Association pour la Protection et la Promotion de l'Indication Géographique Madd de Casamance et à la rédaction du cahier des charges pour l'indication géographique "madd de Casamance". WIPO, 13 p

Hernandez L, Mbodji A, Fournier S, Kanouté PT (2020) Analyse de la spécificité et de la réputation du madd de Casamance et de ses produits transformés. Montpellier SupAgro/ETDS, WIPO, 86 p

Teyssier C, Kanouté PT (2018) Evaluation prospective de l'initiative IG « Madd de Casamance » sur la base méthodologique du Guide FAO/Université de Florence, Italie. Rome, FAO, 39 p

The opinions expressed in this chapter are those of the author(s) and do not necessarily reflect the views of the [NameOfOrganization], its Board of Directors, or the countries they represent.

Chapter 19
Geographical Indication and Its Transforming Role in the Amazon: The Case of Pará State (Brazil)

Paulo de Tarso Anunciação de Melo and Suzana Romeiro Araújo

Abbreviations

CEC	Cation exchange capacity
DO	Designation of Origin
GI	Geographical Indication
INPI	National Institute of Industrial Property
IO	Indication of Origin
IP	Intellectual property
MDIC	Foreign Trade and Services
PO	Products of Origin
SDGs	Sustainable Development Goals
SOM	Soil organic matter
UN	United Nations
WIPO	World Intellectual Property Organization

19.1 Introduction

In Brazil, throughout history, the connection between the territories and the people who live and produce there forms a link that demonstrates how much the true origin of the products has significant importance for society and for the market, because it

P. de Tarso Anunciação de Melo (✉)
Intellectual Property and Innovation Commission of the Brazilian Lawyer' Organization, Belém, Pará, Brazil

S. Romeiro Araújo
Federal Rural University of Amazonia (UFRA), Belém, Pará, Brazil

© The Author(s) 2025
E. Vandecandelaere et al. (eds.), *Worldwide Perspectives on Geographical Indications*, https://doi.org/10.1007/978-3-031-71641-6_19

contributes to protect traditional knowledge and the geographical conditions allied to human intervention that transform the products into something unique.

In this scenario, Geographical Indications appear as an instrument that enables geographical terms linked to recognized territories and their products to be protected from possible falsifications and usurpations, fixing to their true producers the legitimacy to use the geographical name appropriate to the region and the form of making it.

According to the World Intellectual Property Organization (WIPO), Geographical Indications are intellectual property (IP) rights that serve to identify a product originating in a specific geographic region and whose quality, reputation and other characteristics are essentially attributed to this geographical origin.

In accordance with this international definition, several international agreements give normative support to Geographical Indications, particularly the multilateral treaties administered by WIPO, which are the Paris Convention for the Protection of Industrial Property, the Madrid Agreement for the Repression of False or Deceptive Indications of Source on Goods and the Lisbon Agreement for the Protection of Appellations of Origin and their International Registration.

In Brazil, Law No. 9279/96, called the Industrial Property Law, deals with Geographical Indication, specifying two species suitable for recognition, which are:

the Indication of Origin, referring to the geographical name of a country, city, region or locality of its territory, which has become known as a center of extraction, production or manufacture of a certain product or of provision of a certain service, and the Designation of Origin, referring to the geographical name of a country, city, region or locality of its territory, which designates a product or service whose qualities or characteristics are due exclusively or essentially to the geographical environment, including natural and human factors.

Brazilian Geographical Indications are formally recognized through an examination of the application carried out by the National Institute of Industrial Property (INPI), a federal agency linked to the Ministry of Industry, Foreign Trade and Services (MDIC), which has this prerogative to evaluate and grant the requested recognition.

In the year 2022, Brazil reached the number of 100 Geographical Indications, presenting a numerical disparity between the numbers related to the Amazon and the rest of the country (INPI 2022). This can be attributed to the absence of public policies aimed specifically at the recognition and territorial development of GIs in the Amazon. The region's infrastructure faces significant challenges, including difficulties in the management and governance of collective entities and in the distribution of products, exacerbated by long distances and poor internet access in many locations. In addition, Amazonian producers generally have little access to information and examples from elsewhere about the benefits of a GI, which limits their engagement and the ability to seek such recognition. Considering data related to the state of Pará, located in this important region, this locality reaches the amount of only four Geographical Indications during this period, namely *Tomé Açú*, for the cacao product, *Marajó* for the cheese product, *Bragança* for the manioc flour product and *the Andirá-Marau Indigenous Land for the waraná* product and *waraná breads*. Its extensive territorial conditions and the notable production chains that assume a national protagonism, such as that of cocoa (*Theobroma cacao*), açaí (*Euterpe oleracea*) and pineapple (*Ananas comosus*) among others, contrasts with the few official recognitions of Geographical Indications that could add more value and protection of its territory.

According to the International Cocoa Organization (ICCO 2019), 85% of global cocoa bean production is currently concentrated in seven countries: Ivory Coast, Ghana, Ecuador, Cameroon, Nigeria, Indonesia and Brazil. On the other hand, cocoa processing and industrialization predominate in regions that do not produce the fruit, such as Europe (37%) and the USA (8%). Unlike the other major cocoa producing countries, Brazil has a well-established production chain: it produces the fruit, has an industrial park for processing cocoa beans and manufactures chocolate (Conceição et al. 2020). Brazil is also one of the largest pineapple producers in the world, and much of the production comes from Salvaterra, municipality on Marajó Island, in the state of Pará, the state with the highest national production. In 2022, the area harvested for açaí in the state of Pará was 224,044 hectares, with 1,595,455 tons of fruit produced, with an average yield of 7121 kilograms of fruit per hectare (IBGE 2023).

In the present chapter, we analyzed the case of the Pará state, located in the Amazon, that illustrates the contrast between its low recognized GI number until the year 2023 and the GI potentiality of its territory as a result of its unique characteristics, its high values in terms of biodiversity and traditional knowledge, while the region is known for the dynamics of its economy, which thrives in several production chains, which are a reference of development in the national and international scenario. This study is supported by a qualitative approach, with an inductive approach, based on consultations and bibliographic research available in books, scientific journals, laws and norms in general, mainly Law n. 9.279/96 (Industrial Property Law—LPI) and Ordinance/INPI/PR n. 04, of January 12, 2022 (BRASIL 2022). Other laws that were consulted are the Paris Convention for the Protection of Industrial Property—Promulgated by Decree Law No. 75.572, of April 8, 1975 and the Agreement on Trade-Related Aspects of Intellectual Property Rights—TRIPS—As published in the Federal Official Gazette (DOU) of December 31, 1994, Section I, Supplement to No. 248-A.

19.2   Characterization of Pará State

The state of Pará is located in the northern region of Brazil, being one of the 27 federative units. It has a territorial extension of 1,245,870,798 km^2, obtaining the title of second largest in the country. In terms of population, it has an estimated population of 8,777,124 inhabitants, divided into 144 municipalities, with Belém as its capital (Fig. 19.1).

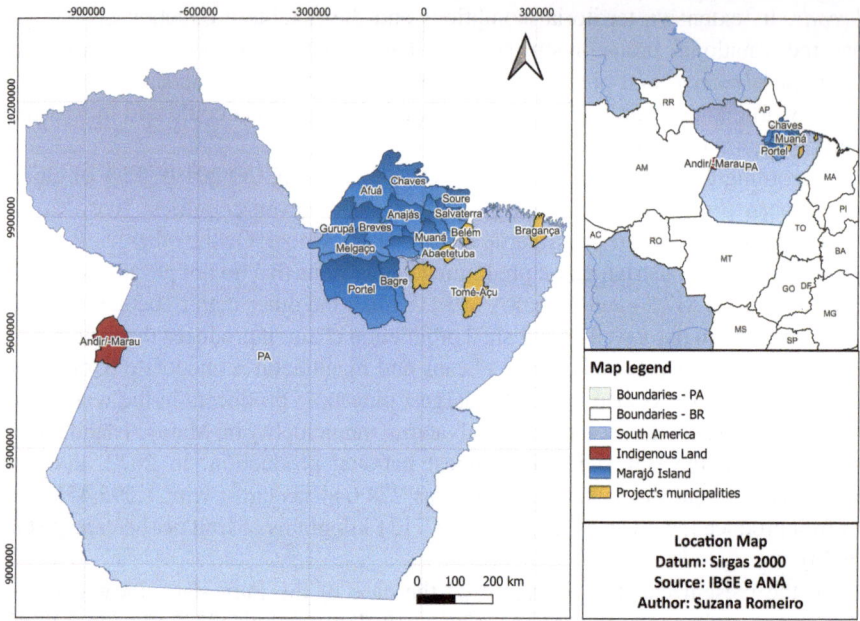

Fig. 19.1 Map of the study area

The Sate of Pará present a plurality of cultures that mix with the various natural conditions, with products reputation and meanings built over time by the connection between place, products and people.

The discussion on the Geographical Indication topic in the Pará state is relatively recent, since the Brazilian legislation that regulates the recognition of a GI dates back to 1996 and, only in 2019, it had its first geographical indication registered with the National Institute of Industrial Property (INPI).

The Geographical Indication can be used as an indicator, so that the development of localities can come from the exploitation of products and services that already exist and are part of everyday life but have not yet been given the necessary look for their history, culture and geographical conditions to be told and exposed in an organized fashion. In the state of Pará, this application is extremely relevant. Traditional products, such as ceramics and Marajoara shirts, as well as açaí, which are deeply rooted in the culture and daily life of Pará, whether in the field of creative economy or in food, can gain visibility and value through GI, boosting the economic and cultural development of local communities.

Producers and consumers realized that the peculiar flavors or qualities of products coming from certain localities gained special attention from the market for their differentiated qualities, in this case neither better nor worse, but typical (Bruch et al. 2009).

Therefore, the state of Pará absorbs in its territory great possibilities for the intellectual property asset geographical indication to act as a factor of transformation for its most diverse localities, situated in an urban area, in a rural area dominated by agribusiness or in a forest preserved by traditional communities.

19.3   The Registered Pará Geographical Indications

Being a great historical and economic reference for the Amazon, the state of Pará has supported the recognition of four geographical indications until the year 2023, namely: cocoa of *Tomé-Açú*, cheese from *Marajó* for the, cassava flour of *Bragança* (*Manihot esculenta Crantz*), which are now registered Indications of Origin, as well as *waraná* (native guarana) and *waraná breads* (guarana stick) from the Andir-á-Marau Indigenous Land registered as Denomination of Origin.

19.3.1   Cocoa of Tomé-Açú

Cocoa of Tomé-Açú GI was recognized in 2019 by INPI. Its notoriety goes back a long way. In 1926, a group of Japanese scientists went on a mission to Pará, with the aim of locating areas in which it was possible to establish agricultural colonies (Silva et al. 2020) and, from them, to boost the economy through the development of crops, as well as modern cultivation practices, and the municipality of Tomé-Açú was one of them.

The first cocoa seeds were introduced to the Tomé-Açu region in 1929 by Japanese immigrants (Silva 2018). Aiming at establishing the cultivation of a perennial species, native to the Amazon rainforest, the region has developed the so-called Tomé-Açú Agroforestry System (*Sistema Agroflorestal de Tomé-Açú*— SAF), in which several species subsist in the same area and the seasonality of products allows for the production of several crops all year round.

The history of the Japanese immigrants who arrived in the Amazon facing and overcoming all kinds of difficulties contributed to the cocoa product becoming notable and allowing the GI to be obtained (Fig. 19.2). Recently, the cocoa kernels from Tomé-Açú were the basis of the official chocolate of the Tokyo Olympic Games in Japan, which brought economic return and was a source of pride for the region.

19.3.2   The Marajó Cheese

The *Marajó cheese* is a product manufactured by hand in the geographical area of the Marajó archipelago (Fig. 19.1), according to the historical and cultural tradition of the island, where the first buffaloes arrived, between the end of the nineteenth century and the beginning of the twentieth century. Over the years and with the

Fig. 19.2 Agroforestry system (SAFs) located in the municipality of Tomé-Açu (**a**); distinctive sign of the Tomé-Açu Geographical Indication, Pará, Brazil (**b**)

increase of the buffalo herd, the product began to be made exclusively with buffalo milk, whose fame spread nationally.

The state of Pará is the largest producer of buffalo in Brazil, with 644,672 head out of a total of 1,598,268 in the whole of Brazil in 2022 (IBGE 2023). Currently, Marajó Island holds the largest herd of buffalo in Brazil, which represents about three times the population of all its municipalities. The animals and the traditional cheese production are the symbol of the region, where the traditional know-how is passed on from generation to generation.

The *Marajó cheese* GI is not only the possibility to combat counterfeits, it also promotes an important local culture, adding the sale of cheese to tourism, which is greatly boosted due to such fame (Fig. 19.3a).

19.3.3   The Bragança Cassava Flour

The *Bragança* cassava flour has an old reputation, from several generations. Cassava crop was already cultivated by the native peoples of the region, being part of the daily menu and the movement of local economy. Throughout the nineteenth century, the region of Bragança had already received European immigrants, as well as Brazilians from other regions. The commercialization of cassava flour, mainly through the railway line on trains that travelled between Bragança and the capital, Belém, allowed this geographical indication to become very famous until the present day.

Bragança flour is recognized as Cultural Heritage of the Pará State and contributes to the regional economy, being found for sale in several cities of the Pará state and sold in other states of Brazil (Fig. 19.3b, c). Cultural heritage of the state is considered to be a collection of movable and immovable assets existing in the country or state, whose conservation is of public interest, whether due to their

Fig. 19.3 *Cheese from Marajó Island* with the distinctive GI sign (**a**); Preparation of Bragança Flour in a traditional oven (**b**); Bragança Flour packaged for sale (**c**); Distinctive sign of the Geographical Indication Terra Indígena Andirá-Marau (**d**)

connection to memorable historical events or their exceptional archaeological, ethnographic, bibliographic, or artistic value. The historical success of the product relies on the commitment of its traditional producers and cooperatives that are starting the preparatory procedures to begin exporting to several countries.

19.3.4 Warana from Andirá-Marau Indigenous Land

The *warana* and *warana* bread of the Andirá-Marau Indigenous Land is on the border between the Brazilian states of Pará and Amazonas (Fig. 19.1). The original guarana produced by the native community experiences several geographical influences, particularly the influence of soils of anthropic origin, including the so-called *Terra Preta de Índio* and *Terra Marrom*. Many soils of the Amazon Region are highly weathered, acidic, with low cation exchange capacity (CEC), low fertility and, consequently, low crop production potential. The *Terra Preta de Índio* and *Terra marron* soils contrast with other Amazonian soils, especially in relation to fertility and resilience. For some authors the *Terra Preta de Índio* formed as a result of anthropogenic concentration of soil nutrients through domestic and agricultural activities during the prehistoric period, including accretion of soil organic matter (SOM) from long-term concentrated settlement and "slash and char" agriculture resulting in a high content of pyrogenic carbon (biochar) through pyrolysis (Glaser and Birk 2012; Glaser and Woods 2004; Lehmann et al. 2003; Steiner et al. 2004). These materials serve as soil conditioners and sequesters of carbon in recalcitrant and reactive forms.

Thus the unique characteristics of the product derive from the indigenous settlements and practices of management of waste, such as charcoal and vegetable ash, animal bones and other remains of kitchens and houses (Eriksson et al. 2016). Anthropic soils allow better growth, yield and quality of guarana, causing its way of making and the result to be unique (Fig. 19.3d).

19.3.5 Towards Further Protected GIs in the State of Pará?

The state of Pará, could support producers private initiatives for the recognition of new Geographical Indications, linking with public policies, and helping disseminate sustainable practices in the localities.

The following names of products are some of the dozens of real possibilities for the recognition of Geographical Indications (Fig. 19.4c):

• Cocoa from the Trans-Amazonian and Baixo Tocantins territories,
• Honey from São João de Pirabas (Fig. 19.4d),
• Chocolate from Ilha do Combú (Fig. 19.4e),
• Andiroba product (*Carapa guianensis*) from Piriquitaquara (Fig. 19.4f),
• Mapará from Cametá (*Hypophthalmus edentatus*),
• Manteiguinha beans (*Vigna unguiculata*) from Santarém,
• Miriti (*Mauritia flexuosa*) toys from Abaetetuba (Fig. 19.4a) and
• Handicrafts, and Pará territory itself for the açaí palm product (*Euterpe oleracea*).

19.4 Importance of GIs for the Pará State

One of the main challenges for the state of Pará is to ensure a right balance between the income production for the Amazonian people from its forests, rivers, mining and agricultural areas and the environmental preservation, which has a profound impact on the planetary ecosystem.

Increasing the number of Geographical Indications in the state of Pará can bring several benefits to the Amazon territory. Geographical Indications could play an important role in promoting sustainable practices, because a GI preserves in a written way the traditional knowledge within the protected territory, and it can encourage the preservation of the natural and cultural resources linked to the production.

In relation with social aspect, geographical indications can contribute to the valorization of local communities and the promotion of regional economic development. GIs could contribute to the preservation of cultural traditions, ancestral knowledge and specific ways of life of the communities involved in the production

Fig. 19.4 *Miriti* toys from the municipality of Abaetetuba (**a**), various products from recognized and potential GIs (**b**), açaí from the state of Pará (**c**), honey from the municipality of São João de Pirabas (**d**); Chocolate from Ilha do Combu (**e**); products derived from Andiroba (Carapa guianensis) from Ilha do Combu (**f**)

of the geographically indicated products. In addition, they can boost local tourism and generate jobs in rural areas (Melo 2022).

In relation with governance principles, geographical indication systems could promote transparency, traceability and compliance with the specifications. By ensuring that the GI product meets specific criteria, such as traditional production methods and good agricultural practices, GI can strengthen consumer confidence and the reputation of the products on the market.

One of the most emblematic examples is the Geographical Indication (GI) of Bragança for the product manioc flour. Historically, it was necessary for flour houses (places where manioc flour is produced) to meet health standards to allow for the free commercialization of the product. After the recognition of the GI and the structuring of its control body, it was possible to discuss and implement specific regulations that meet the reality of the flour houses. This enabled the authorization and operation of these houses throughout the state of Pará, bringing them under both artisanal and industrial production criteria, as well as registering the manioc flour product.

The artisanal certification developed by the Agricultural Defense Agency of the State of Pará aligned the GI products with the United Nations (UN) Sustainable Development Goals (SDGs), meeting economic, social, and environmental demands, helping to conserve Amazonian biodiversity, and mitigating climate change.

In this context, by obtaining a Geographical Indication (GI), local producers gain visibility and a competitive advantage in the market. This can result in an increase in demand for their products, boosting the economic growth of rural communities in the Amazon. Additionally, the Geographical Indication system can enhance the added value of GI products, allowing producers to receive fairer prices for their work.

19.5   Conclusion

Geographical indications play a key role in promoting and preserving the sustainability of the territory of Pará. By highlighting the unique characteristics and traditions of a given region, these indications value local products, encouraging sustainable production and the preservation of natural resources. In the context of Pará, a region rich in biodiversity and culture, geographical indications become even more relevant. They could contribute to the valorization of traditional products, such as açaí, mapará, honey and cocoa, in addition to fostering responsible agricultural and extractive practices. Thus, geographical indications could strengthen territorial identity, boosting sustainable economic development and conservation of natural heritage for future generations.

References

Brasil. Ordinance/INPI/PR n. 04, of January 12, 2022. Estabelece as condições para o registro das Indicações Geográficas, dispõe sobre a recepção e o processamento de pedidos e petições sobre o Manual de Indicações Geográficas. Available https://www.gov.br/inpi/pt-br/servicos/indicacoes-geograficas/arquivos/legislacao-ig/PORT_INPI_PR_04_2022.pdf. Acesso em: 28 set. 2023

Bruch KL, et al. (2009) Indicação Geográficas de produtos agropecuários: Aspectos legais, importância histórica e atual. In: Pimentel, L (Org.). Curso de propriedade intelectual e inovação no agronegócio: Módulo II, indicação geográfica. Brasília: MAPA; Florianópolis: SEaD/UFSC/FAPEU

Conceição, Macedo RDD, Pires MM (2020) Specialization and competitiveness: analysis of Brazilian exports of cocoa beans and products. Revista Mexicana de Ciencias Agrícolas 11(6):1207–1219

Eriksson J, Soderstrom M, Isendahl C (2016) Properties of Amazon DarkEarths at Belterra Plateau, Pará, Brasil. In: Stenborg P (ed) Beyond water, archaeology and envronmetal history of the Amazonian Inland. University of Gothemburg, Ale Tryckteam AB, Bohus

Glaser B, Birk JJ (2012) State of the scientific knowledge on properties and genesis of anthropogenic dark earths in Central Amazonia (Terra Preta de Índio). Geochim Cosmochim Acta 82: 39–51

Glaser B, Woods WI (eds) (2004) Explorations in Amazonian Dark Earths. Springer, Berlin. 212 pp

IBGE—Instituto Brasileiro De Geografia E Estatística. Efetivo de rebanhos. Disponível em: https://sidra.ibge.gov.br/pesquisa/pam/Tabela 3939: Efetivo dos rebanhos, por tipo de rebanho (ibge.gov.br). Acesso em 28, sept, 2023

IBGE—Instituto Brasileiro De Geografia E Estatística. Produção Agrícola Municipal. Disponível em: https://sidra.ibge.gov.br/pesquisa/pam/tabelas. Acesso em: 25, august. 2023

ICCO—International Cocoa Organization. Quarterly Bulletin of Cocoa Statistics, year 2018/2019, 15(3), 2019. Disponível em: https://www.icco.org/about-us/icco-news/415-quarterly-bulletin-of-cocoa-statisticsnovember-2019. Acesso em: 8 mai, 2023

INPI—Instituto Nacional Da Propriedade Industrial. Portaria/INPI/PR n. 4, de 12 de janeiro de 2022. Estabelece as condições para o registro das Indicações Geográficas, dispõe sobre a recepção e o processamento de pedidos e petições e sobre o Manual de Indicações Geográficas. Rio de Janeiro. 2022. Disponível em: https://www.gov.br/inpi/pt-br/servicos/indicacoes-geograficas/arquivos/legislacao-ig/PORT_INPI_PR_04_2022.pdf. Acesso em: 19 set. 2023

Lehmann J, Kern DC, Glaser B, Woods WI (eds) (2003) Amazonian Dark Earths: origin, properties, management. Kluwer, Dordrecht. 524 pp

Melo PAM (2022) Desenvolvimento de marca coletiva para a comunidade extrativista de óleo de Andiroba da Ilha do Combú, Belém, Pará. Dissertação de mestrado (PRFINIT). IFPA, Belém. 74p

Silva BS (2018) Direitos de propriedade no século xx: relação entre japoneses colonos e trabalhadores brasileiros na amazônia. Revista do Departamento de História e do Programa de Pós-Graduação em História do Brasil da UFPI. Teresina 7(1)

Silva AC, Ferreira MC, Silva VP (2020) Indicação Geográfica no Brasil: uma análise do Pará como território amazônico. Revista Brasileira de Gestão e Desenvolvimento Regional 16(3):1–20

Steiner C, Teixeira WG, Lehmann J, Zech W (2004) Microbial response to charcoal amendments of highly weathered soils and Amazonian dark earths in Central Amazonia—preliminary results. In: Glaser B, Woods WI (eds) Amazonian dark earths: explorations in space and time. Springer, New York, pp 195–212

Chapter 20
The Patrimony Blind Spot of Geographical Indication in State-Centered Governance: Mikawa Region Agri-food Products in Japan

Hart N. Feuer and Fatiha Fort

Abbreviations

GI	Geographical Indication
JA	Japan Agricultural Cooperatives
JETRO	Japan External Trade Organization
MAFF	Ministry of Agriculture, Forestry and Fisheries

20.1 Introduction

The global spread of Geographical Indications (GI) policies has provided heritage agri-food producers with an opportunity to promote and safeguard the link between their products and a geographically-specified reputation, but this potential is highly dependent on the institutional structures created to evaluate and certify producers' claims (Marie-Vivien and Biénabe 2017). In East Asia, as in many regions where independent European-style GI policies have been recently adopted, governance of GI has been predominantly state-centered, with governments assertively cultivating, recruiting, and mediating the applicant producer groups (Feuer 2020). In such cases, the priorities of the state, or of the implementing body, displace the global principles of GI and allow more aggressive gatekeeping of the GI landscape (Wang 2008; Thévenod-Mottet 2009; Durand and Fournier 2017). This level of intervention is

H. N. Feuer (✉)
Kyoto University, Kyoto, Japan
e-mail: feuer.hartnadav.4e@kyoto-u.ac.jp

F. Fort
Institut Agro de Montpellier, Montpellier, France

© The Author(s) 2025
E. Vandecandelaere et al. (eds.), *Worldwide Perspectives on Geographical Indications*, https://doi.org/10.1007/978-3-031-71641-6_20

concerning as GI, which has inherited its *raison d'être* from the European approach to terroir, centers on patrimony, diversity, preservation of producer livelihoods, and embodies a producer-driven governance framework (Vandecandelaere et al. 2010; Calboli 2015). In the more state-centered countries found in much of Asia, respecting these principles is not a priority if implementing bodies have divergent mandates or models of public participation (Wang 2008; Lecoent et al. 2010; Augustin-Jean 2012; Pick et al. 2017). Proponents of the *sui generis* GI system, including many stakeholders in Europe who are pushing for the globalization of the European GI approach, brush off such concerns by welcoming the potential diversity of GI models worldwide (Tregear et al. 2007; Feuer 2022). This chapter suggests that such optimistic assessments about the future integrity of the GI concept may be misplaced, as intentional appropriation of the GI scheme by strong governments, rather than contextual diversity, may undermine the global principles of GI.

Japan has a long and rich history of homegrown territorial-focused agri-food certifications that emerged under certain administrative frameworks (local government, tax ministry, patent office, agriculture ministry, etc.) and focused on different outcomes (local tourism, rural development, export, cultural preservation, etc.). This includes hometown certified foods organized at the prefectural level (3E Mark), regional brands organized at municipal or prefectural levels, the Cool Japan mark for export-oriented foods organized by the trade ministry JETRO, and numerous private-label certifications for products of origin. To this existing food certification landscape, Japan has additionally implemented two GI systems in recent decades: a regional collective trademark in 2006 (administered by the Patent Office) and a *sui generis* GI, which was initiated in 1995 for wine and spirits by the National Tax Agency, and for agri-foods in 2015 by the Ministry of Agriculture, Forestry and Fisheries (MAFF). The food certifications preceding GI typically involved government coordination and promotion of producers' activities, so the guiding role and preferences of the state were inherent to these activities. It is important to remember that the GI system joined this regulatory landscape, with many agri-food products having been previously certified under other schemes, and many pre-existing producer groups (particularly those of the Japan Agricultural Cooperatives or JA), pivoting from former schemes to GI. For GI, as a certification with global norms centered around accessibility for producers, grassroots governance, and participatory rural development (Giovannucci et al. 2009; Crescenzi et al. 2022), can these unique values be realized in a context of significant institutional inertia and governmental gatekeeping? Can GI better preserve food heritage or democratize governance? In our assessment of the recent Japanese heritage products, including recently registered GIs, we found that political or administrative barriers in Japan have often precluded some of these potential benefits of GI.

The collective trademark that started in 2006 adopted an independent grassroots approach, allowing producers to independently make submissions and evaluating them based on the global criteria of patent law, including originality, historical provenance, and legitimate representation (Port 2014). The *sui generis* GI system that began in 2015 also refers to such criteria but, as we argue in this paper, can be engaged to prioritize or encourage applicants who deliver governmental goals, such as expansion and trade (rather than, for example, protection of historical patrimony). The potential conflicts and contradictions associated with the business-oriented GI model promulgated by the MAFF have been documented (Galeazzi 2018), particularly in how it impacts the outcome for long-established, so-called "old glory" products (Gugerell et al. 2017; Defrancesco and Kimura 2018). Zappalaglio (2021), in a book about the role of history and provenance in evaluating GI, suggests that history alone is not a sufficient criterion for GI, but rather that history must be substantiated by practices and materials in the present. However, below we characterize how historical products with present-day production practices that match their patrimony, but without obvious political or economic advantages, including export or sectoral expansion, are marginalized by the governance structures of the Japanese variation of *sui generis* GI.

Since the inception of the *sui generis* GI system, the Mikawa region of Japan, a historical territory absorbed into Aichi Prefecture in 1871, has been at the center of numerous conflicts and issues related to the approach of using GIs as a tool for economic expansion. By 2022, it is the only region with a withdrawn GI product and a GI under legal review by intellectual property courts; it also features numerous cases of renowned agri-food products that are, for various reasons, not attracted to the GI model presented by the MAFF. This chapter focuses on three products that characterize the particularities of state-centered GI models, including the politicization of GI selection, ambivalence toward heritage, and emphasis on expansion, export, and productivity. We find that, particularly for "old glory" products, in which weak or antagonistic relations among producers have persisted for generations, the privileging of economic output by an interventionist GI authority will have negative impacts on heritage preservation. Japan's early experiences can serve as a warning sign for the many governments in East Asia and elsewhere who impose a politicized market logic on the unfolding of their GI system.

20.2 Methods

The empirical basis for this chapter centers around the experiences of 3 cases in Japan: *Hatcho Miso, Kokonoe Mirin* and *Nishio Matcha*. The qualitative case study approach is a research methodology that explores a phenomenon in a certain context by engaging a range of data sources and viewpoints to reveal a wide breadth of the phenomenon (Yin 1994; Baxter and Jack 2008). As we are studying the emerging challenges of the GI system in the Japanese context, a multiple case approach is an appropriate strategy as it is employed here to explore: (a) products registered as GIs but in a state of conflict, (b) products withdrawn from GI registration, and (c) very eligible products that are alienated from the GI system. The analytical goal here is synthesis about struggles facing GIs as represented in these cases and verifiable in secondary data, such that each case is not explored in particular depth. An in-depth analysis of the prominent case of *Hatcho Miso* is undertaken by Sekine (this volume).

The primary data is derived from semi-structured interviews among producers of *Hatcho Miso* (4 interviews covering all producers in historic production area) and Kokonoe Mirin (2 interviews covering history, marketing, current production practices) in the Mikawa region in November 2021, as well as *Nishio Matcha* producers (group interview with members of the *Nishio Matcha* Cooperative) and promoters (2 interviews with certifying authorities and promoters, including the Nishio City Tea Association) in 2018.

In the final analysis, the empirical material also includes follow-up data provided by informants in the later months, observations from production areas, marketing documents, legal testimony, and historical artifacts. These data were examined qualitatively using thematic and discursive analysis. Individual accounts were triangulated through reference to publicly available information by various government ministries, archival documents, legal documents, and accounts of the Mikawa region by other scholars, who have evaluated different agri-food products in the preceding 10 years.

20.3 Findings

Given the wealth of empirical material collected and the different data sources used, we present a synthetic analysis showing (1) the relative attractiveness of GI certification (2) the difficulties in governance of products with long heritage and (3) the issues associated with top-down control of product certification processes (i.e. politicization).

20.3.1 Prominent and Historic Heritage Agri-food Products Are Paradoxically Not Attracted by GI Certification

Although the concept of terroir has found fertile ground in most East Asian countries, including Japan, the hesitance to embrace some of the broader principles of *sui generis* GI found in Europe, such as participatory governance, traditional production methods, and cultural preservation, arises from the difficulty of instrumentalizing some aspects of heritage agri-food production to serve contemporary developmental purposes. On the one hand, famous regional foods can serve as markers of national pride and symbols of tradition and the historical continuity of national cuisine. Governments around the world are engaged in various forms of culinary tourism promotion or gastrodiplomacy for these reasons (Zhang 2015; Huysmans 2020). On the other hand, historical foods are often produced using old-fashioned methods (including labor intensive agriculture) that do not lend themselves to modern market conditions, such as standardization, productivity enhancements, and export orientation (Chabrol et al. 2017; Gugerell et al. 2017). It is therefore tempting to use the opportunity presented by GI to support products that are viable for instrumental goals, such as tourism and export, and which already have production systems that are modernized or rationalized (Augustin-Jean and Sekine 2012; Kizos et al. 2017; Galeazzi 2018). Such modern-oriented GIs, such as Kobe Beef or Yubari Melon, are called "Big King" products by Defrancesco and Kimura (2018), underlying their prominence in GI promotion in Japan.

This chapter, meanwhile, focuses on products that are either marginalized by the current GI regulatory framework (discouraged potential applicants) in Japan or those that have been shaped or bullied to achieve instrumental goals through the GI selection process or composition of the product specifications. These case studies from the Mikawa region in Japan are described below.

One prominent case of state intervention through selection politics is *Hatcho Miso*, a soybean miso associated with the namesake municipality (Hatcho district, Okazaki City) in production for more than 400 years. Other producers of a similar red miso are distributed both inside and outside of the Mikawa region, and have both cooperated and conflicted with Okazaki-based producers about production specifications in recent decades. When the MAFF bypassed the Okazaki producers by granting the GI to a prefectural producer group with less strict product specifications, a national dispute about heritage and intellectual property began to unfold. Through its actions (detailed by Sekine, this volume), the MAFF revealed its ambivalence toward anachronistic product specifications, such as fermentation in cedar vats, natural aging, and artisanal workmanship (over automation). There is now a risk that traditional producers will be prevented from using the name *Hatcho Miso* or be forced to legitimate modern production practices by joining a competing producers association.

Meddling public intervention in shaping a GI product is well illustrated by the case of *Nishio Matcha*, a famous green tea powder from the Mikawa region, whose GI mark has since been withdrawn by request of the producers. In this case, public officials coordinated an ahistorical tie-up between processors (large tea powdering companies with significant market presence) and tealeaf producers (farmers) in order to ensure that the final product (matcha) could be traced back to the production area (Nishio). What was intended to raise the prominence of Japanese matcha and create a flagship GI product for Japan thrust tea farmers into an unequal economic partnership with global tea corporations (Sekine 2020). This includes the AIYA Company, a matcha producer founded in 188 that is based in Nishio City and is a familiar entity for tealeaf farmers. Eventually, irreconcilable differences led to the withdrawal of GI and dissolution of the producer group; the tealeaf farmers later applied for a GI for tea leaves only (instead of processed matcha). As with the next case that we profile, GI authorities did not recognize that the historical absence of economic relations between processors and farmers reflected longstanding discordancy that could not be resolved by government action in the framework of GI certification.

The case of *Kokonoe Mirin*, a 250-year-old alcoholic beverage and food ingredient producer, represents the tacit marginalization of historic heritage producers who would seemingly align well with the global norms of GI, but who are not attracted to the GI framework in Japan. Principally, they produce *hon mirin*, a sweet, fermented wine made from rice grown in nearby Mikawa region farms, and which is aged for at least 1 year before it is sold for drinking or cooking. As an institution, they present a storied patrimony that is embodied in historic structures, authentic production practices, embeddedness in local rice agriculture, and engagement in international awards and local agri-tourism. They are also one of the few regional breweries that has not industrialized their production, which narrows the pool of potential members of a GI producer group. As with miso and tea, mirin consumption is declining rapidly in modern Japan, so promotion activities, including certifications, are otherwise welcomed by these producers. Bearing a GI mark to acknowledge their history and certify their ongoing authenticity would appear to be a logical step, but managers at the company claim that a GI mark would provide few meaningful benefits while limiting their dynamism.

The cases of miso, tea, and mirin were chosen for this chapter because they are indicative of a macro-level trend inherent to the Japanese GI system that paradoxically marginalizes some of the most fundamental, daily constituents of the historical Japanese cuisine: tea, miso, and mirin. Within these three categories, there are only two GI teas (one boasting its modern qualities), two misos (one legally contested), and no mirin inscriptions. What this conspicuous absence of traditional ingredients represents is discussed in the following findings.

20.3.2   Strong Governance Is Scarce for Famous Products

The agri-food products studied in this paper are long-established, with a history ranging from 150 years (*Nishio Matcha*) to 250 years (*Kokonoe Mirin*) to more than 650 years (*Hatcho Miso*). It can be assumed that, during the intervening years, the relationships between neighboring producers and/or raw materials providers have undergone many evolutions—competition, cooperation, independence, and everything in between. If we as contemporary researchers encounter weak, strained, or combative relations among local producers, there is likely to be a long-standing basis for this. Creating a context for establishing GI producer groups has succeeded in some cases by relying on pre-existing agricultural institutions, such as the Japan Agricultural Cooperatives (JA), but has been less successful for processed foods like miso, matcha, or mirin. Furthermore, forcible attempts by the MAFF to artificially create producer groups have backfired in the case of *Nishio Matcha and Hatcho Miso*. Without intervention, groups may also not spontaneously assemble, as in the case of *Mikawa mirin*.

In fact, *Kokonoe Mirin*, one of the 7 remaining producers in the region (down from 30 historically) does not have a working relationship with neighboring mirin producers, except for contributing to an annual festival in the region. Even in the context of declining mirin consumption across Japan, Kokonoe does not see other (far more industrialized) producers as suitable partners for promoting their product, which is predominantly artisanal. Divergent views about production scale and quality continue to make collaboration in a producer group—required in GI application under the National Tax Agency—unlikely. Indeed, and as occurred in the case of *Hatcho Miso,* such a divergence of production practices among nearby producers not only undermines cooperative behavior, but can lead to open conflict. As with other processed foods, the fragmented and partially industrialized cohort of historical producers do not necessarily have a basis for mutual cooperation. In short, it is too late in the development of these products to linearly expect artisanality to be a shared goal.

The contested producer landscape for *Hatcho Miso* indicates how the advent of a GI, far from encouraging cooperation, can trigger fraught and contentious relations among producers. Not only have formal conflicts about ownership and fair use of the name "*Hatcho Miso*" existed since the 1980s centered around trademark ownership, but miso producers in the region have long-simmering tension about generic use of the name Hatcho for all kinds of red soybean miso in Aichi prefecture (more detail in Sekine, this volume). New GIs introduced in 2006 and 2015 triggered more outright conflict between newer producers in the Aichi Prefectural Miso and Tamari Soy Sauce Industry Cooperative and the historical producers in Okazaki. Forceful efforts by members outside of Okazaki to use the name *Hatcho Miso* in a collective trademark splintered the prefectural Industry Cooperative and led to the establishment of a competing producer association representing only the historical Okazaki producers. While strict patent laws led the Industry Cooperative to abandon their application for a collective trademark, the 2015 *sui generis* GI law allowed MAFF to impose its preferred outcome by granting the *Hatcho Miso* GI to the Industry Cooperative. This process prompted acrimonious legal action by the Okazaki producers that is still ongoing. The interest of the MAFF to expand the scale of *Hatcho Miso* for economic reasons, export, and inclusivity directly undermined the interests of the most legitimate producers and lowered the standards of product quality that consumers will encounter in the market. Such state-centered actions seemed contrary to the spirit of *sui generis* GI from Europe, and led to many European and multilateral academics and policymakers to denounce the MAFF in filings to the Tokyo High Court of Intellectual Property in 2022.

This struggle echoes that of the now-moribund GI for *Nishio Matcha*, in which the Nishio Tea Cooperative Association (representing historical tealeaf producer) and numerous matcha processors were unable to achieve mutually agreeable conditions in the framework of their producer group. After the dissolution of the GI, the Nishio Tea Cooperative Association was able to retain the right to produce Nishio tea leaves and have thus essentially regained their autonomy. Although an equilibrium has been re-established in the Nishio case, the failure of the producer group is indicative of the limited use of a bottom-up approach and cooperative culture in the context of the Japanese GI policy.

The MAFF (and National Tax Agency for alcohol) should question why so few historically well-regarded producers are interested in GI certification, while contemporary products with less historical pedigree (including grape wine, beef and dairy products), as well as other agri-food categories (fruits, vegetables, seafood) are disproportionately well-represented among GI holders. Continuing on a policy trajectory that marginalizes producers of fundamental ingredients in Japanese cuisine is not likely to be sustainable nor realize the objective of protecting heritage.

20.3.3 Top-Down Control Disadvantages Heritage Producers

In the above cases of Mikawa products, the expansionist market orientation of MAFF's GI policy is a double-edged sword for "old glory" producers of historical agri-food products. Such producers report that they value intellectual property protections and synchronization with global standards, and are very interested in global recognition, but are ambivalent toward efforts to set lower, or more "inclusive" standards that valorize rivals with less patrimony and unique (i.e. strict) product specifications. MAFF can justify privileging inclusivity as an effort to revitalize a declining miso sector or make a rare product (e.g. *Hatcho Miso*) more accessible to consumers, while historical producers may view this as allowing industrial producers to free ride on a long-established reputation. Kokonoe Mirin considers that many neighboring producers have diverged too far from authentic production practices to work with. Similarly, Okazaki Hatcho Miso producers are proud of their uncompromising commitment to authenticity, which even included shutting down during World War II rather than following war-time production orders.

A top-down approach to encouraging cooperation is also uniquely difficult among "old glory" value chains, in which the economic relations between potential partners have likely been fraught or paralyzed for numerous generations. The Nishio Matcha GI, which was engineered by state-centered goals of promoting marquee Japanese matcha regardless of local institutional contradictions, ultimately backfired. For heritage production areas, the creation of comprehensive producer groups may require an empowering, rather than a command-and-control, contribution by authorities (Marie-Vivien et al. 2015). Such gentle relations are more common among underdeveloped production areas, where government support is necessary to initiate dialogue and begin negotiating baseline standards (Feuer 2022), although careful recruitment of producers is no guarantee for establishing a cooperative producer group as expected from European GI experiences.

20.4   Conclusion

A key dimension of European-style GI systems is their associated bottom-up models of collective action. The development of GI in Europe has played an important role in the conservation of local traditional and specialty agricultural products, which also helped to maintain diversified tastes and diets, as well as culinary know-how against a backdrop of generalized agricultural modernization (Barham and Sylvander 2011). For consumers, GI products are meant to be understood as made by small, authentic local producers using traditional production methods. For producers, GI acts both as a marketing tool and intellectual property regime, protecting producers against fraud, counterfeiting and misuse. For example, in 2022, the Court of Justice of the European Union blocked Denmark from labeling cheese as Feta unless it was made in Greece, following a string of cases in the previous decade upholding the Greek origin of Feta against claims from Germany, Bulgaria, Macedonia, and Romania (Yun 2022). However, this historical model of European GI is becoming more pluralized, both through actions within the EU, and in regions that have recently adopted *sui generis* GI systems, including Japan, which do not necessarily ascribe to the 'spirit of GI' centered around patrimony, diversity, and producer-led preservation of agricultural livelihoods.

Globally, GI is facing an institutionalization process that encompasses new purposes like territorial development, biodiversity protection, and sustainability. The recent decisions of European Commission to introduce sustainability criteria in GI product specifications demonstrates that the "model" of *sui generis* GI is fluid. For numerous regions outside Europe, a top-down approach provides the GI system with the potential to fulfill numerous public objectives, such that it also serves as a policy instrument (Marie-Vivien and Biénabe 2017). Our analysis of cases from Japan shows that the historically legitimate regional delimitations and constellations of traditional practices that have defined very old products are marginalized or undervalued by strong GI regulators with an economic expansionist focus. Merit-based or democratic mechanisms for inscribing GIs based on internal motivations, such as pride, fraud-prevention, and global recognition are replaced by strategic concerns reflecting political goals, such as inclusivity, efficiency, expansion, and export. For *Hatcho Miso*, bubbling rivalries concerning authenticity and exclusivity led to dueling GI applications and a deterioration of cooperative behavior. For *Kokonoe Mirin*, longstanding competition between breweries in the region and the lack of differentiation offered by the GI label failed to galvanize mirin producers in the Mikawa region to apply for GI. For former *Nishio Matcha* tealeaf producers, the GI governance structure failed to protect their interests vis-à-vis large processors.

For long-standing producers, hesitance to industrialize represents a raison d'être. Therefore, the model of economic expansion governed by MAFF presents a disadvantageous platform for adjudicating historical rivalries. While fractious cases have been meaningfully arbitrated worldwide, such as in the case of *Aceto Balsamico di Modena* in Italy, this depends on commitment by authorities to protect patrimony over tendency for economic expansion. Such outcomes may also have relevance for European GIs, in which industrial players are beginning to play more influential roles. This represents an increasing trend worldwide, in which the State is taking a more active role in selecting the elements that deserve to be included in the qualification and enhancement processes, on the basis of their heritage value combined with their commercial potential. Heritage enhancement is thus constructed or reconstructed to meet the objectives of modernizing and intensifying production systems (Michon et al. 2016). If the expansion of *sui generis* GI continues on its current trajectory, we can expect a shift in local production systems towards a more intensive, commercialized structure.

References

Augustin-Jean L (2012) Standardization vs products of origins: what kinds of agricultural products have the potential to become a protected geographical indication? In: Augustin-Jean L, Ilbert H, Saavedra-Rivano N (eds) Geographical indications and international agricultural trade: the challenge for Asia. Palgrave Macmillan, London, pp 48–70

Augustin-Jean L, Sekine K (2012) From products of origin to geographical indications in Japan: perspectives on the construction of quality for the emblematic productions of Kobe and Matsusaka beef. In: Augustin-Jean L, Ilbert H, Saavedra-Rivano N (eds) Geographical indications and international agricultural trade: the challenge for Asia. Palgrave Macmillan, London, pp 139–163

Barham E, Sylvander B (2011) Labels of origin for food: local development, global recognition. CABI, Wallingford/Oxfordshire/Cambridge

Baxter P, Jack S (2008) Qualitative case study methodology: study design and implementation for novice researchers. Qual Rep 13:544–559

Calboli I (2015) Of markets, culture, and terroir: the unique economic and culture-related benefits of geographical indications of origin. In: Gervais DJ (ed) International intellectual property: a handbook of contemporary research. Edward Elgar, Cheltenham, pp 433–464

Chabrol D, Mariani M, Sautier D (2017) Establishing geographical indications without state involvement? Learning from case studies in central and West Africa. World Dev 98:68–81. https://doi.org/10.1016/j.worlddev.2015.11.023

Crescenzi R, De Filippis F, Giua M, Vaquero-Piñeiro C (2022) Geographical indications and local development: the strength of territorial embeddedness. Reg Stud 56:381–393. https://doi.org/10.1080/00343404.2021.1946499

Defrancesco E, Kimura J (2018) Are Geographical Indications (GIs) Effective value-adding tools for traditional food? Insights from the new-born Japanese GIs system. Proc Syst Dynam Innov Food Netwr:119–130. https://doi.org/10.18461/PFSD.2018.1808

Durand C, Fournier S (2017) Can geographical indications modernize Indonesian and Vietnamese agriculture? Analyzing the role of national and local governments and producers' strategies. World Dev 98:93–104. https://doi.org/10.1016/j.worlddev.2015.11.022

Feuer HN (2020) Geographical indication out of context and in vogue: the awkward embrace of European heritage agricultural protections in Asia. In: Bonanno A, Sekine K, Feuer HN (eds) Geographical indication and global Agri-food: development and democratization. Routledge, New York, pp 39–53

Feuer HN (2022) Crowding out local initiative in the protection of regional Agri-food specialties: the growing hegemony of sui generis geographical indication in East Asia. Int Sociol. https://doi.org/10.1177/02685809221111940

Galeazzi S (2018) Geographical indications: from tradition to business. The case of Japan and the gradual loss of terroir's fundamental. Masters, Ca' Foscari University

Giovannucci D, Josling T, Kerr W et al (2009) Guide to geographical indications: linking products and their origins. International Trade Centre, Geneva

Gugerell K, Uchiyama Y, Kieninger PR et al (2017) Do historical production practices and culinary heritages really matter? Food with protected geographical indications in Japan and Austria. J Ethnic Foods 4:118–125. https://doi.org/10.1016/j.jef.2017.05.001

Huysmans M (2020) Exporting protection: EU trade agreements, geographical indications, and gastronationalism. Rev Int Polit Econ:1–28. https://doi.org/10.1080/09692290.2020.1844272

Kizos T, Kohsaka R, Penker M et al (2017) The governance of geographical indications: experiences of practical implementation of selected case studies in Austria, Italy, Greece and Japan. BFJ 119:2863–2879. https://doi.org/10.1108/BFJ-01-2017-0037

Lecoent A, Vandecandelaere E, Cadilhon J-J (2010) Quality linked to geographical origin and geographical indications: lessons learned from six case studies in Asia. Food and Agriculture Organization of the United Nations (FAO), Bangkok

Marie-Vivien D, Biénabe E (2017) The multifaceted role of the state in the protection of geograph-ical indications: a worldwide review. World Dev 98:1–11. https://doi.org/10.1016/j.worlddev. 2017.04.035

Marie-Vivien D, Pick B, Dao TA (2015) Geographical indications and trademarks in Vietnam: confusion or real difference? Proceedings of the Second International Conference on Agricul-ture in an Urbanizing Society. Rome, In, pp 151–152

Michon G, Berriane M, Romagny B, Skounti A (2016) Les enjeux de la patrimonialisation dans les terroirs du Maroc. In: Berriane M, Michon G (eds). IRD Éditions, pp 161–179

Pick B, Marie-Vivien D, Kim DB (2017) The use of geographical indications in Vietnam: a promising tool for socioeconomic development? In: Calboli I, Loon N-LW (eds) Geographical indications at the crossroads of trade, development, and culture: focus on Asia-Pacific. Cam-bridge University Press, Cambridge, pp 305–332

Port KL (2014) "Regionally based collective trademark system" in Japan: geographical indicators by a different name or a political misdirection? Cybaris 6:1–56

Sekine K (2020) The impact of geographical indications on the power relations between producers and Agri-food corporations: a case of powdered green tea matcha in Japan. In: Bonanno A, Sekine K, Feuer HN (eds) Geographical indication and global Agri-food: development and democratization. Routledge, New York, pp 54–69

Thévenot L (2009) Governing life by standards: a view from engagements. Soc Stud Sci 39:793–813

Tregear A, Arfini F, Belletti G, Marescotti A (2007) Regional foods and rural development: the role of product qualification. J Rural Stud 23:12–22. https://doi.org/10.1016/j.jrurstud.2006.09.010

Vandecandelaere E, Arfini F, Belletti G, Marescotti A (2010) Linking people, places and products: a guide for promoting quality linked to geographical origin and sustainable geographical indica-tions. UN Food and Agriculture Organization, Rome

Wang G (2008) The definition and the implementation of a system of geographical indications in an economy in transition, the case of China. PhD, Toulouse University

Yin RK (1994) Discovering the future of the case study. Method in evaluation research. Eval Pract 15:283–290

Yun, C (2022, 15 July) 'Feta' is Greek, EU top court says in snub to Denmark, Reuters. https://www.reuters.com/world/europe/feta-is-greek-eu-top-court-says-snub-denmark-2022-07-14/

Zappalaglio A (2021) The transformation of EU geographical indications law: the present, past, and future of the origin link. Routledge, Milton Park/Abingdon/Oxon/New York

Zhang J (2015) The foods of the worlds: mapping and comparing contemporary gastrodiplomacy campaigns. Int J Commun 9:568–591

The opinions expressed in this chapter are those of the author(s) and do not necessarily reflect the views of the [NameOfOrganization], its Board of Directors, or the countries they represent.

Chapter 21
The First Controversy Over GI Registration in Japan: A Case of Hatcho Miso

Kae Sekine

Abbreviations

GI	Geographical Indication
EPA	Economic Partnership Agreement
JPO	Japan Patent Office
MAFF	Ministry of Agriculture, Forestry and Fisheries
METI	Ministry of Economy, Trade and Industry
MIAC	Ministry of Internal Affairs and Communications
PDO	Protected Designation of Origin
PGI	Protected Geographical Indication
TRIPS	Agreement on Trade-Related Aspects of Intellectual Property Rights
TTJ	Terroir and Tradition Japan

21.1 Introduction

While GIs are expected to promote the recognition of the link between producers, territories, and agri-food products (FAO 2010), the outcomes of GI systems depend on the way in which products' quality and code of practices are socially constructed (Barham 2011; Calboli and Loon 2017). Despite the objective of the legislation to protect producers who employ traditional and vernacular production methods, some GIs experienced controversies when powerful stakeholders, such as transnational corporations, co-opted GIs and substituted industrial and modern technology for traditional vernacular methods (Bonanno et al. 2019; Bowen 2015). In these cases, private voluntary quality schemes such as Presidia of Slow Food International are

K. Sekine (✉)
Nagoya, Japan

© The Author(s) 2025
E. Vandecandelaere et al. (eds.), *Worldwide Perspectives on Geographical Indications*, https://doi.org/10.1007/978-3-031-71641-6_21

considered as an alternative to protect the traditional production and processing methods, varieties, and breeds (Fernandez et al. 2020; Parasecoli 2017; Sekine 2020, 2021).

In recent years, several Asian countries integrated GI systems into their national legislations (Augustin-Jean et al. 2012; Bonanno et al. 2019; Calboli and Loon 2017) while these countries' private sectors also promoted territorial branding initiatives. Among them, the Japanese government is actively promoting its GI systems as tools to revitalize the agri-food sector and rural communities following the negative outcomes that economic, social, political, and environmental changes have generated for decades (Augustin-Jean and Sekine 2012; Sekine 2019; Sekine and Bonanno 2018). However, the government approach tends to emphasize these systems' economic role and their contributions in the export of agri-food products rather than their socio-cultural and environmental dimensions. Moreover, Japan faces limits in its territorial development due to diplomatically compromised design of its GI system (See the next section and Sekine 2024).

In December 2017, this government posture caused the first nation-wide controversy over the registration of a GI, namely "*Hatcho Miso*" in Aichi Prefecture. The government approved a code of practice proposed by the Aichi Soy Sauce and Miso Cooperative (hereafter Aichi Miso Coop) that employs modern techniques and an extended area of production. Employing traditional knowledge and limited area of production, the artisanal Hatcho Miso Cooperative (hereafter Hatcho Miso Coop) faced competition with industrial Hatcho Miso labelled as GI and refused to be part of the GI. This case was argued in courts.

Against this backdrop, the objective of this chapter is threefold: (1) to review the GI systems adopted in Japan in comparison with the European system; (2) to illustrate the contradictory process of Japanese GI system employing the case of Hatcho Miso; and (3) to examine the potential roles of other territorial labelling systems, such as Honbano Honmono, that can contribute to the differentiation of quality under GI system.

This study is based on a field survey with 13 semi-structured and open-ended interviews (followed by email and telephone exchanges), conducted from 2015 to 2024 among various stakeholders (Ministry of Agriculture, Forestry and Fisheries (hereafter MAFF), Japan Patent Office (hereafter JPO), Terroir and Tradition Japan (hereafter TTJ), Hatcho Miso Coop, Aichi Miso Coop, and breweries of Miso in Aichi Prefecture) as well as on literature review.

21.2 The Evolution of Japanese GI Systems

While other countries are developing their GI systems under TRIPS Agreement or other international conventions, the Japanese government has also been gradually developing various GI systems (Table 21.1). First, in 1995, when the TRIPS Agreement came into effect, the National Tax Agency revised the Liquor Tax Law

Table 21.1 The GI Systems and a Private Territorial Label in Japan (Oct. 2023)

		GI for Alcohol	Collective Trademark	Geographical Indication	Honbano Honmono
Year		1995	2006	2015	2005
Supervising Organization		National Tax Agency	JPO	MAFF	Honbano Honmono Brand Promotion Organization
Legislation		Liquor Tax Law	Trademark Law	*Sui generis* GI Law	-
Products		Alcohol	All products and services	Agri-food products	Agri-food products
No. of products	Japan	29	776	148	59
	Foreign	0	3	6	0

Source: Author elaboration on data from National Tax Agency (2024), JPO (2024), MAFF (2024), Honbano Honmono Brand Promotion Organization (2024)

and began providing GI protection for sake, distilled spirits, wine, and so on (Sekine 2019). Subsequently, in 2006, the Japan Patent Office, a public organization under the jurisdiction of the Ministry of Economy, Trade and Industry (METI) launched Regional Collective Trademark registration system based on the trademark law, making it possible to register not only agricultural products and foods but also traditional crafts and services such as hot springs. This is in line with countries such as the United States, which seek to protect GIs through the trademark system. Later on, when the Japan-Europe Economic Partnership Agreement (hereafter EPA) negotiations got into full swing, the Japanese government decided to introduce an EU-style *sui generis* GI system and sign on the mutual protection of registered products in response to the EU's request.

In parallel to these public GI systems, a private territorial labeling system "Honbano Honmono'', which means authentic and genuine in Japanese, was established in 2005. It was launched by Japan Food Industry Association as a project subsidized by MAFF, which was unable to legislate the *sui generis* GI system at the time. Since 2016, Honbano Honmono Promotion Organization has operated its evaluation and certification, and TTJ has been in charge of sales promotion, export, and so on.

The Japanese *sui generis* GI system may seem similar to the EU's *sui generis* GI system. However, when comparing the provisions of the law, there are major differences (Table 21.2). On the one hand, the article 5 of EU Regulation 1151/2012 stipulates the establishment of two types of GIs, Protected Designation of Origin (hereafter PDO) and Protected Geographical Indication (hereafter PGI). For instance, in the case of processed foods, the former requires that the ingredients are produced within the certified region, but the latter allows ingredients produced outside the certified region (within the EU). On the other hand, such a provision is not specified in Article 2, Paragraph 2 in Japan's GI Law. In other words, the food processed with imported ingredients can be issued as GIs in Japan. In contrast,

Table 21.2 A Comparison of GIs and Honbano Honmono

	Year	Requirement	
		All stages of production, processing, and preparation must take place in the region	At least one of the stages of production, processing or preparation takes place in the region
European GI	1992	PDO	PGI
Honbano Honmono	2005	Type I	Type II
Japanese GI	2015	–	GI

Source: Author elaboration based on European Commission (2023), MAFF (2024), Honbano Honmono Brand Promotion Organization (2024)

Honbano Honmono employs a certification scheme that distinguishes Type I and Type II which are equivalent to PDO and PGI under European GI legislation.

The reason why MAFF did not adopt the scheme equivalent to PDO was that (1) Japan's agricultural and food industries are highly dependent on imported raw materials and feed, and therefore few agri-food products could be registered as PDO; (2) the idea was to make the GI System as simple as consumers did not get confused. However, as there are many miso and pickles made with locally produced ingredients in Japan, as well as livestock products that are pastured locally, and there is no consumers' confusion due to two-GI categories in the EU, the rationality of this decision can be alternatively explained. In other words, there was a political consideration for countries and stakeholders that export raw materials and feed to Japan. The business circles of the US and Australia expressed their concerns about the possible negative effects of *Japanese sui generis* GI system on international trade and therefore their businesses (The Study Group of Protected Geographical Indication Systems 2012). Based on their request, the Study Group of Protected Geographical Indication Systems, that was organized by MAFF, proposed to establish only PGI equivalent GI category in Japanese GI System.

By not introducing PDO in Japan, the possibility of using the GI system as an opportunity to rebuild ties between local agriculture and the local food industry and restore local economic circulation has been greatly undermined (Sekine 2024). The establishment of PDO in Japanese GI legislation remains a challenge in its possible amendment in the future.

21.3 The First Controversy: A Case of Hatcho Miso

21.3.1 Agriculture and Food Culture in Aichi Prefecture

The Japanese GI system lacking the EU's PDO equivalent category led to the first nation-wide controversy of GI registration in the country, namely the case of the

Hatcho Miso GI in Aichi Prefecture (see also Chap. 37). Aichi Prefecture has a thriving manufacturing industry represented by Toyota Motor Corporation, but it is also one of the leading agricultural prefectures in the country, boasting the 8th largest agricultural output in the country in 2018 (Aichi Prefecture 2020). In addition, the dishes cooked with Hatcho Miso are known as local traditional and typical food culture.

Hatcho Miso is a traditional food that has been produced since the Edo period (1603–1867) in Hatcho Town, Okazaki City (formerly Hatcho Village) in the West Mikawa Region of Aichi Prefecture. The former Hatcho Village was a crossroad for water and land transportation where the Tokaido road and Yahagi Rivers intersect, and therefore was a convenient location for collecting soybeans and salt, the raw materials for *Hatcho Miso*. There is also a record that *Hatcho Miso* was carried along with the Mikawa samurai group that supported the shogunate Ieyasu Tokugawa.

While around 80% of miso produced in Japan is rice miso, it could be spoiled in the warm and humid climate in Hatcho Village (Miso Health Promotion Committee 2001). Therefore, soybean *koji*, soybean malt, was invented instead of rice *koji*, mixed with salt and water, fermented, and aged in wooden vats. Miso has been produced employing a unique method of aging for a long period of time. Then the production of soybean miso extended to the Chita Peninsula in the Owari Region of Aichi Prefecture, where the brewing industry was flourishing and then spread to the adjacent Gifu and Mie prefectures. Today, the soybean miso is mainly produced and consumed in these prefectures in Tokai Region.

21.3.2   The Controversy Over GI Registration of Hatcho Miso

Since the *sui generis* GI legislation has been launched, there are two parties that applied for GI registration of "*Hatcho Miso*", namely Hatcho Miso Coop and Aichi Miso Coop, both located in Aichi Prefecture, in 2015. However, the controversy over the use of the name "*Hatcho Miso*" dates back to the 1980s (Table 21.3). In the 1980s, a lawsuit was filed regarding trademark registration, and the Tokyo High Court ruled that the name "*Hatcho Miso*" was a common name for a type of soybean miso produced mainly in Okazaki City. In 2006, under the Regional Collective Trademark registration system, both Hatcho Miso Coop and Aichi Miso Coop filed applications for "*Hatcho Miso*" and "*Aichi Hatcho Miso*" respectively, with JPO. At that time, the applications were not registered as an agreement had not been reached between the two parties, and both applications were withdrawn.

Under these circumstances, between the Hatcho Miso Coop and Aichi Miso Coop there was a significant divergence of contentions over the right to use the name "*Hatcho Miso*" and the methods of *Hatcho Miso* production. While Hatcho Miso Coop consists of two long-established miso breweries in Hatcho Town, Okazaki City (Limited Partnership Company Hatcho Miso, commonly known as Kakukyu (hereafter Kakukyu) since 1645 and Maruya Hatcho Miso Co., Ltd. (hereafter Maruya) since 1337), Aichi Miso Coop includes 37 member brewers, six of whom

Table 21.3 The History of the Controversies over the Name of Hatcho Miso

Dates	Events
1983	Kakukyu applied for a trademark "Hatcho Miso Limited Partnership" which is the company's name, but it was rejected by JPO.
1989	Tokyo High Court ruled that "Hatcho Miso" was a common name.
Apr. 13, 2005	Establishment of Hatcho Miso Coop.
Feb. 2006 Mar. Apr.	Aichi Miso Coop's application for a trademark "Hatcho Miso" was rejected by JPO. Hatcho Miso Coop applied for a Collective Trademark "Hatcho Miso" to JPO and withdrew later. Aichi Miso Coop applied for a Collective Trademark "Aichi Hatcho Miso" to JPO and withdrew later.
Mar. 2009	Hatcho Miso Coop withdrew from Aichi Miso Coop and Japan Federation of Miso Manufactures.
Jun. 1, 2015 Jun. 24	Hatcho Miso Coop applied for registration of GI "Hatcho Miso." Aichi Miso Coop applied for registration of GI "Hatcho Miso."
Jun. 15, 2017 Dec. 15	Hatcho Miso Coop withdrew GI application. GI "Hatcho Miso" applied by Aichi Miso Coop was registered.
Jan. 2018 Mar. 14 Mar. 26 May 29	Mass media widely reported on the GI registration of "Hatcho Miso." Hatcho Miso Coop filed an appeal request to MIAC. The mayor and city council president of Okazaki City handed opinion letters to the Deputy Minister of MAFF. An association of Okazaki citizens was established and started signature campaign to support Hatcho Miso Coop.
2019 Sep. 27	MAFF and Hatcho Miso Coop made vindications at MIAC. The Administrative Appeal Committee in MIAC reported that MAFF's decision was inappropriate.
Mar. 19, 2020 Mar. – Dec.	MAFF established a third-party committee regarding GI registration of "Hatcho Miso." The meetings of the committee were held four times.
Mar. 12, 2021 Mar. 19 Sep. 17	The report by the third-party committee was published. MAFF announced that it rejected the Hatcho Miso Coop's request for cancellation of the GI registration. Maruya filed a lawsuit with the Tokyo District Court against MAFF to cancel the registration of GI "Hatcho Miso."
Jun. 28, 2022 Jul. 8	Tokyo District Court dismissed Maruya's lawsuit. Maruya appealed the Tokyo High Court.
Jan. 27, 2023 Mar. 8 Mar. 16	Okazaki City renamed Hatcho Town employing an old *kanji* character so that the spelling of geographical name became same with Hatcho Miso in Japanese. The Tokyo High Court dismissed Maruya's lawsuit. Maruya filed a lawsuit with the Supreme Court against MAFF to cancel the registration of GI "Hatcho Miso."
Mar. 6, 2024	The Supreme Court dismissed Maruya's lawsuit.

Source: The author developed the table based on the interviews and newspapers
Notes: JPO is Japan Patent Office. MAFF is Ministry of Agriculture, Forestry and Fisheries. MIAC is Ministry of Internal Affairs and Communications

Table 21.4 The Comparison of the Specifications of Hatcho Miso

Applicants		Hatcho Miso Coop	Aichi Miso Coop
Number of Concerned Miso Brewers		2	6
Delimited Areas		Hatcho Town in Okazaki City, Aichi Prefecture (former Hatcho Village)	Aichi Prefecture
Production Methods	Ingredients[a]	Soybeans (domestic or import), salt	Soybeans (domestic or import), salt
	Food Additives	Not allowed	Alcohol is allowed
	Miso Ball Size	Size of Fist	Diameter >20 mm Length >50 mm
	Container	Wooden Vats	Brewing Tanks (Stainless Tanks are allowed)
	Weight on Container	Native river stones laid forming a pyramid on the lid of wooden vats (3t/vat)	No specification on material or laying forms
	Temperature Control	Not allowed (only natural brewing is allowed)	Temperature control by machine is allowed
	Aging	More than 24 months (2 summers and 2 winters)	More than 3 months When temperature is controlled, more than 10 months over 25°C
Features	Appearance	Shiny blackish brown Amino acid crystals can be found	Dark reddish-brown
	Taste Aroma	Astringent, bitter, sour, sweet, seasoned salty, and umami with a rich and unique aroma	Moderate acidity, umami, bitterness, and astringency

Source: Author elaboration based on the GI specifications and interviews
Note: [a]Based on Food Labeling Law, water, and malt (*koji*) are not indicated on the product's label

produce soybean miso (Nakari Co., Ltd. since 1896, Noda Miso Co. Ltd. since 1928, Morita Co. Ltd. since 1665, Ichibiki Co. Ltd. since 1772, Nakamo Co. Ltd. since 1830, and Satoh Co. Ltd. since 1874).

Hatcho Miso Coop claimed that the origin of Hatcho Miso is Hatcho Town and therefore only the soybean miso produced in this delimited area should be named "Hatcho Miso" as well as it should be produced under traditional methods (Table 21.4). However, Aichi Miso Coop that has large food industry as its member companies stated that all soybean miso with certain features produced in Aichi Prefecture could be named "Hatcho Miso" and modern technologies must be allowed to its production (Table 21.4). In 2009, the former finally separated from the latter and also withdrew from the national organization of miso producers (Table 21.3).

When GI registration began in June 2015, Hatcho Miso Coop immediately applied for GI "*Hatcho Miso*," followed by Aichi Miso Coop (Table 21.3). However, the production standards (specifications, management operational regulations of production process, and so on) applied by both Coops were significantly different

Fig. 21.1 Hatcho Miso in a Wooden Vat at Maruya. (Source: Maruya)

(Table 21.4). *Hatcho Miso* of Hatcho Miso Coop is produced only in Hatcho Town in Okazaki City, made from only soybeans and salt, and has strict standards that require it to be naturally brewed in wooden vats for more than 2 years. In addition, the weights of the wooden vats are natural stones from the local Yahagi River, and skilled craftsmen stack them in a conical shape using a construction method derived from the stone walls of Okazaki Castle (Fig. 21.1). In other words, *Hatcho Miso* of Hatcho Miso Coop is the culmination of traditional methods, history, and craftsmanship.

On the other hand, the latter has expanded its production area to cover all of Aichi Prefecture. Although it is not specified in the production standards, alcohol as food additive is permitted. Furthermore, in addition to wooden vats, stainless steel tanks are also permitted. Moreover, modern manufacturing methods that speed up the brewing period by adjusting the temperature are permitted. In other words, the Aichi Miso Coop adopts manufacturing methods similar to that of generic soybean miso, rather than traditional manufacturing methods.[1]

These differences in production standards are thought to have a major impact on the quality of miso. However, the special sub-committee on processed foods of

---

[1] A member of the third-party committee that examined the case of GI registration of *"Hatcho Miso"* also pointed out that the currently registered production standards of GI *Hatcho Miso* cannot be distinguished from those for general soybean miso (The Third-Party Committee of *"Hatcho Miso"* GI Registration 2020). However, this statement was deleted from the final report of the committee (The Third-Party Committee of *"Hatcho Miso"* GI Registration 2021).

MAFF's GI Academic Expert Committee has determined that there was "no significant'' difference between the two without conducting any additional scientific or organoleptic examination. Furthermore, MAFF required Hatcho Miso Coop to enlarge the delimited area from Hatcho Town to the entire Aichi Prefecture, stating that the production area and the use of the name "*Hatcho Miso*" have been already expanded in the early Showa Era (the 1920–1930s) and it would be advantageous to do so for a provision to the promising export market.

When Hatcho Miso Coop informed MAFF that it could not comply with the request, MAFF notified Hatcho Miso Coop that it would be difficult to register "Hatcho Miso" as a GI in 2017 (Table 21.3). As Hatcho Miso Coop recognized that the situation was similar to that of Collective Trademark in 2006, it withdrew its application for GI. However, MAFF subsequently approved Aichi Miso Coop's application for GI registration in December 2017. The MAFF's decision was based on the intention of accelerating the GI registration of "Hatcho Miso" in line with the schedule of the Japan-Europe EPA that was concluded in December 2017, and adding it to the list of Japanese GI products protected within the EU. In addition, as there was an increase in applications for trademark registration of Japanese product names in China and other countries, MAFF might have been in a hurry to make an action in protecting the name "Hatcho Miso". As a result, due to Aichi Miso Coop's GI registration, an unusual situation has arisen in which Hatcho Miso, manufactured by the long-established breweries in Hatcho Town employing traditional methods, has become prohibited from being named and labeled "*Hatcho Miso*" under the GI legislation. This goes against the purpose of the GI Law, which is to protect the names of local traditional products, methods, producers, and ultimately consumers.

In March 2018, Hatcho Miso Coop filed an administrative appeal request to the Ministry of Internal Affairs and Communications (hereafter MIAC), and the Okazaki City Council and the mayor also requested that the Ministry of Agriculture, Forestry and Fisheries provides guidance and coordination to encourage consensus building among stakeholders (Table 21.3). Furthermore, in May 2018, local residents and other interested parties began signing a petition asking MAFF to review Aichi Miso Coop's GI registration.[2] In September 2019, MIAC's Administrative Appeal Board determined that the GI registration of Aichi Miso Coop by MAFF was not appropriate. Then a third-party committee consisting of members selected by MAFF was established to examine the case. However, the committee's final report in March 2021 confirmed the claims of MAFF and supported GI registration of Aichi Miso Coop's "*Hatcho Miso*", contrary to the conclusion of the Administrative Appeal Board of MIAC.

Based on this report, MAFF issued a ruling in March 2021 to dismiss the Hatcho Miso Coop's request to cancel the registration of GI *Hatcho Miso*. However, in September 2021, Maruya filed a lawsuit with the Tokyo District Court asking MAFF to cancel the registration of GI *Hatcho Miso* (Table 21.3). Though Maruya wished to

---

[2] 2 As of April 2022, the number of signatures had exceeded 100,000.

question Japanese GI system, its lawsuit was dismissed in June 2022. After Maruya's appeal to the Tokyo High Court in July 2022 and its dismissal in March 2023, it filed an appeal to the Supreme Court in the same month. However, the Supreme Court dismissed Maruya's lawsuit on March 6, 2024. Following the decision of the Supreme Court, the Hatcho Miso Coop applied to MAFF for registration as a GI *Hatcho Miso* brewers' organization on July 30, 2024.

21.4 Conclusion

Employing the case of GI registration of "*Hatcho Miso*", this chapter reviewed the evolution of GI systems and illustrated the first controversy of GI registration in Japan. The GI system could be a powerful means of contributing to regional sustainable development by protecting traditional agri-food production systems that have been marginalized under the globalization and industrialization of agri-food systems. However, there are cases in which food industry bring modern technology, improved varieties and breeds, and cost squeezing logics into GI production standards, jeopardizing small and medium-sized producers and businesses and traditional methods, knowledge, varieties, and breeds. The case of *Hatcho Miso* shows that the alienation of traditional manufacturers is arising under the Japanese GI system.

Furthermore, a serious flaw in Japan's GI system is the lack of category equivalent to the EU's PDO. As a result, miso made with locally produced traditional soybeans varieties and that made with imported soybeans (including genetically modified soybeans) can be registered as GIs in Japan. Under such a GI system, it is difficult to encourage collaboration between local farmers and local food manufacturers. This is in contrast to the private territorial labeling Honbano Honmono, which registers Hatcho Miso Coop's "Hatcho Miso made from Mikawa soybeans" as Type I (PDO equivalent). It shows that it has more potential to differentiate the quality and authenticity of agri-food products than the existing GI system in Japan. The roots of the controversy over GI registration of "*Hatcho Miso*" can be found in the government's neoliberal policy, which prioritizes export expansion rather than the protection of local original and traditional products and the promotion of regional economic circulation, as well as Japan's weak diplomatic position.

At the same time, citizens' movements that directly challenge the government's GI system have emerged. The main actors are traditional food producers, local residents, local governments, and consumers pursuing authentic agriculture and food. For example, the Palsystem Consumers' Cooperative Union has vigorously questioned the GI registration system and called for a revision of the current system. Journalists who reported the complex issue of the GI controversy also played a role in stimulating nationwide debate. The controversy over GI registration of "*Hatcho Miso*" is not only a significant concern for local stakeholders, but it also questions what kind of agri-food systems and ultimately what kind of future society we want to build. Although the potential of the GI system for sustainable and fair development contains contradictions, these awoken actors can be a hope for the future of the region.

Acknowledgments The author appreciates the cooperation of all informants that made this publication possible.

Conflicts of Interest The author declares no conflict of interest.

Funding This work was supported by JSPS KAKENHI Grant-in-Aid for Early-Career Scientists, Grant Number 18K14542.

References

Aichi Prefecture (2020) Easy-to-understand agriculture in Aichi. Aichi Prefecture (in Japanese)
Augustin-Jean L, Sekine K (2012) From products of origin to geographical indications in Japan: perspectives on the construction of quality for the emblematic productions of Kobe and Matsusaka beef. In: Augustin-Jean L, Ilbert H, Saavedra-Rivano N (eds) Geographical indications and international agricultural trade: the challenges for Asia. Palgrave Macmillan, New York, pp 139–163
Augustin-Jean L, Ilbert H, Saavedra-Rivano N (eds) (2012) Geographical indications and international agricultural trade: the challenges for Asia. Palgrave Macmillan, New York
Barham E (2011) In: Sylvander B (ed) Labels of origin for food: local development, global recognition. CAB International, Wallingford
Bonanno A, Sekine K, Feuer HN (eds) (2019) Geographical indications and global Agri-food: development and democratization. Routledge (Earthscan Food and Agriculture)
Bowen S (2015) Divided spirits: tequila, Mezcal, and the politics of production. University of California Press, Oakland
Calboli I, Loon N-LW (2017) Geographical indications at the crossroads of trade, development, and culture: focus on Asia-Pacific. Cambridge University Press, Cambridge
European Commission (2023) Geographical indications and quality schemes explained. Retrieved from https://agriculture.ec.europa.eu/farming/geographical-indications-and-quality-schemes/geographical-indications-and-quality-schemes-explained_en#pdo on October 12, 2023
FAO (2010) Linking people, place and products: a guide for promoting quality linked to geographical origin and sustainable geographical indications. FAO, Rome
Fernandez A, Liu B, Galante AP, Slattery S, Sekine K, Ponzio R, Palandri C, Pantzer Y, Barletta MT, Martin G (2020) Globally important agricultural heritage systems, geographical indications and slow food presidia. FAO Technical Note, Rome
Honbano Honmono Brand Promotion Organization (2024) Honbano Honmono. Retrieved from https://honbamon.com/ on December 18, 2024
JPO (2024) The number of applied and registered collective trademarks. Retrieved from https://www.jpo.go.jp/system/trademark/gaiyo/chidan/document/index/ranking.pdf on December 18, 2024
MAFF (2024) Geographical indication system. Retrieved at https://www.maff.go.jp/j/shokusan/gi_act/ on December 18, 2024
Miso Health Promotion Committee (2001) Miso Culture Magazine. Japan Federation of Miso Manufacturers and Central Miso Research Institute (in Japanese)
National Tax Agency (2024) The list of registered GI alcohols. Retrieved from https://www.nta.go.jp/taxes/sake/hyoji/chiri/ichiran.htm on December 18, 2024

Parasecoli F (2017) Knowing where it comes from: labeling traditional foods to compete in a global market. University of Iowa Press, Iowa

Sekine K (2019) The impact of geographical indications on the power relations between producers and agri-food corporations: a case of powdered green tea "Matcha" in Japan. In: Bonanno A, Sekine K, Feuer HN (eds) Geographical indications and global Agri-food: development and democratization. Routledge (Earthscan Food and Agriculture), pp 54–69

Sekine, K (2020) New endeavors of Italian tomatoes industry at employing agri-food labeling systems: contributions to sustainable development goals. Veg Inform, 190: 61–70 (in Japanese)

Sekine K (2021) Roles and challenges of Agri-food labelling systems towards sustainable society: a case of Sorana beans in Tuscany, Italy. J Ritsumeikan Gastron Arts Sci 3:89–104. (in Japanese)

Sekine K (2024) The potential and challenges of geographical indications in building regional economic circulation: a case of Hatcho miso in Aichi Prefecture, Japan. Aichi Gakuin Econ Rev 11(2):22–38. (in Japanese)

Sekine K, Bonanno A (2018) Geographical indication and resistance in global Agri-food: the case of miso in Japan. In: Bonanno A, Wolf SA (eds) Resistance to the neoliberal agri-food regime: a critical analysis. Routledge, New York, pp 106–119

The Study Group of Protected Geographical Indication Systems (2012) Draft report of the study group of protected geographical indication systems. Study group of protected geographical indication systems (in Japanese)

Third-Party Committee of "Hatcho Miso" GI Registration (2020) The Minutes of the 4th Meeting on December 8, 2020. Retrieved from https://www.maff.go.jp/j/kanbo/tizai/brand/gi_iinkai/attach/pdf/8-iinkai-6.pdf on October 12, 2023 (in Japanese)

Third-Party Committee of "Hatcho Miso" GI Registration (2021) Final report on March 12, 2021. Retrieved from https://www.maff.go.jp/j/kanbo/tizai/brand/gi_iinkai/attach/pdf/8-iinkai-8.pdf on October 12, 2023 (in Japanese)

Chapter 22
Comparison Between Geographical Indication Indigenous Rice in India and Thailand: Regulations and Practices

Orachos Napasintuwong, Chitra Parayil, and A. M. Radhika

Abbreviations

CB	Certification Body
COP	Code of Practice
GAP	Good Agricultural Practice
GI	Geographical Indication
KSYMP	Khao Sangyod Muang Phatthalung
PGI	Protected Geographical Indication

22.1   Introduction

India and Thailand are two important exporters of rice and possess several premium quality rices such as *Basmati* and *Hom Mali*. As the competition in the world market of quality rice is increasing with emerging competitors, to distinguish the products from competitors, Geographical Indication (GI) is one of the ways to protect the property rights of rice production. Both countries are endowed with rich biodiversity and have high potentials to benefit from GI registration. The concept of GI in these

O. Napasintuwong (✉)
Department of Agricultural and Resource Economics, Faculty of Economics, Kasetsart University, Bangkok, Thailand
e-mail: orachos.n@ku.ac.th

C. Parayil
Department of Agricultural Economics, College of Agriculture, Vellanikkara, Thrissur, Kerala, India

A. M. Radhika
Department of Agricultural Economics, School of Agricultural Sciences, Amrita Vishwa Vidyapeetham, Coimbatore, Tamil Nadu, India

© The Author(s) 2025
E. Vandecandelaere et al. (eds.), *Worldwide Perspectives on Geographical Indications*, https://doi.org/10.1007/978-3-031-71641-6_22

299

countries is tightly interwoven with traditions, practices and know-how of rural lives. The strategy of building an image of quality for a class of products made in a certain area can help indigenous agricultural products achieve consumer acceptance quickly and can also help resource poor farmers command premium price. India and Thailand initiated GI laws that came into force in 1999 and 2003, respectively. As of September 2023, India has registered 17 GI rices while Thailand has registered 21 GI rices in the domestic market. While Thailand has successfully registered two of its local rices in the European Union (EU) market, namely *Hom Mali Thung Kula Rong-Hai* in 2013 and *Khao Sangyod Muang Phatthalung* (KSYMP) in 2016, India submitted only one, that is Basmati rice in 2018.

Previous studies have shown that the beneficial effects of GI depend strongly on the quality of the supply chain governance and on the elements of the code of practices (Radhika et al. 2021a). However, the benefits of complying to GI standards on farmers is unclear due to cost associated with GI production (Cei et al. 2018; Török et al. 2020). Nevertheless, the GI registration improves profitability of farmers in the GI protected area (Petruang and Napasintuwong 2022). To understand the context of the importance of GI protection on the livelihood of farmers and the market situations in these two countries, this chapter compares GI system and of two GI rice cases. In Thailand, KSYMP is selected as it is the first rice to be registered as GI protected product in Thailand and also GI protected in Malaysia, Indonesia and Protected Geographical Indication (PGI) rice outside of the EU territory registered by the European Commission. *Sangyod rice* is originally from Phatthalung province. It is popularly cultivated and consumed in Southern Thailand, but the registration of GI KSYMP is confined to Phatthalung province of Thailand. *Navara rice* is selected for the India case for its first recognized GI rice in India. Navara rice is a medicinal rice and is one of the native genetic resources of Kerala, famed for its use in Ayurveda. Both cases are indigenous rice historically reputable in the geographical location of registered GI.

The following sections provide details of GI system and discuss the value chain of *Khao Sangyod Muang Phatthalung* and *Navara rice* followed by a discussion of the comparison between the two which leads to conclusion and policy recommendations.

22.2   Thailand: Khao Sangyod Muang Phatthalung
(ข้าวสังข์หยดเมืองพัทลุง)

22.2.1   Description of the Case

Khao Sangyod Muang Phatthalung (KSYMP) (Single Document PGI *Khao Sangyod Muang Phatthalung*, European Union 2016).

KSYMP is produced from Sangyod Phatthalung variety. It can be paddy rice, brown rice and semi-milled rice. KSYMP is the first GI rice product in Thailand. The registration of GI protection of KSYMP by Thailand Department of Property Right was given to the Rice Department on June 23, 2006, and by the European Union as PGI protection on May 27, 2016 (European Union 2016; Department of Intellectual Property 2006).

Fig. 22.1 Sangyod rice variety. (Source: Author)

22.2.1.1   Rice Variety: Sangyod

Sangyod Phatthalung is an indigenous rice variety originally grown in Phatthalung Province. In 1988, the Sangyod rice was selected by the Phatthalung Rice Research Center. Sangyod was registered as a recommended rice variety on April 23, 2007 (Rice Department 2016). Sangyod is a photoperiod-sensitive, non-glutinous, light, small, lean and gently fragrant rice (Fig. 22.1). It is characterized by its red pericarp, soft texture when cooked and light fragrant. The taste is slightly sweet with mildly aromatic fragrance.

22.2.1.2   Geographical Location: Phatthalung Province

Phatthalung is the southern province of Thailand (Fig. 22.2). It is a lowland plain created by the deposits from the flooded lake during the monsoon season.

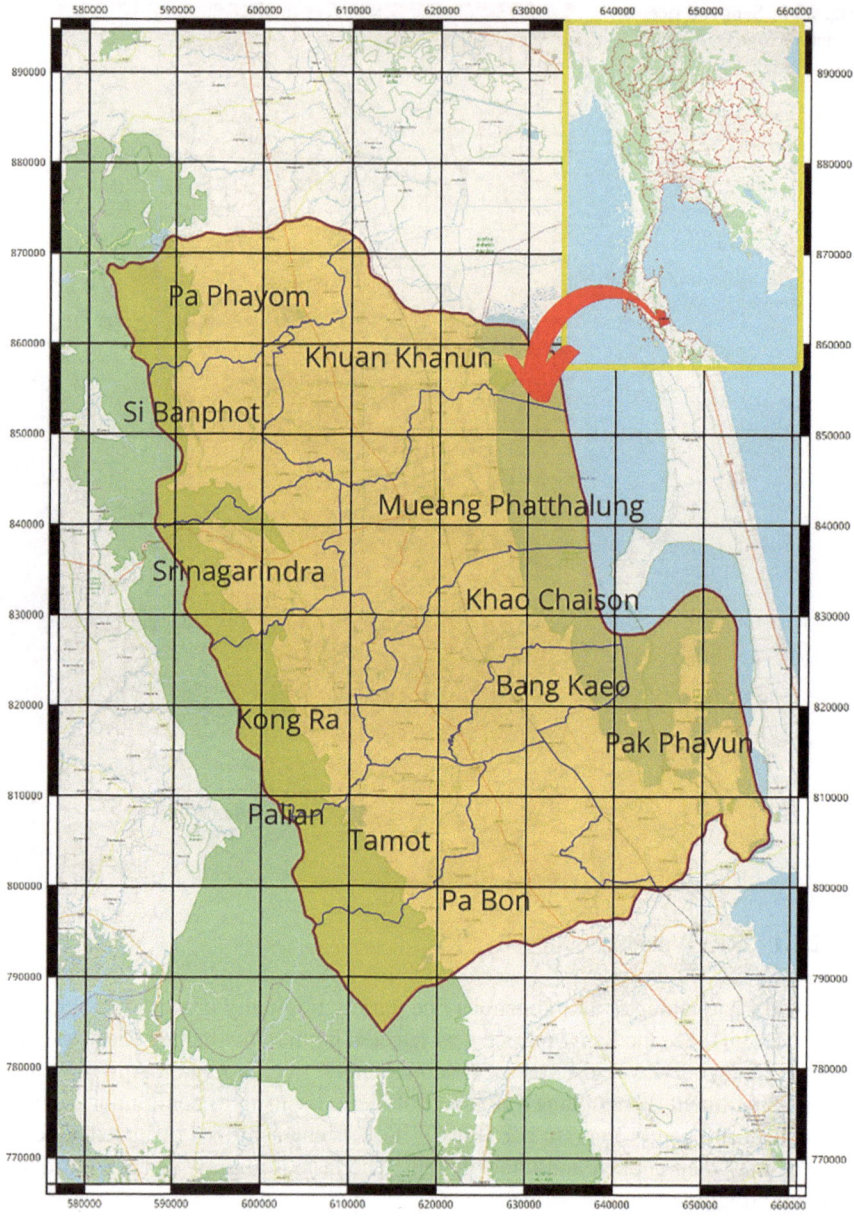

Fig. 22.2 Map of Phatthalung province. (Source: Author)

Phatthalung province has two climatic influences, both oceanic and continental. The continental influence is responsible for the distinctive dry season characterized by long hours of photoperiod.

22.2.1.3   Distinct Characteristics and Relation to Phatthalung Province

Local people in Phatthalung province prefer to consume granule and hard (high amylose content) rice such as Sangyod variety. Traditionally Sangyod rice is reserved to be given as a gift to respected senior people and cooked for special persons, such as for the royal guests, official foreign visitors, on special occasions, in religious ceremonies or traditional festivals (Single Document PGI *Khao Sangyod Muang Phatthalung*, European Union 2016). This tradition has created a reputation for KSYMP to be known as the rice for special persons and occasions in Phatthalung.

The nature of Phatthalung province especially the lowland plain provides several nutrients to the soil. The drainage capacity of the soil, the water storage and redistribution capacity, and the climate in Phatthalung allows KSYMP to have special characteristics of slender grain shape and being gently fragrant (Single Document PGI Khao Sangyod Muang Phatthalung, European Union 2016).

22.2.2   Technical Specifications of Khao Sangyod Muang Phatthalung

To ensure the quality of the rice, not only that KSYMP must be sown, grown, harvested, milled, packaged and labelled in the province of Phatthalung, the specifications of GI KSYMP are summarized in Table 22.1.

There are three control systems of GI rice in Thailand: self-, internal-, and external (Napasintuwong 2019). Self-control system is where groups of farmers and processors follow a Code of Practice (COP) or production manual . This self-control is needed for the internal control system. The internal control system is required for the authorization to use Thai GI symbol. Producers must follow the COP complying with the Thai Agricultural Standard for Geographical Indications on Rice: TAS 4005–2014[1] (National Bureau of Agricultural Commodity and Food Standard 2014) and in KSYMP case, the provincial GI certification committee[2]

---

[1]TAS 4005–2014 specifies requirements for planting, processing, and labeling and usage of GI symbol for paddy rice, brown rice, germinated brown rice, semi-polished rice, traditional parboiled rice and white rice that has registered geographical indication rice in Thailand.

[2]The current provincial GI certification committee consists of 13 members including Director of Phatthalung provincial commerce office as a chair, Head of the Business promotion and marketing division of the provincial commerce office as a secretary, Director of provincial agricultural extension office, Director of provincial industrial office, Director of provincial cooperative office, Director Phatthalung rice research canter, Director Phatthalung rice seed center, President of KSYMP producer association, Dean of Faculty of technology and community development of Thaksin University, etc.

Table 22.1 Technical specifications of *Khao Sangyod Muang Phatthalung*

Territory	
Geographical area	Phatthalung province
Variety	Sangyod Phatthalung variety
Farming practices	
Seed	Seed must be of Sangyod Phatthalung variety and produced in Phatthalung province. The seeds should be produced by Thailand Rice Department such as the Phatthalung Rice research Centre and the Phatthalung Rice seed Centre. Seeds from other sources such as farmers' organizations, must be approved by Thailand Rice Department. Saved seeds from approved Sangyod Phatthalung variety are allowed, but the seeds must be planted within 3 years after obtaining them from the source.
Soil preparation	Includes first rough plough, second plough, to appropriately control the number of volunteer rice plants (rice plants from seeds remaining in the field from the previous season) and off-type rice (rice plants of other varieties)
Soil fertilization and improvement	Regular basis
Rice plant care	Implementing appropriate water management, removing the off-type rice in the tillering stage, reproductive phase and maturation stage, appropriately terminating pests
Cultivation	Must be done during the southern wet season (august–October) after the ripening phase. It is a tradition of Phatthalung province to cultivate a bit over ripening rice
Harvest	The harvest is fixed during ripening stage between December and February according to the climate and the maturity of the rice.
Processing	
Grain moisture	Moisture content 14% or less.
Packaging	Must be packaged and labelled in Phatthalung province. Repackaging is not allowed.

Source: Thai Department of Intellectual Property (2006), Single Document PGI *Khao Sangyod Muang Phatthalung*, European Union (2016)

appointed by Phatthalung Provincial Governor inspects the operations of farmers, millers, and distributors (what is called the internal control in Thailand).

The external control system is required for PGI products under European Union regulations. The Certification Body (CB), here Bioagricert performs external control on behalf of the Department of Intellectual Property (DIP). The DIP is responsible for the validation of specifications and inspection methods e.g. production manual and approving and supervising the CB. The accreditation body, specifically National Bureau of Agricultural Commodity and Food Standard (ACFS) has the right to accredit CB according specific requirements laid down by DIP. The authorization for Thai GI c symbol is valid for 2 years.

22.2.3   Value Chain of Sangyod Phatthalung Rice in Thailand

The value chain of Sangyod Phatthalung rice is displayed in Fig. 22.3. There are two groups of Sangyod Muang Phatthalung rice farmers: one is those who comply with the KSYMP GI specification and generally are those who are compliant with the Thai Agricultural Standard for Geographical Indications on Rice and are authorized to use GI symbol. GI farmers are typically members of farmers' groups that received certification for either Thai Good Agricultural Practice (GAP) or organic standards. The other is those who are not compliant with the KSYMP GI specification. Both groups can acquire Sangyod Phatthalung rice seeds from Phatthalung Rice Research Centre and the Phatthalung Rice Seed Centre. Farmers can also acquire Sangyod Phatthalung rice seeds from other approved seed suppliers such as approved farmers' organizations. They can also save seeds, but the seeds must be planted within 3 years after obtaining them from the source.

In 2022/23, the planting area of Sangyod Phatthalung rice in Phatthalung province is about 139 hectares producing 556 tons of Sanyod Phatthalung rice and only 38.5 hectares (2.78%) or 96.4 tons (2.77%) is certified GI KSYMP. From 1727 households cultivating Sangyod Phatthalung rice, only 34 households (2%) are certified GI KSYMP farmers (Phatthalung Rice Research Center 2023) and can use the GI national logo.

After the harvest, approved GI KSYMP farmers sell their products to their member farmers' groups to be processed by authorized GI KSYMP millers into two typical products: unpolished rice or semi-milled rice. About 20% of farmers process themselves the product (using miller's services) and sell it directly to

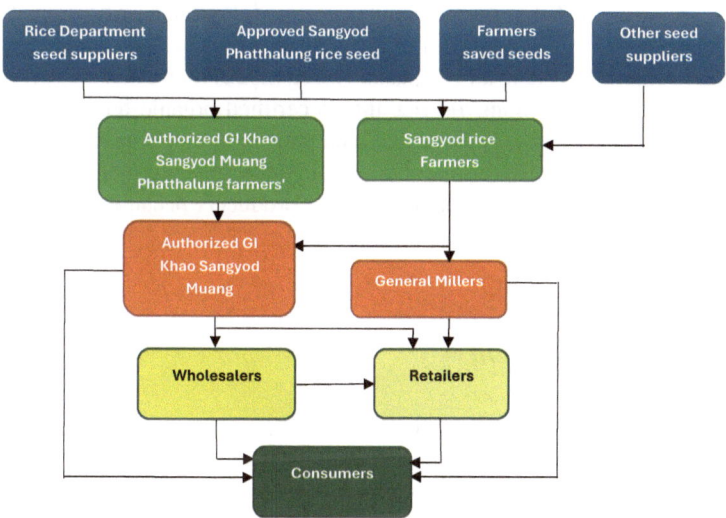

Fig. 22.3 Value chain of Sangyod Phatthalung rice

consumers. Other 70% sell it to authorized GI millers or to farmers' groups (Regional Office of Agricultural Economics 9 2021). These farmers' groups or millers almost always perform processing and marketing activities. Authorized farmers' groups may sell their products to authorized GI millers or collect processed rice, package it with GI symbol, and sell it to wholesalers, retailers or directly to consumers. Often, the authorized farmers' groups create their own brands and use a Thai GI as a marketing strategy. The unauthorized GI rice from general Sangyod Phatthalung rice farmers is sold to either general millers or authorized GI millers. This rice may use the GI name, KSYMP, but not authorized to use the GI symbol unless registered members are compliance with the production manual and follow the GI control plan. Some farmers can also process products at small local millers and keep them for household consumptions and/or sell them to retailers or directly to consumers in the local markets.

The approval for the use of the EU PGI and the associated EU PGI logo is much more costly than the approval for the use of the Thai GI symbol due to inspection fee of external control that is mandatory for GIs in the EU while in Thailand only internal control is possible. In 2017, the cost of annual certification service by certification body was about 1000 euros for a producer excluding fees for certification of products and the use of certification seal. The approval for Thai GI although is far less expensive, it is only good for the domestic market. At present, nearly all the Thai GI inspection fee is subsidized by the DIP through Phatthalung Provincial Commerce Office. In 2023, there are four GI KSYMP authorized farmers' groups. There were 24, 23, and 11 authorized farmers groups in 2018, 2019 and 2022, respectively. The number of authorized groups declined due to market and economic situation during the COVID-19 pandemic (Department of Intellectual Property 2023).

Sangyod Phatthalung rice products are found in four standards: organic, GAP, GI (that requires either GAP or organic) and uncertified. Producers who are authorized to use GI symbol (Fig. 22.4) on their products received higher price than the product without GI symbol. Petruang and Napasintuwong (2022) found that at the farm-level in Phatthalung province, Sangyod rice that is certified organic fetched the highest price followed by GI and GAP while the unauthorized product received the lowest price. Nevertheless, Farmers in Phatthalung province generate more profit from GI registration even if they are not authorized GI producers because the rice and the name KSYMP has some value.

22.2.4   Market Situation

Even though KSYMP is registered as GI or PGI protection in other countries, KSYMP has not reached the foreign markets. Most of the consumers remain in Southern region where their palatable preference is hard rice.

Farmers or farmers groups that process rice themselves use milling services and sell Sangyod Phtthalung rice directly to consumers through community markets,

A) Thai Symbol for GI   **B) European Union Symbol for Protected**
                                      Geographical Indication (PGI)

Fig. 22.4 Symbols of GI (**a**) Thai Symbol for GI (**b**) European Union Symbol for Protected Geographical Indication (PGI)

social media or to customers who visited the farms. Millers sell processed rice to wholesalers or retailers or directly to consumers through local shops, rice exhibition, agricultural exhibition or other events organized by the government (provincial or national level) (Regional Office of Agricultural Economics 9 2021). The average retail price of in December 2019 in Phatthalung provincial administrative shop of authorized GI KSYMP was about 85 THB/kg, about 13% higher than unauthorized Sangyod Phatthalung rice (75THB/kg).

22.3   India: *Navara Rice* (നവര)

22.3.1   *Description of Case*

Navara, known as the sastika rice is a special and unique variety of rice that is primarily grown in southern Indian state of Kerala. India's Geographical Indications of goods (Registration and Protection) Act, 1999, is a *sui-generis* legislation that aligns with its international obligation under TRIPS agreement to protect and promote product associated with specific geographical regions (Government of India 1999). Navara rice received GI protection in 2007, and *Navara Rice* Farmers Society, Chittur, Palakkad is the registered proprietor of Navara rice. As per the norms of GI recognition any farmer from any part of Kerala can produce and market *Navara rice*.

22.3.1.1   Rice Variety: Navara

Navara is a short duration rice variety that is traditionally cultivated as an organic crop to maintain the medicinal value (Fig. 22.5). There are two varieties of Navara:

Fig. 22.5 Navara rice
variety. (Source: Unny
2024)

the black color glumed and golden yellow color glumed, but the processed rice is
purple in color in both cases. It typically matures in 60 days making it fast growing
compared to other varieties. Navara is not typically grown as a staple variety but
rather it is often used in the preparations of ayurvedic medicines, health supplements
It has a nutty, earthy flavor and supersedes even virtuous brown rice in nutritional
value.

Researchers suggest that *Navara rice* is a good source of potassium, phospho-
rous, magnesium, sodium and calcium and is comprised of antioxidants, polyphe-
nols, flavonoids such as tricin, oryzanol, and proanthocyanidins that help to relieve
from obesity, diabetes, neurodegenerative disorders, and heart disease (Kowsalya
et al. 2022).

22.3.1.2   Geographic Location: Kerala State

Rice cultivation in Kerala dates back to 3000 B.C. (Manilal 1990) and is known for
its unique varieties and traditional farming practices. Rice is grown in widely
different regions from low lying areas below sea level extended to high attitudes
of the western ghats. Navara originating in Kerala State is unique with a given
quality, reputation and characteristic essentially attributable to its cultivation in a
specified soil and under and ambient weather condition coupled with a unique labour
intensive, organic and environmentally friendly manner that has traditional and
historical significance (Intellectual Property India 2004) (Fig. 22.6).

Parliamentary Constituencies
Kerala

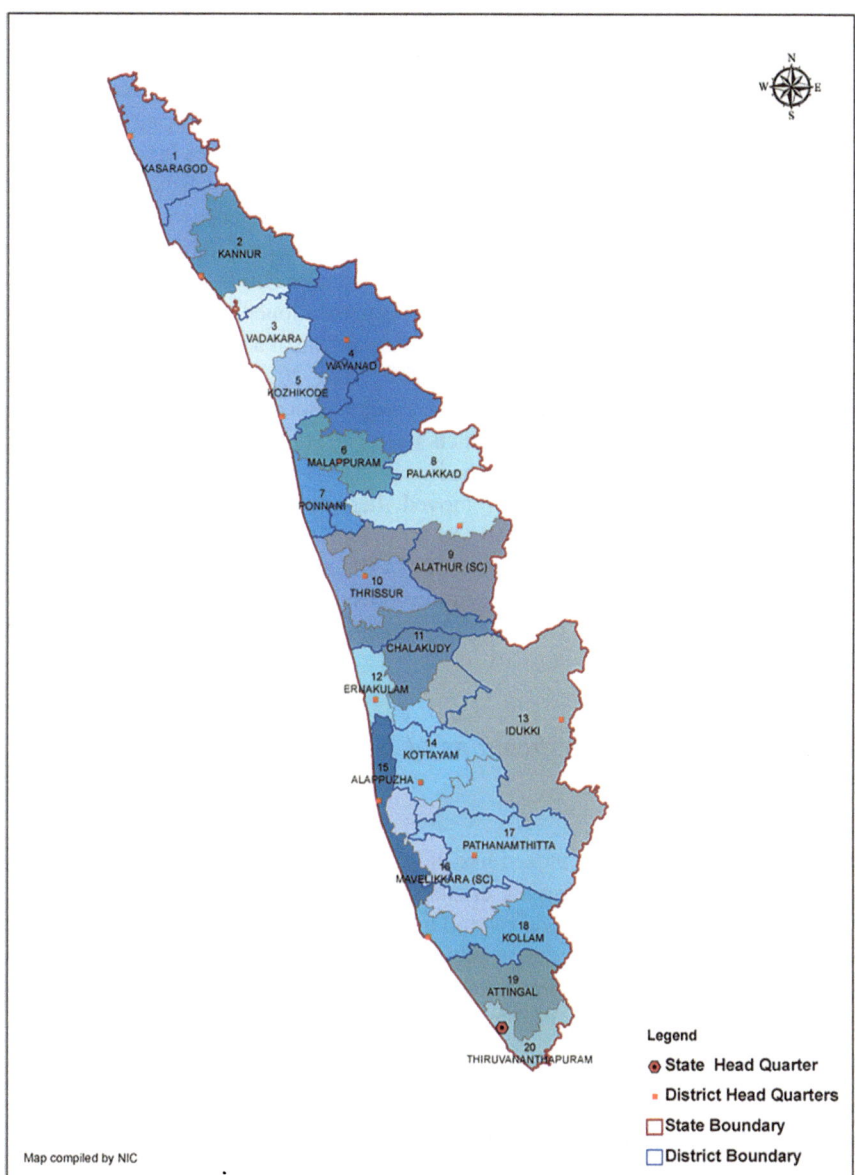

Fig. 22.6 Map of Kerala State. (Source: Election Department of Kerala 2024)

22.3.1.3 Distinct Characteristics and Relation to Kerala State

Kerala's rich tradition of folk medicine and healing practices is deeply rooted in its abundant biodiversity of medicinal plants. Navara is an important red rice (vreehi) variety cultivated in Kerala for more than 2500 years (Gopi and Manjula 2018; Variar 1985). "*Navara kizhi*" and "*Navaratheppu*" are two significant components of ayurvedic treatments, both of which are therapeutic techniques involving the use of Navara (John and Nizar 1998). It is a highly nutritious and easily digestible rice characterised with medium sized red colour grains. The reference to Navara is found in ancient ayurvedic scriptures of Ayurveda like 'Susrutha Samhitha', which dates back approximately to 400–200 B.C. and in 'Ashtanga Hrudaya' which dates back approximately to 600 A.D. This rice has been mentioned in 'Sushruta Samhita' dating back to 2200–2400 years (400–200 BC), as 'Sastika rice'- a rice that matures in 60 days (Intellectual Property India 2004; Unny 2023). This historical legacy signifies the deep-rooted traditions and knowledge associated with Navara in Kerala.

There are several value-added products from Navara rice such as rice powder and rice flakes. The products use unpolished rice. This is because 'bran' has the nutritional constituents and when bran is removed, these nutritional constituents would be lost. Rice powder, used as baby food and for preparation of typical Kerala breakfast dishes such as *Puttu, Pathiri, Kozhukatta* has been branded and marketed. Rice flakes as breakfast cereal has also been introduced in the market. These rice flakes can also be used as snacks (Trust for Advancement of Agricultural Sciences 2023).

22.3.2 Technical Specifications of Navara—The Grain of Kerala

The cultivation of Navara is indeed like those of other rice varieties in many aspects, with primary distinction being the emphasis on organic farming methods towards maintaining medicinal value and overall quality. By avoiding chemicals, pesticides and fertilizers, organic cultivation helps preserve the therapeutic potential of the variety and contributes towards long term sustainability and environmental consciousness of Navara variety (Table 22.2).

22.3.3 Value Chain of Navara Rice in India

Navara rice farmers can be categorised into two groups, GI certified farmers and Navara rice farmers (non-GI certified) (Fig. 22.7). The foundation of geographical Indication (GI) law rests on two key principles: registration of a product as a GI through collective rights and the subsequent acknowledgement of an individual as an authorized user after registration. Despite Kerala taking a leading role in the

Table 22.2 Technical specifications of Navara—The Grain of Kerala

Territory	
Geographical area	Kerala State
Variety	Navara-two ecotypes black colour glumed and golden yellow colour glumed
Farming practices	
Seed	Seed of Navara has short life cycle. After 5–6 months of storage the viability of the seed reduces considerably. Ensure that the seed for further multiplication is pure and must be grown in the state of Kerala. Seeds are soaked in water for 12 hours and then taken out and kept in gunny bags tightly tied, covered with hay, and kept for 3–4 days till sprouts emerge. In the meantime, moisture in the gunny is maintained by sprinkling water 2–3 times a day over the hay cover.
Soil preparation	Navara rice is grown in well prepared fields or wetlands. This seed is sown directly in puddled plots.
Soil fertilization and improvement	The only specific condition for cultivation is that it must be organic (chemical-free).
Rice plant care	Farmers must adhere to organic practice while controlling weeds and pests.
Cultivation	Cultivation practices from sowing to harvest are like other rice varieties. Three crops can be grown during a year depending on the climatic condition. This rice variety can be grown under low land and upland situation. The seeds are directly sowed in the main field.
Harvest	Navara being a short duration crop is harvested manually in 60–90 days. Scriptures prescribe Navara rice to be kept at least for a year as paddy before conversion to rice for better results.
Processing n/a	

Source: Government of India (2007), Intellectual Property India (2004)

registration of new geographical indications, the count of authorized users for a registered GI is surprisingly minimal. In case of *Navara rice*, the scenario of authorized registration doesn't match with the volume of Navara rice produced and marketed from the state (Pant 2015).Over the last few years, the landscape of protection has visualised a host of changes to address the challenges faced by genuine producers towards authorised user registration. Streamling the procedures for both GI and authorized user registration along with reduction in registration and renewal fees for authorized user registration in recent years has motivated communities to protect their products under these laws. As a result, now the authorised user registration has slowly gained speed and the registered users of Navara is 33 as on July 11.2023. Those 33 farmers are considered as Navara authorised users, according to self-imposed quality monitoring mechanism. Producers who belong to the geographical region from where the unique good originates following same variety, similar practices, therefore complying with the GI specification, and using the name Navara to sell the goods are marked as Navara rice farmers (producers who has not completed user registration).

Certified GI farmers sell their produce to either millers or to consumers directly. The Malayalam month of *Karkadakam* (August–September) is the month during

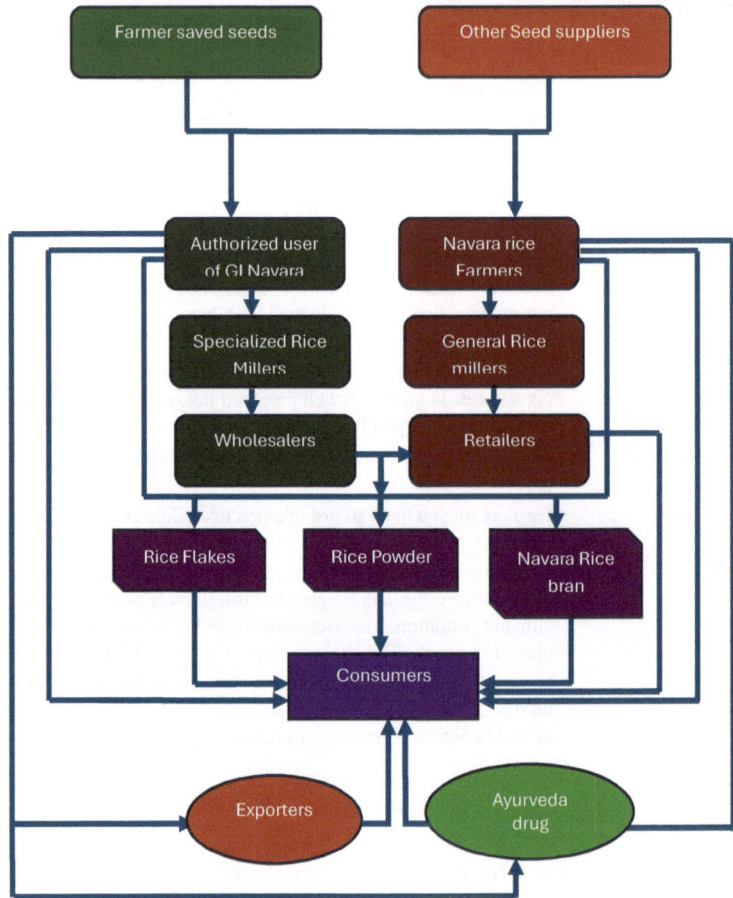

Fig. 22.7 Value chain for Navara rice in Kerala

which the consumption rises due to the medicinal properties. The rice is used both by the ayurvedic companies for their treatment purposes, (*Navara Kizhi and Navara Thepu*) as well as it is consumed directly by the consumers as porridge during the above month. It was estimated that about 15% of farmers process the product (using miller's services) and sell it directly to consumers. Other 85% is sold to the Ayurvedic pharmaceutical companies who are the main stakeholders in the value chain. Often, the certified farmers' groups will create their own brand and use a GI logo (Fig. 22.8) which is granted as a certification mark for all GI registered products in India (as a selling strategy. On the other hand, the small scale Navara farmers using traditional techniques and producing Navara rice for self or local consumption are marketing through general millers, wholesalers and retailers. They are deprived of this entitlement better price discovery as they lack direct access to the supply chain (Radhika and Raju 2021).

Fig. 22.8 Geographical Indication of India logo, registered as an Indian GI certification mark of Department for Promotion of Industry and Internal Trade (Source: Government of India 2019)

अतुल्य भारत की अमूल्य निधि

Invaluable Treasures of Increadible India

In the past 50 years, Navara rice has come to the verge of extinction. The vagaries of monsoon and pest attack are the major hurdles limiting the production of *Navara rice*. In 2022/23, the planting area of Navara rice in Kerala came down drastically to about 50 hectares with a production of about 25 tons (Regional Agricultural Research Station 2022). Despite investments in registration and marketing, GI has not yielded results for small and marginal farmers due to higher cost of production, lower yield, limited access to supply chain, lack of branding and lower profit margins. As organic certification is costly for small and marginal farmers, the absence of organic certification and branding is a common problem (Prabhu 2021). Given the limited no of completed user registrations, it becomes challenging to differentiate genuine producers from others. (Pant 2015); (Radhika et al. 2021b). The COVID-19 pandemic has also affected the supply of the crop drastically.

22.4 GI Impacts on Khao Sangyod Muang Phatthalung and Navara Rice

The motivation of both GI rice cases are to ensure the authenticity and quality of indigenous varieties specific to the origin and to provide a quality guarantee for consumers seeking high quality, traceable and traditional products. GIs are also important for local producers, helping them to use qualitative strategies and increase opportunities in existing and new markets. In Thailand case, the government plays a significant role in the registration and implementation of GI certification.

The rice variety itself is selected and registered as recommended variety by the Rice Department and sources of seed as one of the critical control points of the GI product demand to be from or approved by the Rice Department. In the case of *Nava rice* of India, the registry of Navara only mentions that the varieties are local to Navara and the sources of seed are not mandatory controlled. Furthermore, the registration of GI KSYMP was submitted by the public authority, namely the Rice Department. The production of seeds, the technical supports and other support programs are profoundly provided by the local government, and the GI inspection fee is also subsidized by the government The export promotion of KSYMP is evidently by the government as noticed by GI registration in foreign markets. On the other hand, the Indian government's support to *Navara rice* is insignificant from the registration to production and marketing of *Navara rice* products.

The volumes of certified products in both cases are very small although the rice is native and are commonly cultivated in the geographical locations (Table 22.3). Despite a strong government support to KSYMP in Thailand's case, the benefits to GI rice growers and retailers are not substantial as compared to India's case. KSYMP received very small price premium at the farm level (2%) and retail level (13%) while Navara received a significant price premium both at the farm level (50%) and at retail level (57%).

The commonality in both cases is that the production must be organic or at least meet GAP standard in Thailand's case. The reasons that Thailand GI rice does not significantly receive price premium could be because GAP is not well-recognized but is more common for GI certified products while organic rice of both cases received a significant price premium. India case of Navara rice also includes substantial element of local wisdom in cultivation practice specific to the location while Thailand case of KSYMP does not require any special practice except for the authenticity of the registered variety. As Thailand KSYMP has been registered as GI in the foreign markets, and with government supports to promote KSYMP in the international markets, the export markets are very small and almost unobserved. On the other hand, Navara rice has been exported to several countries mainly to the United Arab Emirates and other European countries. This could also be attributable to the demand for organic products and stronger reputation of India for indigenous rice which gives Navara rice more opportunity to reach high-end markets abroad.

Another similarity is the effort to add more value to the rice by processing into food (and non-food) The ayurvedic oil from authentic Navara husk is an example. This may not be specific to GI products as the registry of both cases do not cover processed products. However, as the rice became more reputable due to the GI registration, it gives the variety more recognition and can be a marketing strategy for the processed products using registered GI names.

22.5 Conclusion and Suggestions

As rice is the important product for Thailand and India, the GI protection gives the countries the opportunity to protect their products from international competitors. The GI systems in both countries are somewhat different but have a common ground

Table 22.3 GI system and its impacts on *Khkao Sangyod Muang Phatthalulng* and *Navara rice*

	Khao Sangyod Muang Phatthalung	Navara—The Grain of Kerala
Geographical area	Phatthalung	Kerala
Rice type	Gently fragrant	Medicinal
Technical practices		
Seed sources	Rice Department agencies in Phatthalung province or authorized suppliers	Unspecified
Plant protection measures	Organic or good agricultural practice methods	Organic method
Market situation		
Farmgate price (US$/kg)	0.38–0.64	2.4–4.81
Farmgate price premium	2.8%	50.1%
% of certified products	2%	Less than 1%
Retail price	85 THB/kg (2.70 US$/kg)	400 INR/kg (4.79 US$/kg)
Retail price premium	13%	57%
Value added products (yes/no) if yes, examples	Yes Cookies, noodles, coffee mix, cosmetics	Yes Rice flakes, rice powder, Navara rice bran
Export market (yes/no)	No record	Yes
Regulatory practices		
Registered proprietor	Rice Department, Ministry of Agriculture	Navara Rice farmers society, Chittur, Kerala
Government support for registration	Yes	No
Agency of registration	Dept of intellectual property	GI registry
Does technical specification include traditional knowledge/practice that are required for quality?	No, only specified variety and geographical area	Yes
Sustainability in terms of seed availability and quality?	Somewhat	Somewhat
Control system (self-, internal (group), external (third party)	Self, internal, external	Public vigilance

Note: Data from 2020

to recognize the importance of indigenous varieties and local knowledge especially local people more opportunity to gain benefits from the recognition of the quality products by the market. In India, local wisdom and the local people play more significant role in the control in the production of GI products while in Thailand, the government give a strong support to the GI system. The benefits as observed from the price premium and the potential to export market, however, is greater in the case of India.

This suggests that GI as collectively owned and managed by the GI group is effective in generating the benefits when the registration of GI products essentially come from the people who own the property right of the knowledge in producing them. The GI group who owns proprietary of the GI registry is the key element in the success of its existence which allows the proprietor to manage and play a crucial role in controlling the value chain stakeholders and reflecting collective efforts in protecting the knowledge and products. The benefits in practice can be more realized when the GI proprietors are passionately monitor and control the quality of the products to ensure the best and distinct quality comes from their geographical location—the true spirit of GI protection.

Acknowledgement Parts of the data and information in this chapter come from the research supported by Indian Council of Social Science Research and Kasetsart University Graduate School.

References

Cei L, Defrancesco E, Stefani G (2018) From geographical indications to rural development: a review of the economic effects of European Union policy. Sustainability (Switzerland) 10(10). https://doi.org/10.3390/su10103745

Department of Election of Kerala (2024) Kerala district map. Retrieved from http://webfile.ceo.kerala.gov.in/maps.html

Department of Intellectual Property (2006) Announcement of the Department of Intellectual Property Regarding registration of geographical indication "Khao Sangyod Muang Phatthalung". Registration number Sor Chor 49100011 (in Thai)

Department of Intellectual Property (2023) List of authorized users of the Thai Geographical Indication logo, S.C. 49100011, Khao Sangyod Mueang Phatthalung

Gopi G, Manjula M (2018) Speciality rice biodiversity of Kerala: need for incentivising conservation in the era of changing climate. Curr Sci 114(5):997–1006

Government of India (1999) The geographical indications of goods (Registration & Protection) Act, 1999. Retrieved from https://ipindia.gov.in/act-1999.htm

Government of India (2007) Geographical Indications Journal (No:17). Retrieved from https://search.ipindia.gov.in/GIRPublic/Application/ViewDocument

Government of India (2019) Public notice: guidelines for permitting the use of Geographical Indication (GI) Logo and Tagline-seeking stakeholders' comments-regarding. Retrieved from https://ipindia.gov.in/writereaddata/Portal/News/536_1_GI_Guideline__Finalised_..pdf

Intellectual Property India (2004) Geographical indications registry. Navara rice. Retrieved from https://search.ipindia.gov.in/GIRPublic/Application/Details/17

John KJ, Nizar MA (1998) Collection of rice germplasm from Malabar, Kerala. Indian J Plant Genet Resour 11(2):173–181

Kerala Agricultural University (2022) Annual progress report (2021–2022), Regional agricultural research station Pattambi, Unpublished manuscript

Kowsalya P, Sharanyakanth PS, Mahendran R (2022) Traditional rice varieties: a comprehensive review on its nutritional, medicinal, therapeutic and health benefit potential. J Food Compos Anal 114:104742

Manilal KS (1990) Ethnobotany of the rices of Malabar. In: Contribution to ethnobotany of India. Botanical Survey of India, pp 243–253

Napasintuwong O (2019) PGI Hom Mali Thung Kula Rong-Hai Rice in Thailand. In: Arfini F, Bellassen V (eds) Sustainability of European food quality schemes. Springer, pp 87–109. https://doi.org/10.1007/978-3-030-27508-2

National Bureau of Agricultural Commodity and Food Standard (2014) Thai agricultural standard TAS 4005-2014: geographical indication rice. Ministry of Agriculture and cooperatives (in Thai)

Petruang N, Napasintuwong O (2022) Economic sustainability of geographical indication indigenous rice: the case of Khao Sangyod Muang Phatthalung, Thailand. Asian J Agric Rural Dev 12(2):104–112. https://doi.org/10.55493/5005.v12i2.4467

Phatthalung Rice Research Center (2023) Area, output and number of households of Sangyod Rice production in Phatthalung Province. Expert interview. Unpublished data

Prabhu B (2021) Organic farming in Kerala. Int J Creat Res Thoughts 9(3):76–80

Radhika AM, Raju RK (2021) Rice Gis in Kerala: gap in desired and achieved outcomes. J Intellect Prop Rights 26:83–91

Radhika AM, Jesy TK, Rajesh KR (2021a) Geographical indications as a strategy for market enhancement-lessons from rice Gis in Kerala. J World Intellect Prop 24:221–236

Radhika AM, Thomas KJ, Raju RK (2021b) Geographical indications as a strategy for market enhancement-lessons from rice GIs in Kerala. J World Intellect Prop 24:221–236

Regional Office of Agricultural Economics 9 (2021) Khao Sangyod Muang Phatthalung development guidelines, Agricultural economic research no. 120. Office of Agricultural Economics, Ministry of Agriculture and Cooperatives. (in Thai)

Rice Department (2016) Sang Yod Phatthalung. Rice Knowledge Bank. Retrieved from https://www.ricethailand.go.th/rkb3/title-index.php-file=content.php&id=93.htm

Ruchi Pant (2015) Protecting and promoting traditional knowledge in India: what role for geographical indications? IIED working paper. Retrieved from International Institute for Environment and Development, London. https://www.iied.org/sites/default/files/pdfs/migrate/16576IIED.pdf

Single Document PGI Khao Sangyod Muang Phatthalung, European Union (2016) Publication of an application pursuant to Article 50(2)(a) of Regulation (EU) No 1151/2012 of the European Parliament and of the Council on quality schemes for agricultural products and foodstuffs (2016/C 188/08). EUR-Lex Access to European Union law. Retrieved from https://eur-lex.europa.eu/legal-content/EN/ALL/?uri=CELEX%3A52016XC0527%2803%29

Török Á, Jantyik L, Maró ZM, Moir HVJ (2020) Understanding the real-world impact of geographical indications: a critical review of the empirical economic literature. Sustainability (Switzerland) 12(22):1–24. https://doi.org/10.3390/su12229434

Unny PN (2023) Navara Rice—a success story. Trust for Advancement of Agricultural Sciences, New Delhi

Unny PN (2024) Photograph of the Navara variety, June 1, 2024

Variar PK (1985) The ayurvedic heritage of Kerala, ancient. Sci Life 1:54–64

The opinions expressed in this chapter are those of the author(s) and do not necessarily reflect the views of the [NameOfOrganization], its Board of Directors, or the countries they represent.

Part IV
Geographical Indications and Sustainability

Chapter 23
The Potential of Geographical Indications (GI) to Enhance Sustainable Development Goals (SDG) in Japan, with GI *Mishima Potato* as a Case Study

Junko Kimura and Cyrille Rigolot

Acronyms

GI	Geographical Indication
SDGs	Sustainable Development Goals
MAFF	Ministry of Agriculture, Forestry and Fisheries
EPA	Economic Partnership Agreement
MDGs	Millennium Development Goals
JA	Japan Agricultural Cooperatives
UN	United Nations

23.1 Introduction

Geographical indications (GIs) correspond to the labelling of products referring to their geographical origins. GIs have been described as a promising device to foster sustainable rural development (FAO-EBRD 2020). In both scientific literature and political discussions, two contrasted conceptions of GIs have been intensively debated: On the one hand, GI as a "trademark" correspond to a collective property, using a registered geographical name (US position); On the other hand, "sui generis" GI refers to a delimited and protected area, associated to a stronger state intervention

J. Kimura (✉)
Faculty of Business Administration, Hosei University, Tokyo, Japan
e-mail: kimura@hosei.ac.jp

C. Rigolot
France's National Research Institute for Agriculture, Food and Environment (INRAE), Clermont-Ferrand, France

© The Author(s) 2025
E. Vandecandelaere et al. (eds.), *Worldwide Perspectives on Geographical Indications*, https://doi.org/10.1007/978-3-031-71641-6_23

(UE position). GI are increasingly used worldwide in both developed and developing countries, variably referring to "trademark" or "sui generis" conceptions (Marie-Vivien and Biénabe 2017; Feuer 2019). The case of Japan is particularly interesting. After firstly adopting the "trademark" system in 2006, the country changed its perspective in 2015 to adopt a *"sui generis"* conception (Sekine and Bonanno 2017; Baumert 2019). Subsequently, the number of product certifications in Japan has been growing very fast (in January 2021, after five years and a half, 105 products are registered). Certified products in Japan have some specifies compared to other countries. Particularly, there is a high proportion of fresh non-processed products (kaki, pears, etc.) and bovine meats (nine certified products in January 2021) (Baumert 2019). This very fast expansion is partly due to a Japan–EU Economic Partnership Agreement (EPA) that came into force on February 2019, and which specified mutual protection of Geographical Indication (GI) products (Huysmans 2020; Morisaki and Suda 2017). These recent political changes can be interpreted has part of what Bestor (2014) calls Japan's "gastrodiplomacy", i.e., the efforts "to promote, protect, and prove the essence of culinary authenticity, internationally and domestically". From these different perspectives, Japan offers a unique case of a recent introduction of *sui generis* GI followed by a very fast expansion in a developed country, with a rich and specific food culture (Kimura 2015; Morisaki and Suda 2017).

Concomitantly with the inception of a *sui generis* GI system in Japan, 2015 was also the first time in history that world leaders unanimously agreed on a common vision for the future of humanity: The 2030 Agenda for Sustainable Development (UN 2020). After an exceptional deliberative process, the agenda articulated a universal and integrated plan for action, through a set of 17 Sustainable Development Goals (SDG) and 169 targets (Kimura 2015). The 17 SDGs integrate all three interrelated dimensions of sustainable development (economic, ecological, social): #1: No Poverty; #2: Zero Hunger; #3: Good Health and Well-being; #4: Quality Education; #5: Gender Equality; #6: Clean Water and Sanitation; #7: Affordable and Clean Energy; #8: Decent Work and Economic Growth; #9: Industry, Innovation and Infrastructure; #10: Reduced Inequality; #11: Sustainable Cities and Communities; #12: Responsible Consumption and Production; #13: Climate Action; #14: Life Below Water; #15: Life on Land; #16: Peace and Justice Strong Institutions; #17: Partnerships to achieve the Goal (UN 2020). SDGs were created to overcome the limitations of pre-existing Millennium Development Goals (MDGs), with some major inflections in the sustainability debates:

(1) The idea of the SDGs is to consider a global cooperation system, rather than the support from the North to the South that the MDGs advocated (130 out of 169 targets relate to developed countries); (2) SDGs try to address issues comprehensively. They are not a solution to one problem, but a system that responds to issues of economic, social, political, peace and security, environment, gender, etc., as a whole; (3) Civil society and corporate activities, not the state, have come to the fore as the main actors of the international community as expert groups and NGOs played an important role in the negotiation process of SDGs (Takahashi 2016).

To enhance the contribution of food systems to SDGs, Caron et al. (2018) insist on the key role of territorial approaches and the importance of "vibrant rural territories" (Caron et al. 2018). From this perspective, GIs seem to have an interesting potential for sustainability, as they are typically associated to a territorial approach (by definition). However, to date, most studies on the link between GIs and rural development have given emphasis to the economic dimensions. For example, in a worldwide comparative study in nine countries in four continents, FAO-EBRD (2020) show that GIs generate positive economic impacts in all nine case studies, in terms of price, income for producers and market access (FAO-EBRD 2020). These positive economic effects have been well explained in the literature by several factors, such as specific value-chain governance and institutional frameworks in GI systems (Belletti and Marescotti 2011; Barjolle and Jeanneaux 2012). In another recent important book with worldwide case studies, Bonanno et al. (2019) indicate that GIs might have rather mixed effects, depending on local contexts and many factors and drivers, especially when considering issues of social equity and power relationships (Bonanno et al. 2019). In both these references, the environmental dimension of sustainability is often considered as a background element, not directly assessed. Particularly, FAO-EBRD (2020) suggest that because of the localized resources, for long term viability, GI actors should rationally preserve the resources, which advocates the need for a sustainability strategy (FAO-EBRD 2020). However, as shown by Baritaux et al. (2016), the relationships between localized food systems and environmental performances is more complex as it may seem, depending on the multiple possible configurations of "ecological embeddedness" (Baritaux et al. 2016). FAO-EBRD (2020) conclude their study by stressing the importance of considering possible trade-offs between economic development, environmental preservation and social welfare, as a perspective (FAO-EBRD 2020). Recently, some GIs studies consider sustainability in a broader sense, such as in the European project Strenght2Food (Arfini and Bellassen 2019) and several papers in this special issue (Marescotti et al. 2020; Owen et al. 2020; Millet et al. 2020). To our knowledge, most of these studies are based in Europe, where GIs have been established for quite a long time (Arfini and Bellassen 2019; Marescotti et al. 2020; Owen et al. 2020). These European studies are performed in the context of a recent inflection in EU and national policies toward greener or more agroecological practices (i.e., "Farm to Fork", green deal, etc.), which gives an important role to food certification (Marescotti et al. 2020; Owen et al. 2020; Millet et al. 2020).

In Japan, the recent and rapid development of GIs has generated a significant academic interest and a number of reports and studies. Most of existing studies are based on economic and social perspectives, corresponding to a limited number of SDGs. We found no study considering the full set of SDGs. We present a survey performed in the *Mishima potato* GIs case study, taking explicitly the SDG as underlying analysis framework. Although limited in scope, this survey provides original insights on the potential positive contribution of GIs to SDGs in Japan. Finally, these insights are relativized in light of the existing literature, in order to identify some general perspectives.

23.2 Materials and Methods

Mishima Bareisho potato is a variety of May Queen produced in Mishima City and the Kannami-cho area of the Tagata District in Shizuoka Prefecture, which is located about 130 km south of Tokyo, at the western foot of Mount Fuji (Fig. 23.1). The potato is characterized by a "beautiful glossy surface with no skin scratches, a creamy texture which does not easily disintegrate like that of the May Queen, and soft, flaky mouthfeel with a sweet flavor like that of the Danshaku (baron) variety" (information website on Japan's geographical indications 2020, https://gi-act.maff. go.jp/en/outline.html). *Mishima Bareisho* has been granted the 18th GI products certification under the new *sui generis* system in Japan in 2016. There are several local human and natural factors contributing to *Mishima Bareisho* distinctiveness. For example, the potatoes are harvested carefully one by one so as not to damage the potato skin. They are then stored by drying in a cool, dark place with good

Fig. 23.1 Two GI products' (fish and potato) collaboration croquette. (Source: The Tagonoura Fishery Cooperative Association)

ventilation to mature for 1–2 weeks after harvesting. This gives the potatoes sweetness and a flakey texture, and a longer storage life. In regard to the natural factors, the volcanic soil in the production area has excellent water permeability, breathability and water retention, and the south-facing slopes of the fields in the area facilitate the long daylight hours and good drainage (Information Website on Japan's Geographical Indication 2020). In November 2017, the GI registration application group, JA (Fuji Izu), developed a GI collaboration croquette in partnership with another GI registration group, (The Tagonoura Fishery Cooperative Association, Tagonoura Shirasu (whitebait) GI), in the same prefecture. The fishery cooperative as producers' organization collaborated with the agricultural cooperative, producers' organization of Mishima potato, to develop a new product, although they are in different industrial sectors. The croquette is called "Tamiko-chan", (TA for TAgonoura shirasu, MI for MIshima potato, and KO for CROquette in Japanese). The croquettes is made by frying baby sardines and potatoes. This product generated confidence and pride among the local community. Generally speaking, collaboration between the agricultural and fishery sectors is difficult. This project, instead, was realized because the effect of the GI registration was to increase the motivation to utilize local GI products.

From 2016 to 2018, just before and after the GI product registration, one of the authors (Junko Kimura) conducted interviews and fieldworks based on participatory observation in the Mishima Bareisho area. In total, 19 stakeholders, who play important and active roles in food value chain of the GI, have been interviewed. These include (i) farmers; (ii) a manager of the fruit and vegetable local market; (iii) one processing company (which produces Mishima croquette); (iv) two retail stores; (v four local institutions (Japan Agricultural Cooperative (JA), the Mishima municipality, the Mishima Tourism Association, the Mishima Chamber of Commerce and Industry; (vi) four restaurants and cafés (including one bakery, one national chain café developing Mishima croquette burgers only for the store, and one restaurant using zero-kilometer vegetables)}. The Mishima local fruit and vegetables market is located in the middle of the production area. Potatoes are distributed to the processing company 9 km away in a plain field. Many of the cafés, retail stores, and HoReCa (hotels, restaurants and caterings) are located in the western part of the city near the Mishima train station. The farms and the fruit and vegetable market are located in the eastern rural area of the city, and the processing company, café, restaurants, and retail stores are located on the western urban side of the city (Fig. 23.1). Interviews were semi structured, with questions covering stakeholders' relationships and all aspects of *Mishima potato* and GI certification value creation by stakeholders' activities. The authors applied an interpretive approach, and all interviews were recorded and documented for interpretation. The relationships between stakeholders are presented in Fig. 23.2. Finally, the SDG framework has been used to identify all potential contributions of *Mishima potato* GI to sustainability of product origin. The potential contributions were divided into the production stage and the transformation and commercialization stages, and related to corresponding individual SDGs as proposed by the UN (2020).

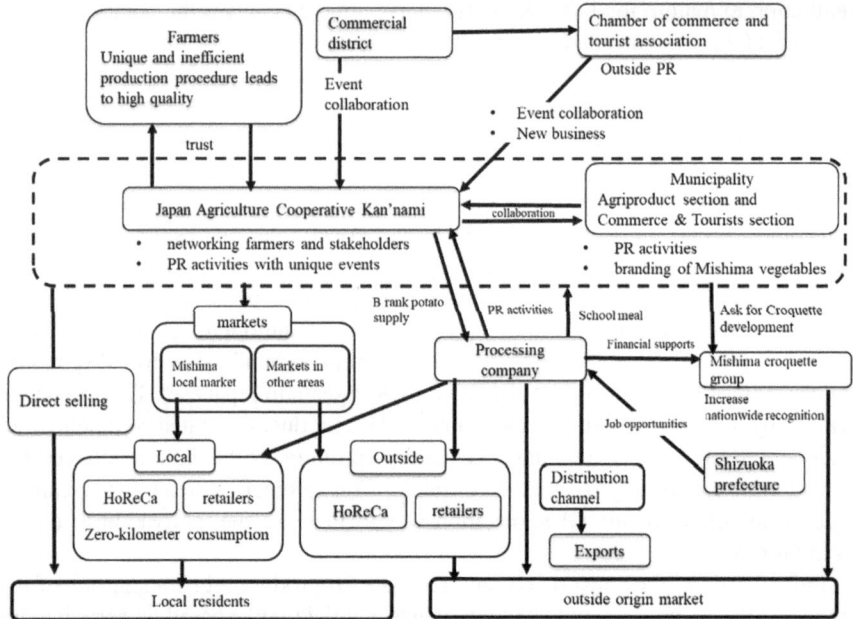

Fig. 23.2 Networks and interactions between stakeholders (from the authors based on interviews)

23.3 Results and Discussion: The Contribution of the Japan GI *Mishima Bareisho Potato* to SDGs

The contribution of the *Mishima potato* GI to sustainability is presented in Table 23.1 for both the production stage and for the transformation and commercialization stages. The table illustrates some potential contributions of GI to a diversity of Sustainable Development Goals (SDG), beyond those directly related to food (such as #2: Zero Hunger) or those considered in government reports (#8: De- cent Work and Economic Growth) or in some previous research studies (#10: Reduced inequalities). We identify original contributions to at least nine other SDGs, at both the production stage, and the transformation and commercialization stages. At the production stage (Table 23.1), stakeholders indicate that the *Mishima potato* contributes to the employment of women and disabled people). For example, NPO Libera Human Support supports disabled people with Asperger's, who are not good at communicating with others, grow potatoes in the field and are able to become economically and socially independent (SDG #1 and #5). In regard to the social dimension, education farms are used to promote the product to children as the next generation of consumers (SDG #4). Young farmers are forming the group "Nomins (literally means peasants)" and working together to make higher quality products by exchanging information and encouraging one another (SDG #8). As the *Mishima potato* is air dried in the production process, it has become a potato with higher

Table 23.1 Potential contribution of *Mishima potato* to SDGs at the production, transformation and commercialization stages

SDGs goals	Production stage	Transformation and commercialization stages
#1 No Poverty	Young generation newly become farmers	Local employment
	Disabled employment	
#3 Good health and well-being	Consumers more conscious on zero-kilometer vegetables and its freshness	Product development solving metabolic disease (non- fried croquette)
#4 Quality Education	Educational farm for the next generation (harvest experience)	School meal menu of GI product and children learn terroir-based products
#5 Gender Equality	GI registration encourage female farmers	
#7 Affordable and clean energy	Small environmental load in traditional production process including harvest by hands without using machines and dry by wind for 1–2 weeks	Zero kilometers due to local production for local consumption
#12 Responsible Consumption and production		Zero waste by utilizing B-class products
#8 Decent work and economic growth	Mutual cooperation in a young farmers' group NOMINS	Local tourism resources (National croquette contest)
	Highest may queen potato in Japan	
#15 Biodiversity	Motivate farmers to preserve local specialties	Land use in mountainous and cold regions
#17 Partnerships	JA works networking with different subjects and stakeholders (other GIs)	Buy raw materials from farmers without lowering prices

nutritional value, potentially useful for nutrition issues of the elderly (SDG #3). Agriculture in mountainous areas is hard work, but in Mishima, potato GI contributes to maintain young farmers, both those who inherited from fathers and those newly entered into agriculture. All lands are still cultivated, which maintains the landscape and sound soil (SDG #15), and prevents the invasion of wild boar and deer. Moreover, as potatoes are grown in mountainous areas, farmers cannot use machines but only their hands to harvest including digging the soil. This does not consume petroleum resources nor pollute the air (SDG #7 and #13).

At the transformation stage (Table 23.1), the processing company which produces Mishima croquettes employs local people. Croquettes must always be made from *Mishima potatoes*. Other ingredients can be decided freely by each croquette producer. Each croquette has a different taste and shape, and tourists can enjoy the differences between them. The role of JA as the *Mishima potato* Producers Association is to help producers sell low grade Mishima potatoes at a profitable price without discarding them. By purchasing lower grade potatoes without lowering the price from the farmers, it contributes to economical support for the farmers (SDG#1). The company retails their product to public schools for lunches and teaches food balance and safety while the products become familiar with children

(SDG#4). It has also developed new non-fried products to meet consumer needs (SDG#8). The local restaurant offers a menu with Mishma croquette and other local meats and vegetables. The National Croquette Festival held by the Mishima Croquette Association and the Mishima Croquette Contest held by the Mishima municipality are attracting tourists to visit the region. In regard to the environment dimension, interviewed stakeholders highlight that processing within the production area realizes "zero kilometers" and reduces the environmental impact. Moreover, processing lower grade potatoes and making croquette by processing company, bakery and restaurant have achieved zero waste (SDG #7 and #12).

23.4 Conclusion

To some extent, the *Mishima potato* can be seen as a relative "success story" in the context of Japan, where GI registration has fostered positive relationships between stakeholders. In our understanding, four key factors can explain this relative "success story": (1) Bottom-up endogenous development; (2) Collaborative activities and close networks within and outside the value-chains (for example, with the municipality); (3) Altruistic attitude; (4) Innovation and open mind. As a particularly interesting illustration, according to one interviewed stakeholder, "Tamiko-chan" croquette is the very first GI collaboration product in Japan (with the *Mishima potato* and *Tagonoura Shirasu* fish as raw materials). The Mishima case study could be inspirational for other GI products and more generally for GI development policy in Japan. After the adaption of the *sui generis* GI system in 2015, many products certifications have been in fact directly solicited by MAFF, following rather a "top-down" approach. A consequence is that the first GIs products have often been the "easier" to certify, such as raw or minimally processed products, rather than more traditional products, such as sake, tea or soy sauce (Baumert 2019). Moreover, MAFF emphasizes elimination of counterfeit products, expansion of transactions, increase of bearers and price increase. These are very important effects, but our study shows that GIs can contribute more generally to SDGs when following a terroir-oriented approach, as proposed by (Casabianca et al. 2005) in a European context. In order to develop terroir-oriented agriculture that can contribute to SDGs, it is necessary to create an organizational system in which all relevant parties involved in the product can cooperate and create value. In this perspective, the concept of "social capital" can be used as a theoretical framework to better understand the capacities of GI value-chains to adapt to sustainability challenges (Rigolot 2016). Particularly, it is essential to understand how trust between various actors can develop, particularly in regard to the trust of consumers toward food certification labels such as GIs (Rupprecht et al. 2020).

An obvious limitation of the proposed case study is that it focuses on the positive contributions of the *Mishima potato* to SDGs, from the perspective of local stakeholders. In order to avoid a criticism that the positive effects are not systematic, a quantitative survey would be conducted with stakeholders of different GI products to statistically and objectively measure both positive and negative effects and to generalize the conclusions. In fact, it is important to stress that the case study has to be considered in combination with the proposed overview of the effects of GIs inception in Japan, based on both official reports and academic studies, to get a fuller and more nuanced picture (Sekine and Bonanno 2017; Sekine 2019). Clearly, not all certified GIs in Japan contribute equally to SDGs, and it is important to consider also potentially negative contributions (to our knowledge, there is no important controversies or environmental issues in this case study). To get a fuller and more "objective" picture, other methodologies and frameworks could be used, like the SAFA methodology (Sustainability Assessment of Food and Agriculture Systems) developed by the FAO (2013) (Arfini and Bellassen 2019; FAO 2013).

As a conclusion, the fast expansion of *sui generis* GIs since 2015 can be seen as a significant and ambitious evolution in Japanese agricultural development policy. On the other hand, the observed "rush" in products' registration might also raise perplexity as regard the meaning of GIs in Japan and their sustainability outcomes. The *Mishima potato* case study illustrates how the close connections of GI products to their local environment (natural and socio-cultural) can translate into positive contributions to several SDGs. To enhance the potential of GIs for sustainable development (in synergy with other agricultural policies), the SDG framework can be considered as a useful tool, among others, to support decision making and to align local action with the context of global priorities.

Funding This work was supported by JSPS KAKENHI Grant Number 19H01544 and 19KT0014. The contribution of CR is partly integrated in ANR GIngKo Project.

References

Arfini F, Bellassen V (eds) (2019) Sustainability of European food quality schemes: multi-performance, structure, and governance of PDO, PGI, and organic agri-food systems. Springer, New York

Baritaux V, Houdart M, Boutonnet JP, Chazoule C, Corniaux C, Fleury P, Tourrand JF (2016) Ecological embeddedness in animal food systems (re-)localisation: a comparative analysis of initiatives in France, Morocco and Senegal. J Rural Stud 2016(43):13–26. https://doi.org/10.1016/j.jrurstud.2015.11.009

Barjolle D, Jeanneaux P (2012) Raising rivals' costs strategy and localised agro-food systems in Europe. 2019. Int J Food Syst Dyn 3:11–21

Baumert, N. Les indications géographiques alimentaires made in Japan (2019) Une nouvelle orientation géopolitique et une évolution des critères de définition de la qualité. Ebisu Études Jpn 56, 163–189

Belletti G, Marescotti A (2011) Monitoring and evaluating the effects of the protection of geographical indications. A methodological proposal. In: The effects of protecting geographical indications ways and means of their evaluation. Swiss Federal Institute of Intellectual Property, Bern, pp 31–122

Bestor T (2014) Most f (l) avored nation status: the gastrodiplomacy of Japan's global promotion of cuisine. Public Dipl Mag 11:57–61

Bonanno A, Sekine K, Feuer HN (eds) (2019) Geographical indication and global agri-food: development and democratization. Routledge, New York

Caron P, de Loma-Osorio GF, Nabarro D, Hainzelin E, Guillou M, Andersen I, Bwalya M (2018) Food systems for sustainable development: proposals for a profound four-part transformation. Agron Sustain Dev 2018(38):41. https://doi.org/10.1007/s13593-018-0519-1

Casabianca F, Sylvander B, Noel Y, Beranger C, Coulon JB, Roncin F (2005) Terroir et Typicité: Deux concepts clés des Appellations d'Origine Contrôlée, Essai de définitions scientifiques et opérationnelles. In: Colloque International de Restitution des Travaux de Recherches sur les Indications et Appellations D'origine Géographiques. INAO, Paris, pp 199–213

FAO (2013) SAFA. Sustainability Assessment of Food and Agriculture systems indicators. FAO, Rome. ISBN 978-92-5-108486-1

FAO-EBRD (2020) Strengthening sustainable food systems through geographical indications: evidence from 9 worldwide case studies. J Sustain Res 4

Feuer HN (2019) Geographical indications out of context and in vogue: the awkward embrace of European heritage agricultural protections in Asia. In: Geographical indication and global agri-food: development and democratization. Routledge, London, pp 39–53

Huysmans M (2020) Exporting protection: EU trade agreements, geographical indications, and gastronationalism. Rev Int Polit Econ:1–28

Information Website on Japan's Geographical Indication (2020) Available online https://gi-act.maff.go.jp/en/register/entry/18.html. Accessed 5 July 2020

Kimura J (2015) The act on protection of the names of specific agricultural, forestry and fishery products and foodstuffs (Geographical Indication (GI) Act). In: World food culture encyclopedia. Maruzen Publishing Company, Tokyo. (in Japanese)

Marescotti A, Quiñones-Ruiz XF, Edelmann H, Belletti G, Broscha K, Altenbuchner C, Scaramuzzi S (2020) Are protected geographical indications evolving due to environmentally related justifications? An analysis of amendments in the fruit and vegetable sector in the European Union. Sustainability 12:3571

Marie-Vivien D, Biénabe E (2017) The multifaceted role of the state in the protection of geographical indications: a worldwide review. World Dev 98:1–11

Millet M, Keast V, Gonano S, Casabianca F (2020) Product qualification as a means of identifying sustainability pathways for place-based agri-food systems: The case of the GI Corsican grapefruit (France). Sustainability 12:7148

Morisaki M, Suda F (2017) Patrimonialisation of foods and agriculture in Japan. In: Colloque SFER. Université de Reims Champagne Ardenne, Reims

Owen L, Udall D, Franklin A, Kneafsey M (2020) Place-based pathways to sustainability: Exploring alignment between geographical indications and the concept of agroecology territories in Wales. Sustainability 12:4890

Rigolot C (2016) The social capital of value chains: A key dimension of their adaptive capacities. Illustration with the "comet" cheese value chain. Cah Agric 25:45007

Rupprecht CD, Fujiyoshi L, McGreevy SR, Tayasu I (2020) Trust me? Consumer trust in expert information on food product labels. Food Chem Toxicol 137:111170

Sekine K (2019) The impact of geographical indications on the power relations between producers and agri-food corporations. In: Geographical indication and global agri-food: development and democratization. Routledge, London, pp 54–69

Sekine K, Bonanno A (2017) Geographical indication and resistance in global agri-food: the case of miso in Japan. In: Resistance to the neoliberal agri-food regime. Routledge, New York, pp 106–119

Takahashi K (2016) Ideological background and real challenges of sustainable development goals: development, environment and security. Interdiscip J World Peace Educ 2016(47):1–8. (in Japanese)

UN (2020) Transforming our world: The 2030 Agenda for Sustainable Development. 2015. Available online: http://www.un.org/ga/search/view_doc.asp?symbol=A/RES/70/1&Lang=E. Accessed on 10 June 2020

Chapter 24
The Teachings of the *Bouhezza Cheese* GI in Algeria Through the Perception of the Actors on the Economic, Social, Environmental and Cultural Effects

Samir Messaili

Acronyms

AB	Organic Agriculture
AO	Appellation of Origin
CNL	National Labelling Committee
DZA	Algerian Dinars.
GI	Geographical Indication
IMESSENDA	Association holder of GI Bouhezza cheese
INAPI	Algerian National Institute of Industrial Property.
INATAA	National Institute of Food, Nutrition and Agro-food Technologies.
MADR	Ministry of Agriculture and Rural Development
Outre	Goat skin specially prepared to serve as a container and maturing chamber for Bouhezza cheese.
VAOG	Appellation wines of guaranteed origin
Wilaya	Administrative division in Algeria (prefecture or region) composed of several municipalities.

24.1 Introduction

The recognition of the quality of agricultural products or of agricultural origin by acquiring distinctive quality signs such as the geographical indication (GI) is a recent process in Algeria. After the promulgation of law n° 08–16 on agricultural

S. Messaili (✉)
IMESSENDA association for the promotion and protection of the "Bouhezza cheese" dénomination, Oum El Bouaghi, Algeria

© The Author(s) 2025
E. Vandecandelaere et al. (eds.), *Worldwide Perspectives on Geographical Indications*, https://doi.org/10.1007/978-3-031-71641-6_24

orientation and publication of the executive decree 13–260 of 07 July 2013 setting the quality system of agricultural products or of agricultural origin; four quality signs are available in the Algerian system of recognition of the quality of agricultural products or of agricultural origin, namely the GI, the Appellation of Origin (AO), the organic agriculture (AB) and the agricultural quality labels.

It should also be noted that seven areas of Algeria have benefited from appellations Wines of Guaranteed Origin Appellation-VAOG within the framework of ordinance 76–65 of 16/07/1976 relating to appellations of origin, it concerns wines of origin; seven VAOGs have been registered (within the framework of the Lisbon Arrangement), namely: The *Dahra hills*, the *hills of Mascara*, the *hills of Tlemcen*, the *hills of Zaccar*, *Ain-Bessem Bouira*, the *Tessala Mountains* and *Medea*.

To date, three GIs are recognised by the Algerian Ministry of Agriculture: the *Deglet Nour date* of *Tolga* in the wilaya of Biskra in 2016, the *dried fig of Béni-Maouche* in the wilaya of Bejaia in 2016, and *Bouhezza cheese* in the wilaya of Oum el Bouaghi in 2020. Many other products are in the process of identification or dossier construction (for example, the *Sig table olive* in the wilaya of Mascara, the *Oulhassa white onion* in the wilaya of Ain Temouchent, and the *Mésserghine Clementine* in the wilaya of Oran, etc. No study has been conducted in Algeria on the effects of GI recognition, which limits the availability of data to justify their dissemination in Algeria. A collaboration was therefore envisaged with the association IMESSENDA, holder of the GI *Bouhezza cheese* to provide initial data.

This chapter will attempt to answer several questions related to the perceived effects of the registration of the GI *Bouhezza cheese*, in particular, those related to economic, social, cultural and environmental aspects.

24.2   The Process of Recognition of the GI *Bouhezza Cheese*

The *"Bouhezza cheese"* is a traditional fermented cheese whose manufacture is typical from the regions of Eastern Algeria (Oum el Bouaghi, Khenchela, Batna, Tebessa) (see Picture 24.1), once known for a significant practice of extensive goat and sheep farming. Indeed, originally, the "Bouhezza" was the product of the transformation of goat and sheep milk. However the current trend seems to be moving towards the use of cow's milk.

The name Bouhezza of the cheese means according to the testimonies collected from the elders, the fact that the skin, containing the *Bouhezza cheese* is always hanging (traditionally on a wooden tripod) (see Picture 24.2) and the word HEZ or MEHZOUZ which means "lifted", hence the name Bouhezza. *Bouhezza cheese* is widely known in the area, especially in Oum el Bouaghi, where it is widely consumed and marketed. It is a product that is loved to be offered in rural families and that people come to get even from abroad.

The association IMESSENDA (which takes its name from the Amazigh word IMESSENDA which means tripod), was specially constituted in 2017 to organise the producers of this cheese and carry the application file for registration of the GI with the national labelling committee (CNL) at the Algerian Ministry of Agriculture

Picture 24.1 *Outre* hanging on the traditional tripod (Location Sidi-R'ghiss Mountain in Oum El Bouaghi). (Source: author)

Picture 24.2 *Bouhezza cheese* Outres at different stages of maturation. (Source: author)

and Rural Development. Composed of 18 members (cheese artisans, cattle and sheep breeders, academics…) including four women, it has carried the labelling process as a voluntary action in close collaboration with the University of Constantine. The collaboration began in 2013, the product specifications were drawn up to be submitted to the national labelling committee; it particularly records:

- The applicant of the sign and its representativeness and its eligibility.
- A detailed description of the product.
- Its method of obtaining.
- The geographical area of its production.
- The link to the terroir.
- The control plan.

The GI application was submitted to the CNL in February 2018; the CNL appointed a specialised sub-committee composed of academics, specialised technical institutes, members of the association, to study and amend the specifications before submitting a final version to the CNL for deliberation. For *Bouhezza cheese*, the GI was awarded by the CNL on 06 December 2019, then registered in the National Register of Recognition of the quality of Agricultural products or of Agricultural origin of the MADR (Ministerial Order of 30 August 2020 granting the distinctive sign "geographical indication" for the recognition of the quality of the agricultural product "*Bouhezza Cheese*"). The registration at the Algerian National Institute of Industrial Property (INAPI) was carried out on 25 January 2021.

24.3   Materials and Methods

The study on the perception of effects was carried out among the members of the IMESSENDA association during the month of December 2021 in Oum El Bouaghi. The method adopted is the "focus group". The panel is a group of ten (10) people, corresponding to two thirds of the members of the association), the cheese producers and the preparers of skins for the making of wineskins from different localities of the geographical area. The technique used is the "focus group by questioning". During the semi-directive interview of 3 h, the panel answered a series of questions previously recorded in an "interview guide". The focus group followed the following steps:

1. Installation, presentation and introduction.
2. Interview guide.
3. Animation, relaunch and moderation of debates.
4. Establishment of the group sheet.
5. Analysis of results.

Five themes or axes were submitted to the group in the form of questions, namely:

- The registration of the GI.
- The future missions of the association after the registration of the GI, especially those related to the management of the Label.
- The commercial aspects of the products and the economic fallout.
- The reputation of the association and its members and the cheese since the registration of the GI.
- Sustainable development and environmental aspects after registration of the GI. A group sheet was produced after the closure of the "focus group".

24.4 Results of the Study

For each of the axes, the group's responses were synthesised in Table 24.1, and the following analysis can be made.

The analysis of the results leads to highlight important aspects in terms of rural development.

24.5 Importance of the Registration of the GI *"Bouhezza Cheese"*

The acquisition of the quality sign GI of *Bouhezza cheese* was a process:

- Long (about 5 years);
- Technical; which required a lot of skills from the association and its partners (especially university);
- Costly; in promotion, events, missions and others...

But the satisfaction of success outweighs everything. The opinions of members who participated in the survey are unanimous on this aspect. The recognition of the quality of the cheese as an Algerian product of the terroir and its labelling as GI by the state and its registration by the INAPI and its protection is a great satisfaction and a great honour.

The task was laborious for this group of modest producers to manage the different aspects of the registration procedure and raise the necessary funds (the association did not benefit from public funds).

This saga was enriching (unanimity of responses) and for many survey participants, they are ready to start another registration procedure for another product (the association plans to register another traditional cheese called "MEDGHISSA".

Table 24.1 Summary of focus group questions

Themes	Questions	Answers
1. Registration of the GI "*Bouhezza cheese*"	Are you satisfied with the registration of the GI?	Yes, unanimous personal and collective satisfaction, because the challenge was significant and the cheese deserved recognition and protection.
	Was it easy?	No, it was difficult and very technical and costly. The help of the team from the University of Constantine was very useful as well as the financial contributions from the association's members and sponsors.
	If you had to do it again, would you?	Yes, because other cheeses that we produce deserve recognition (like MEDGHISSA cheese).
2. Future missions of the association after registration of the GI, especially those related to the management of the Label.	How do you apprehend this change of missions?	With concern, given the lack of training and experience.
	Are you capable of managing this mission?	Yes, with assistance and learning, as it is a necessity for the protection and sustainability of the label.
3. Commercial aspects of the products and the economic fallout.	Has the price of your cheese increased?	Yes, the price of Kilogram of *Bouhezza cheese* has increased more than 2.7 times on average: from 400 to 500 DZA (2.6 to 3.3 €) before GI to 1000 to 1500 DZA (6.7 to 10 €) after GI.
	Has the demand increased?	Yes the demand has increased in volumes, before the GI, the cheese maker launched the production of a cheese bag per month at most, after the GI, several bags are launched each month.
	Have you acquired new customers?	New customers and new territories have been acquired gradually (especially the central and western provinces of the country, which are discovering the product).
	Are you considering more investments?	Yes, investments are essential (infrastructure, packaging, training...)

(continued)

Table 24.1 (continued)

Themes	Questions	Answers
4. Reputation of the association and its members and the cheese since the registration of the GI.	Do you feel a change in the appreciation of your surroundings for what you do?	Collective positive observation on the social status acquired from the surroundings (esteem, encouragement...), the cheese that was falling into oblivion comes back to the forefront and becomes emblematic for the region.
	Do you feel an identity appropriation from your surroundings for what you have managed to do?	The appropriation of success, the promotion of the product and the region are especially visible during cultural events.
5. Sustainable development and environmental aspects after registration of the GI.	Since the registration of your GI, what does environment and sustainable development mean to you and why?	The group members are predominantly more sensitive to the preservation of the environment and the rational use of natural resources, the history of their product and the traditional practices they preserve are a testament to this respect for the environment and the rational use of resources.

24.5.1 Future Missions of the Association After Registration of the GI, Especially Those Related to the Management of the Label

The Ministerial Decree of 30 August 2020, awarding the distinctive "geographical indication" sign for the recognition of the quality of the agricultural product "*Bouhezza Cheese*", has, among other things, mandated the association to manage the GI, its new missions for the association in addition to those provided for by its approval of constitution raise some fears among the members surveyed in relation to the lack of experience and skills in the field, and have expressed the need for external assistance in training or assistance.

24.5.2 Commercial Aspects of Products and Economic Impacts

It is indisputable for the surveyed group that the GI *Bouhezza cheese* and the media coverage that accompanied it, benefits the cheese producers and, has a direct impact on the price of the cheese, the new markets and on demand.

Cheese lovers especially, and consumers of local and typical products in general are the first demanding customers (knowledgeable consumers who are ready to pay the fair price and with regular consumed quantities) and also the new consumers who discover the cheese, especially in the centre and west of the country.

Thus, and in this perspective, a group of producers organised themselves into a traditional cheese-making cooperative in 2022 to meet this demand and make the cheese available to large-scale distribution.

24.5.3   Reputation of the Association and Its Members and the Cheese Since the GI Registration

The group unanimously agrees that the GI has improved the personal and collective social status of the cheesemakers and that many economic and cultural activities and events solicit and associate the *Bouhezza cheese* and the association during local and national thematic events (annual *Bouhezza cheese* festival in Oum el Bouaghi, international exhibition of agricultural productions and services, international exhibition of agri-food sciences, television reports and programmes...)

The association and its members, are privileged partners in all local and national thematic events on local products and are often asked to testify during national days (ex, like the day of 12 January, called YENAR Amazigh New Year's Day).

24.5.3.1   Sustainable Development and Environmental Aspects After GI Registration

Production according to the specification which are subject to self-checks, internal checks and external checks, in addition to the concern to do well and the duty to preserve and protect resources and to present to the consumer a good and authentic product, make producers more responsible for issues of resource sustainability, product traceability, intellectual and industrial property rights and also environmental aspects related to waste management.

24.6   Conclusion

The study has allowed us to answer the questions we asked ourselves about the effects of the registration of the GI *Bouhezza cheese*, as perceived and testified by the producer members of the association carrying the approach. The financial benefits (increase in the price of cheese, increase in the volumes sold and the number of customers and acquisition of new territories) are confirmed, the need to invest for better work is confirmed. The appropriation by the population of the region of the recognition of the cheese *Bouhezza* under GI and the status attributed to the cheese and the association are also confirmed, they are highlighted during all thematic events (economic, cultural, scientific...). The members of the association are more sensitive to environmental problems and notions of sustainable development. The labelling process being recent in Algeria, more in-depth studies must be carried out by researchers on labelling and its short, medium and long term effects.

This study is a first on the subject in Algeria. It remains limited, both in time and space, it was carried out within a young association of 18 members and on a product newly labelled, but it is surely the door open to researchers to study the valorisation approaches of local products through quality distinctive signs including GIs.

Acknowledgements I would like to thank on my personal behalf as well as on behalf of my colleagues from the *IMESSENDA* association for the promotion and protection of the *Bouhezza cheese* denomination, all those who have supported the association in the labelling process of its cheese and who contribute to the promotion of Algerian local products; especially;

To the members of the association *IMESSENDA* for the promotion and protection of the *Bouhezza cheese* denomination *Bouhezza* from Oum El Bouaghi. Algeria for their efforts and perseverance in promoting and protecting the *Bouhezza cheese Bouhezza* since 2014.

To the researchers and students of the National Institute of Food, Nutrition and Agri-food Technologies of the University of Constantine, Institute of Nutrition, Food and Agri-food Technologies (INATAA). for their coordination efforts and support to the *IMESSENDA* association.

Chapter 25
Sustainability Strategy for GIs; A Bottom-Up and Participatory Approach for GI Sustainability

Emilie Vandecandelaere, Luis F. Samper, Florence Tartanac, and Massimo Vittori

Abbreviations

GI	Geographical indications
GRI	Global Report Initiative
IP	Intellectual Property
SAFA	Sustainability Assessment of Food and Agriculture Systems
SDG	Sustainable Development Goal
SSGI	Sustainability Strategy for GI
UNCTAD	United Nations Conference on Trade and Development

25.1 Introduction: The Need for a Tool in the Hands of Producers

The 2030 Agenda for Sustainable Development makes an ambitious and crucial call for the transformation of agrifood systems. The United Nations Food Systems Summit of 2021 recognized the importance of territorial approaches to achieve this transformation. One of these approaches is the development and implementation

E. Vandecandelaere (✉) · F. Tartanac
Food and Agriculture Organization of the United Nations (FAO), Rome, Italy
e-mail: emilie.vandecandelaere@fao.org

L. F. Samper
4.0 Brands, Bogota, Colombia

M. Vittori
oriGIn, Geneva, Switzerland

E. Vandecandelaere et al. (eds.), *Worldwide Perspectives on Geographical Indications*, https://doi.org/10.1007/978-3-031-71641-6_25

of Geographical Indications (GIs).[1] As highlighted in previous publications (FAO 2009; FAO and European Bank for Reconstruction and Development (EBRD) 2018), GI processes can enhance sustainability, understood as the combination of three pillars (economic, environmental and social sustainability). In particular, GI systems can increase incomes for all actors in local value chains, including small-scale producers (farmers and processors), thus improving their livelihoods and boosting their resilience. In addition, increased incomes allow GI producers to invest in GI systems to strengthen their performances in these three pillars.

However, the literature on GIs is not unanimous about their positive impacts and controversial cases might also exist when specific sub-categories of any individual pillar are emphasized. Specific performance may depend indeed on the conditions of establishment and management of the GIs (FAO-EBRD 2018). In addition, in many cases GI producers themselves may not be aware of the impacts of their system on the local economy, environment and community, nor on the potential of their GI as a tool to boost local sustainability.

From the other extreme of value-chains, sustainability is in many cases a market access condition of downstream market actors and civil society. Sustainability is therefore a key product attribute to consider (Samper, Quiñones, 2017). This demand driven sustainability has led to many frameworks and standards proposed, if not imposed, to upstream segments to document conformity with these standards. In addition, the public sector is also becoming more sensitive to the need to enhance sustainability in general, and in the agrifood sector in particular, to address important global challenges of the 2030 Agenda and specific challenges such as climate change, imposing additional requirements in destination markets.

The visions and approaches at global or national levels show important limits in practice. Providing general rules or requisites can't specifically address the particularities of different territorial sustainability issues and may not contribute to a true engagement of local communities. In the case of GI systems, evaluations built on general frameworks can also miss their benefits and special contributions to different aspects of sustainability compared to other standards or approaches.

The objective of this chapter is to present an original framework jointly developed by FAO and oriGIn, the Sustainability Strategy for GI (SSGI), which aims at supporting GI producers through participative process and alliances to develop and implement a tailored strategy to improve the sustainability of their GI systems.

---

[1] Geographical indications are place names (in some countries also words associated with a place) used to identify the origin and quality, reputation or other characteristics of products. They can become collective intellectual property instruments, always linked to specific territories of origin.

25.2   GI Sustainability and the Role of GI Organizations

By definition GIs are not sustainable food production systems and should not be viewed as a sustainability standard. However, the processes for their establishment and management can provide the ground to preserve the local resources that make GI products unique and special, while adding values to those involved in their production and commercialization (FAO-EBRD 2018). GI systems can also become collective intellectual property rights, jointly managed by GI producers (i.e. farmers and processors involved in the production of GI products) and/or other institutions. In the methodology of the virtuous origin-quality circle (FAO 2009), a GI is conceived as a tool for local sustainability as far as the local GI community -in particular GI producers- are at the center of decision making and are involved in the establishment, management and evaluation of their GIs system, being ready to adjust their self-imposed rules and modalities as necessary. From this viewpoint, GIs can be used both as a local development and sustainability tool for their territories.

A key condition of success is to provide GI producers with the necessary empowerment to navigate their sustainability challenges in four dimensions: economic, social, environmental and governance (FAO 2023). The nature of the GI product and its link to a specific origin represents a driver to address and measure progress on the challenges associated with each of the first three (usual) pillars of sustainability. As GI systems are intrinsically attached to a specific territory and are collective in nature, these challenges must resonate with local communities where GI producers are located and, at the same time, the collective initiatives and actions undertaken to confront them may have a larger impact than addressing them through individual efforts.

Collective efforts, however, require collective decisions and their implementation, hence the SSGI is considering a fourth pillar on governance as used in the Sustainability Assessment of Food and Agriculture Systems (SAFA) framework (FAO 2014). Maintaining an effective governance of GI systems and their ability to build alliances requires credibility and legitimacy among internal and external stakeholders.

Thus, GI systems sustainability performance needs to be assessed and monitored under four pillars:

- the economic pillar is positively impacted by Intellectual Property (IP) protection through GIs. Added value generation and distribution along the value chain must positively impact local actors, and GI coordination generates positive effects associated with GI protection related to market positioning or complementary activities in the territory;
- for the social pillar, the importance of human resources is highlighted together with local know how and the way they are recognized in the GI product specifications (e.g. specific role of certain producer categories, self-esteem...) that can lead to the preservation and promotion of local social capital;

- similarly, the identification of the roles of the <u>natural resources</u> for the specific quality of the GI product and the need to contribute to the preservation of the local environment (as the production can't delocalize, the resources have to be preserved);
- <u>governance</u> is at the heart of any GI organization that represents collective GI producer interests. The local resources and public (reputation, landscape, biodiversity, local development, etc.) goods implicit in any GI system must be managed with a collective view; the good practices for sustainability, when collectively applied, maximize scaling up and sustainability impact at the territory level.

It is now clear that competition alone is not enough to achieve local development nor to address the need to efficiently handle finite public resources and collective reputation. The implicit "tragedy of the Commons" (Hardin 1968) in the scramble for finite collective resources associated with competitive behavior needs to be addressed with collective rules that, without restricting competition to rule-abiding GI producers, enforce common rules that favor the conservation of resources and the benefits of collective action. Local governance and regulation principles allow indeed for a sustainable use of limited common resources (Ostrom 2015).GI systems have often been recognized as fertile for co-opetition (Bengtsson and Kock 2000; Dentoni et al. 2013), the combination of healthy competition between firms combined with fruitful cooperation around collective assets which are numerous in the case of GI systems, and include at first the name and reputation. This "cooperative" environment can be more easily attained if developed collective rules are relevant to the local context. From this viewpoint, GI organizations should facilitate cooperation among members and other actors of the territory, which are particularly relevant and necessary when dealing with complex issue sustainability topics.

Cooperation is therefore crucial to address sustainability in evolving contexts. Credible and representative GI producer organizations are in a strong position to develop alliances, which are essential to tackle challenges that go beyond their financial capacity. Also, in an increasingly complex world, challenging sustainability issues may need to be addressed through science and innovation initiatives that also require enlisting different types of allies.

The FAO and oriGIn collaboration, which started in 2017, recognizes these collective sustainability challenges and seeks to provide guidelines and tools to GI organizations to identify their local sustainability challenges through a bottom-up approach that can empower producers in the evolving sustainability conversation. This main aim is achieved through a guided methodology that allows GI organizations and GI producers to elaborate and implement a strategy to improve their sustainability performance in all four dimensions by identifying stakeholders and potential allies with common interests.

25.3 SSGI Objectives and Approach

The SSGI objectives are twofold: enhance the levels of sustainability in the GI local systems with proper monitoring, and empower GI organizations and producers in the development of a tailored sustainability roadmap that can give ground to relevant alliances. Consequently, important principles and operational objectives are structured within the SSGI framework (Vandecandelaere et al. 2021) which are here summarized:

- an adapted approach to reflect GI contribution to sustainability:

    - the nature of GI has to be well reflected in the approach, through an approach which is voluntary, participative and a place-based;
    - identification of all relevant indicators to cover specific topics (not always found in generic sustainability frameworks) for example the redistribution of added value along the value-chain, the contribution to landscape, the cultural identity, and key collective governance elements, considered as the fourth sustainability pillar;

- a credible approach allowing reporting:

    - review of existing frameworks at global and sectorial levels (86 sources of which 36 have fed directly the database), to identify and characterize (or adapt) relevant indicators for GIs, and allowing reporting with references to other important frameworks(e.g. SDGs, GRI...), As part of this effort new indicators have also been included in a 442 strong indicator database from which GI organizations can choose from for benchmarking and monitoring purposes (Vandecandelaere et al. 2021);
    - various levels of peer reviews and expert contributions: a task force of scientific and practitioners including GI organization representatives has followed the development of the SSGI framework from its beginning; the database of sustainability indicators relevant for GIs and the guide have been carefully reviewed by a group of experts with different backgrounds;

- empowerment of GI groups and organizations in the global and their local contexts:

    - providing guidance and concrete tools for any GI organizations (whatever the capacities and previous experience) to develop its own sustainability roadmap;
    - providing guidance in the identification of specific issues and vulnerabilities and the definition of priorities together with key external stakeholders;
    - recognizing influent economic players by working with concepts, topic and metric frameworks consistent with those they use (SDG's, GRI, UNCTAD-FAO, etc.);
    - increasing capacities to provide information required to access new markets or comply with evolving regulations, to better communicate and report on progress made;

– enhancing their legitimacy and relevance in the territories and industries by creating opportunities for dialogue and the creation of alliances for sustainability with the public and private sectors.

In the SSGI approach, each GI organization is accompanied to develop its own sustainability roadmap, in a way that it also represents a roadmap for alliances. Although the approach is primarily designed for GI producer groups, it can be used by any groups engaged in the promotion and representation of a specific product, and can be useful for producers interested in the registration of a GI.

25.4 The Process Towards a GI Sustainability Roadmap

25.4.1 The Roadmap Designing Actors

Most GI organizations have a top decision-making body (a GI board or board of directors), as well as a collective body that represents the organization's members, such as a general assembly of GI producers. The GI board is therefore the most appropriate and efficient decision-making body to overview the process of developing a GI sustainability roadmap, throughout the necessary frequent consultations with the members. The key person to execute all the activities necessary is called the GI practitioner and is appointed by the GI board. He/she can be a GI board member, a staff member, an external consultant or similar figure reporting to the board. The GI practitioner does not need to be a sustainability expert but to allocate the necessary time to leading the process and ensuring stakeholders' full cooperation.

25.4.2 Governance and Engagement

The SSGI builds on two crucial concepts. The first is governance, which is an important aspect of any organization. Good governance provides credibility with consumers and regulators and helps build and maintain alliances. Credibility is based on many factors, including the organization legitimacy, transparency or participation, which allows it to make decisions that are accepted by stakeholders. As most GI organizations around the world have rather limited resources, their institutional strength is their main asset. Governance topics are therefore particularly important for GI systems; they are the basis of their credibility and their ability to build long-term partnerships. Since governance is a key dimension for GI system sustainability, some governance topics that provide credibility and legitimacy of GI systems are compulsory in the selection of priorities by the GI organization.

The other key concept is engagement. GI organizations can't identify priorities in all sustainability pillars nor implement significant actions to enhance the sustainability performance by themselves. They may need to partner with various local, regional national or even international stakeholders. Engaging with key stakeholders and understanding their own priorities is a critical part of the process of developing a GI sustainability roadmap. Engagement is the process by which an organization identifies and selects stakeholders who have a significant interest or are strategically aligned with the organization's sustainability goals. Once selected, stakeholders are engaged by soliciting their views and analysing their challenges concerning the GI system (stakeholder consultation); then, stakeholders should be kept involved in the formulation and implementation of initiatives and actions that address common priorities (cooperation, alliance building or implementation engagement).

25.4.3   The SSGI Practical Guidance Along: Eight Steps and Three Phases

The GI practitioner is guided all along the journey through a detailed guide and its associated toolkit (an excel file containing programs) that provide all necessary inputs to facilitate the identification, selection and reporting processes.

The step-wise process is organized in three phases: prioritization, assessment and improvement. These phases are organized in 8 steps (see Table 25.1). Each step leads to a specific output which is supported in the guidelines with detailed information, examples, templates. The associated toolkit also facilitates the work, presentations and reporting for GI practitioners, by providing tables to fill with the information collected, automatic programs to generate the results in reporting tables and radar graphs.

25.4.4   Timing

The prioritization process can be expected to take approximately 8–12 weeks, assuming that the GI board is available to review and approve certain steps, information can be obtained when required and GI stakeholders can be consulted promptly. This period includes the time devoted to stakeholder research before interviews, and to the analysis of the different sustainability reports from the industry or value chain actors, territory development plans, etc. Well-prepared prioritization interviews with key stakeholders can take up to 2 hours. Depending on the number of GI producers involved, additional time may be required for the communication and validation processes.

The expected duration of the assessment phase will depend mostly on the availability of information and the indicators selected. If the information is readily available, it can take 4 weeks to build the baseline, with an additional 2 weeks to share the results with key stakeholders.

Table 25.1 The 3 phases and 8 steps

Phase	Steps	Activities
PRIORITIZATION	1. Preparing for the engagement of stakeholders	Understanding the specific GI system Developing a stakeholder consultation plan
	2. Conducting stakeholder consultations	Familiarization with the SSGI framework for sustainability topics Preparing the engagement of all stakeholder groups Implementing the consultation plan for internal stakeholders Implementing the consultation plan for external stakeholders
	3. Prioritization of topics	Selecting a target number of GI system sustainability topics Analysis of the results of consultation process by the GI practitioner Selection of priority topics by the GI board *Note: Use of the SSGI framework by organizations that have already engaged in a prioritization exercise*
	Communicating with and engaging GI stakeholders	The importance of communicating about the prioritization process Communicating with internal stakeholders Communicating with external stakeholders
ASSESSMENT	4. Selection of the GI sustainability indicators	Preparing the list of relevant GI sustainability indicators. Assessing the feasibility of indicators. Ensuring a balanced selection of indicators. Validation of the list of preselected indicators by the GI board.
	5. Monitoring frequency and designing responsible actors	Organizing the sources of information and methodology. Establishing the baseline assessment plan. Calculating the baseline. Assessing performance and gap analysis.
	Communication	Communication to stakeholders and possible allies to prepare for improvement.
IMPROVEMENT	6. Developing the improvement plan	Assessing existing initiatives and identifying possible roles for the GI organization Drafting an improvement plan Engaging with stakeholders towards improvement

(continued)

Table 25.1 (continued)

Phase	Steps	Activities
	7. Implementation of the improvement plan: Iterative monitoring and evaluation	Monitoring and evaluation Reporting A continuous pathway: Evolution of the roadmap along the way
	8. Communicating to ensure continuous engagement	Communication basics Internal communication External communication

Building the improvement phase (not the implementation as such) requires the active participation of the GI board and the staff members of GI organizations. Determining gaps and defining goals may take up to 4 weeks, including validation of proposals. For some priorities (in particular governance priorities and other priorities led by the GI organization itself), an improvement plan can be defined within a few weeks. For those priorities where the GI organization chooses to partner with other actors, improvement plans with concrete sustainability initiatives may take longer to develop. Actions may include active and structured dialogues with potential allies, government actors and other GI stakeholders. Generally, 12 to 16 weeks should be dedicated to defining initiatives and plans for most topics. Regular evaluation and communication throughout the implementation of the roadmap is highly recommended.

25.5 Conclusion

The SSGI framework was developed with the involvement of many experts during several years and led to the publication of a guide and its related toolkit including a database: *"Developing a roadmap towards increased sustainability in geographical indication systems; Practical guidelines for producer organizations to identify priorities, assess performance and improve the sustainability of their geographical indication systems"* (FAO & oriGIn, 2024).

Then the SSGI provides an innovative approach to sustainability assessment and targeted activities, due to a number of its characteristics:

- it is a framework designed for actions and alliances,
- designed for a place-based approach, which ensure local issues are addressed by the relevant actors in the community,
- thereby contributing to empowering local actors, in particular GI producers and their organizations.

This approach can be used by other types of actors, including local authorities, trade associations, NGO and academics, to perform sustainability assessment based on existing priorities or priorities to be identified with a participative approach. Other types of actors that can use this framework can be producer associations, cooperatives or other types of producer groups, or trade / sector association, at regional, national or even international level. Beneficiaries can also share experience and results of their sustainability journey with FAO and oriGIn. This will help generate more knowledge in terms of best practices as well as recommendations and indicators targeted to specific GI sectors.

For more information:

The guidelines and toolkit can be downloaded on FAO or oriGIn websites at: https://
openknowledge.fao.org/items/2c234a56-163c-4006-85bc-e23ce2d328a5
www.origin-gi.com/web_articles/sustainability

References

Bengtsson M, Kock S (2000) "Coopetition" in business networks. To cooperate and compete simultaneously. Ind Mark Manag 29:411–426

Dentoni D, Tonsor GT, Calantone R, Peterson HC (2013) Brand coopetition with geographical indications: which information does lead to brand differentiation? New Medit 12(4):14–27

FAO (2014) SAFA sustainability assessment of food and agriculture systems guidelines version 3.0. FAO, Rome

FAO (2009) Linking people, places and products. A guide for promoting quality linked to geographical origin and sustainable geographical indications, by E. Vandecandelaere, F. Arfini, G. Belletti & A. Marescotti. Rome.

FAO (2023) Using geographical indications to improve sustainability—lessons learned from 15 years of FAO work on geographical indications, Rome. https://doi.org/10.4060/cc3891en

FAO & oriGIn (2024) Developing a roadmap towards increased sustainability in geographical indication systems—Practical guidelines for producer organizations to identify priorities, assess performance and improve the sustainability of their geographical indication systems, Rome. Available online: https://doi.org/10.4060/cc9122en Hardin 1968. The tragedy of the Commons Science, 162: 1243–1248

FAO, The European Bank for Reconstruction and Development (EBRD) (2018) Strengthening sustainable food systems through geographical indications: an analysis of economic impacts. In: Vandecandelaere E, Teyssier C, Barjolle D, Jeanneaux P, Fournier S, Beucherie O (eds). FAO/EBRD, Rome/London, p 135. Available online: www.fao.org/3/a-i8737en.pdf

Hardin G (1968) The tragedy of the commons. Science 162(3859):1243–1248

Ostrom E (2015) Governing the commons: the evolution of institutions for collective action. Canto classics, vol 2015. Cambridge University Press, pp i–iv

Samper LF, Quiñones-Ruiz XF (2017) Towards a balanced sustainability vision for the coffee industry. Resources 6:17

Vandecandelaere E, Samper LF, Rey A, Daza A, Mejía P, Tartanac F, Vittori M (2021) The geographical indication pathway to sustainability: a framework to assess and monitor the contributions of geographical indications to sustainability through a participatory process. Sustain For 13:7535. https://doi.org/10.3390/su13147535

The opinions expressed in this chapter are those of the author(s) and do not necessarily reflect the views of the [NameOfOrganization], its Board of Directors, or the countries they represent.

Chapter 26
Terroir-Based Geographical Indications in the Face of Climate Change: The Narrow Path of a Strategic Reinterpretation of the Link to Origin

Claire Bernard-Mongin

Acronyms

GI	Geographical Indications
IPR	Intellectual Property Rights
AP	Appellation of Origin
INAO	National Institute of Origin and Quality
PDO	Protected Designations of Origin
PGI	Protected Geographical Indications
CO	Control Organisation

26.1 Introduction

Climate change is gradually becoming an essential, even existential, issue for geographical indications (GI). A GI is an intellectual property right that reserves for a collective of producers the exclusive use of a name due to the unique and specific aspect of the product linked to its origin. The administration of proof of the link to the place is at the heart of the justification of the instrument. However, the effects of climate disruption are changing the characteristics of the terroir (precipitation, temperatures, etc.) affecting productivity and/or the final quality of the product. This results in increasingly numerous and recurrent requests for derogation from the specifications. The model is pushed to its limits (Clark and Kerr 2017).

C. Bernard-Mongin (✉)
CIRAD, UMR INNOVATION, INNOVATION, Univ Montpellier, CIRAD, INRAE, Institut Agro, Montpellier, France
e-mail: claire.bernard-mongin@cirad.fr

© The Author(s) 2025
E. Vandecandelaere et al. (eds.), *Worldwide Perspectives on Geographical Indications*, https://doi.org/10.1007/978-3-031-71641-6_26

Faced with the acceleration and increased amplitude of these upheavals, producer collectives adapt, innovate. This situation refers to what the field of strategic management problematises as a "tension between perseverance and flexibility" (Gersick 1994), which characterises the dynamics of organisational adaptation. In this respect, the notion of adoption is particularly used in management sciences to restore the dynamics and processes of interpretation, negotiation and construction of meaning that take place around the instrument, which actors use to develop and reinvent the collective action model that suits the situation (Segrestin 2004). It seems interesting to consider this challenge to GIs by climate change in the light of this managerial problematisation. In the current dynamics of adoption of geographical indications, can the link to the terroir, the heart of the "historical promise" of GIs, be a support for collective learning and of innovations in the face of the need for adaptation to climate change?

We propose to conduct our reflection around the European system *sui generis* of registration and protection of GIs, envisaged as an organisational support for learning and collective innovations (Le Masson et al. 2006; Moisdon 1997). Initially, in a descriptive and synchronic manner, we will detail the different dimensions as they have stabilised in the current system in force. Secondly, in an analytical and diachronic manner, we will revisit the dynamics of endogenous adoption (Grimand 2012) of this European system of origin protection, from a definition of its *adaptive attributes* (Ansari et al. 2010). We will then specify the dimensions of the instrument put under tension during this process, particularly in relation to the link to the terroir. Thirdly, we will consider the effects of climate change as an exogenous disruption of the GI appropriation process, and propose an exploratory discussion on emerging strategies and new learning modalities around a reinterpretation of the link to origin under the constraint of adaptation to climate change.

26.2   Geographical Indications, a Support for Learning and Collective Innovations

The geographical indication makes the recognition of the particular, local, identified conditions of production, the heart of a quality or a unique reputation to protect and promote. The European Commission and the Member States have made this legal tool an economic lever for enhancing a part of their agricultural, food and wine productions on domestic markets and for export. In order to better understand how this legal-economic tool interacts with its users to become a support for innovations and collective learning, we propose to describe it, with Hatchuel and Weil (1992, pp. 122–126) then David (1996) following three dimensions considered dynamically and interdependently: the " *formal substrate* " carrying a " *managerial philosophy* " and " *a simplified vision of organisational relations* ".

The managerial philosophy refers to the " *system of concepts that designates the objects and the objectives forming the targets of a rationalisation* " (Ibid, pp. 124).

In the managerial philosophy carried by geographical indications, origin is at the heart of a double legal and economic rationalisation of the link to place. Legally, the progressive incorporation of GIs into the regime of intellectual property rights (IPR) and the consequent recognition of a differentiated treatment is based on the assertion that a distinctive and unique link exists between a certain category of products and their area of origin. In the European system of registration and protection of GIs, it is therefore from the administration of the proof of this particular link that legal protection is granted to applicants. This link to the origin is also subject to an economic rationalisation as it is at the heart of a mechanism for creating particular value. From the revelation of attributes linked to origin, the product can stand out on the markets, thus justifying the premium price displayed (European Commission, 2021). In an ideal-typical operation, this rent linked to the origin can then be redistributed to all private operators involved in the production of the product, and reinvested in the mechanism of collective quality regulation. This "virtuous circle of quality" has the effect of creating an incentive to maintain local production conditions and to generate positive externalities at the territorial level (Vandecandelaere et al., 2009).

While these rationalisations are evolving and contested (Allaire et al. 2005; Sylvander et al. 2006), crossed by controversies and divergent interests, they nevertheless find points of momentary stabilisations, which is the case today at the European level, in the Regulation (EU) No. 2024/1143[1] on geographical indications for wine, spirit drinks and agricultural products, as well as traditional specialties guaranteed and optional quality terms for agricultural products—as well as all the delegated regulations that specify their correct application. These texts and their concrete apparatus (e.g. specifications, control plan, European register of geographical indications, etc.) constitute the technical substrate, that is to say the set of artefacts on which the instrument relies to function. It is notably through the specifications specific to each product that the administration of the proof of the link to the place is made—a proof that is then guaranteed by the control plan and the product's traceability monitoring system, throughout its manufacturing chain. To these localised artefacts, resulting from often long collective learnings, stabilisers of local compromises (Bernard-Mongin et al. 2021; Millet 2019; Quiñones-Ruiz et al. 2016) are added national and European artefacts, which ensure the registration and national and then European recognition and protection of GIs. The minimal European foundation is composed of the single registration (or modification) procedure, finalised by publication in the Official Journal of the European Union and entered in the European register of GIs on the basis of a synthetic description of the product and the link to the place, called the "single document". These common European rules are also transposed into the law of the Member States, and broken down into rules and registration (or modification) procedures at the national level,

---

[1] This new regulation amends regulations (EU) No 1308/2013, (EU) 2019/787 and (EU) 2019/1753 and repeals regulation (EU) No 1151/2012.

which in turn define the content requirements of the specifications, control plan, but also requirements on the nature or organisation of the producers carrying the GI.

Finally, these managerial artefacts also carry a simplified vision of organisational relationships, defining " *a scene whose characters come to explain the roles that a small number of actors, summarily, or even caricaturally defined* " (Hatchuel and Weil 1992, pp. 125). Thus, a restrictive scene is organised, which revolves around these artefacts, composed of "Producer Groups", that is to say the set of producers who organise collectively to apply for the recognition of their exclusive right to use the GI for the product which they justify by a specific link to the origin by a set of specifications and an associated control plan. The producer group is also involved in the defence and protection actions of the product and the terroir and assumes a role of promotion and valorisation of the product (Réviron and Chappuis 2011). The "National Competent Authorities" are responsible for the official approval of the specifications and the associated control plan. They must ensure the conformity of the practices of the producer group to the specifications. They can delegate the inspections of this conformity to a public control body or to an independent private operator called "Certification Body", which intervenes against payment by the producer group. Where national legal regulation so provide, the certification body is itself controlled on its competence to certify conformity to the standards and rules of GIs by an accrediting body, according to the standard currently in force (ISO/EC 17065:2012). Consumers, finally, are the final recipients of the innovation, through the act of purchase, they reveal their preference and their consent (to pay) for a specific quality guaranteed by the GI certification system.

This first key to reading GIs allows us to consider them as legal and commercial instruments, organisational supports for innovations or collective learnings embedded in the sense that the European system *sui generis* provides for derogatory routes to the global free-trade commercial regime, linked to production spaces anchored in their biophysical and cultural-historical environment ("terroir-niche") (Belmin et al. 2018). These unique and differentiated arbitrations, are thus the result of local compromises, of learning processes stabilised in specifications and their control plan.

26.3   Dynamics of Appropriations of Geographical Indications

We now propose to revisit the dynamics of adoption of the European GI system by following how the link to the origin, the heart of the historical promise of the instrument, has evolved. The dynamics of adoption test the "game" possible between the constitutive dimensions of the instrument (artefacts, managerial philosophy, simplified organisational relations) by making it evolve in a space defined by two main dimensions: fidelity and extensiveness (Ansari et al. 2010). In other words, *in the ways in which diffusing practices are implemented*, the nature of the initial version is more or less respected (substitution, modification or hybridisation of the managerial philosophy and/or simplified organisational relations) and the degree of implementation of the innovation is more or less high (more or less significant omission of certain elements of the technical substrate). The plasticity of this learning process is a function of the "adaptive attributes" *(key affordances)* of the innovation (ibid.): the *interpretive viability*, the *divisibility* and the *complexity*. The interpretive viability refers to the latitude of reinterpretation allowed by the instrument supporting innovations. Divisibility refers to the possibility of appropriating the innovation independently of scale. The complexity indicates a difficulty perceived by the actors to use and understand the innovation due to numerous grey areas and uncertainties. If the attributes are given by the material and cognitive dimension of the instrumentation, they are also constructed by the practices, uses and interpretations that are made of it. It is by describing the evolutions in time and space of the innovation that one can appreciate the tension between "perseverance and flexibility".

The concept of GI in Europe is derived from that of appellation of origin (AO), this one carried by the southern Mediterranean countries such as France, Italy, Spain or Portugal. Initially limited to wines and spirits, the appellation of origin was then extended to agri-food products in the 1990s (Bienaymé 1995), which explains the way in which the link to the origin crystallised initially around the notion of terroir in the viticultural sense. As Barham (2003) deciphers, " *Historically, terroir refers to an area or terrain, usually rather small, whose soil and microclimate impart distinctive qualities to food products. The word is particularly closely associated with the production of wine* ". In this understanding, the terroir pre-exists the product and the collective of producers who only reveal its potentialities through their know-how (Bérard and Marchenay 2004). This encoding of the attributes of a product linked to its origin defines an "essentialist" type relationship between the product and its territory of origin.

However, during the Europeanisation (then internationalisation) of GIs, the terroir paradigm had to compose with another understanding of the protection of origin based on the concept of reputation. Gangjee (2017) thus speaks of the "European compromise", to designate the harmonised regime at the beginning of the 1990s which cohabits under Regulation (EU) No 2081/92, geographical indications " *whose quality or characteristics are due essentially or exclusively to the geographical environment including the natural and human factors, and whose production, processing and elaboration take place in the delimited geographical area* ", but which recognises and protects just as well products whose "*a certain quality, reputation or other characteristic can be attributed to this geographical origin and whose production and/or the transformation and/or processing takes place in the defined geographical area*".

This compromise was maintained (and strengthened) in the successive regulations of 2006 and 2012, leading to two different denominations: Protected Designations of Origin (PDO) for GIs whose uniqueness is given by a close link with a specific terroir (GI-terroir); and Protected Geographical Indications (PGI) for GIs based on a reputation or know-how (GI-reputation). A recent study outlining the state of the art on GIs in Europe, confirms from a chronological analysis how PGIs have become the predominant mode of registration and protection from the mid-1990s, to represent, since 2012, nearly 67% of registration requests against 33% for PDOs (Zappalaglio et al. 2022). This shift from the concept of terroir towards a simplified relationship with origin was reinforced by the standardisation of the GI control system initiated by Regulation (EU) No 520/2006. This latter, by specifying the nature of the controls, standardises a model of certification by third-party bodies approved (by the competent national authority) and accredited according to the European certification standard. In France for example, this precision of the technical substrate of the European GI system has significantly altered the balance of "simplified organisational relations" that existed before, by giving a more important role to private Control Bodies (CBs) in the evaluation of product conformity. Bodies that operate from a standardised evaluative reference, whereas previously the public authority in charge of control (i.e. the National Institute of Origin and Quality (INAO)) operated with localised, territorialised references, due to its historical presence in production areas (Marie-Vivien et al. 2017).

This reinterpretation of the endogenous adoption dynamics of GIs shows that the interpretative flexibility of the instrument has allowed the original management philosophy to be adapted, to allow for wider dissemination. This was possible by loosening the relationship with the terroir, and allowing a broader interpretation of the link to origin, also understood as a historical (know-how, tradition) or reputational link to the territory, freed from its biophysical or ecosystemic dimension. However, it is observed that the *sui generis* registration and protection regime for GIs in the European Union has maintained all of its architecture[2] . In other words, this model of origin protection does not easily accommodate partial or "low dosage" adoption, and requires full implementation to function, or even "something more is necessary" (Gangjee 2015). This is explained by the fact that the simplified model of the organisational relations of the GI does not saturate either the meaning or the form of the local arrangements necessary for the GI to perform: by leaving grey areas, and interpretative margins it allows for ad-hoc organisational combinations or assemblies, specific to a given situation. For example, the status of the "group of producer applicants" for the GI, as defined by the European Regulation, is broad and does not predispose its legal form. Thus, several organisational realities coexist under the status of applicant: profit-making or non-profit organisations (e.g. Italian Consorzio (cooperative form) or French defence and management organisation (ODG) (associative form)), local authorities in some cases, several groups of producers in the case of a cross-border geographical area. The instrument also requires collective coordination that actively involves the "shareholders" directly involved in the stages of production, valorisation and protection of the product, but also the stakeholders in

---

[2]In comparison, it should be noted that the Agreement on Trade-Related Aspects of Intellectual Property Rights (TRIPS, 1992) within the framework of the World Trade Organisation (WTO) and the Geneva Act (2015) of the Lisbon Agreement within the framework of the World Intellectual Property Organisation (WIPO) both recognise a plurality in the modes of organisation of the defence of intellectual property related to origin.

the process, facilitators of its creation and implementation (local authorities, international NGOs, intergovernmental organisations, sector organisations, research structures, etc.). This complexity in turn reinforces the dimension "tailor-made" for the dynamics of GI adoption, by maximising the interpretative flexibility of the instrument, while maintaining the entire technical substrate in a rather rigid manner, as it guarantees the final effectiveness of the instrument, mainly on its legal dimensions (protection of the name, fight against counterfeiting) (EUIPO 2016) and economic (added value, market share gain) (European Commission 2021).

26.4 Emerging Strategic Perspectives for "Terroir-Based GIs" in the Face of Climate Change

Climate disruption now exogenously seizes this adoption process, necessitating strategic reflection on the tool in times of crisis. However, as Clark and Kerr (2017) rightly analyse, not all GIs are questioned with the same intensity. It is essentially the "terroir-based GIs" (*terroir based*), which are particularly affected, as the effects of climate disruption alter the very characteristics of the terroir (rainfall, sunshine, temperatures, etc.) in an unpredictable, jerky manner, reducing productivity or affecting the final quality of the product. The increase in the frequency of these effects makes a conservative stance and strict adherence to the specifications difficult in the long term. Producers' collectives adapt, innovate (new varieties or breeds, modification of the cultural calendar, feeding outside the area, correction of the organoleptic qualities of the product, etc.) at the risk of stretching the link to the place. The terroir-based GI model is therefore pushed to its limits: " *As a result, pressure for altering the legal specification of terroir may arise* " (Ibid). These elements have been incorporated into the ongoing reflection on the different adaptation strategies in the French wine industry, which envisage a difficult future for GIs whose link to origin is essentially based on terroir (Ollat et al. 2021). Would it then be possible to envisage a third way forward for terroir-based GIs (Touzard and Ollat 2022) and propose an interpretation of origin that is not fixed but adaptive? More precisely, it would be the effort to adapt to the terroir and its characteristics— translated by an adaptive co-management of territorial resources—that would (re-)establish the link to the place. This proposal opens the way to a redefinition of origin, in a procedural—and no longer essentialist—conception of the relationship between the product and its terroir.

In an exploratory manner, we can then sketch out elements of necessary rearrangements for the development of this third way. On the managerial philosophy, first, this adaptative perspective of origin-based quality proposes a reinterpretation of the link to the place under the sign of coevolution under climate constraint, of production practices and terroir. It requires putting the act of production at the heart of the agroecosystem, more systematically highlighting the link of the product to the biophysical environment and the specific resources it mobilises, and organising the recognition of evolving attributes that are different from the static attributes traditionally highlighted in the specifications (nature, climate, breed, raw materials). The terroir would be tinged with experimental agroecology, and terroir-based GIs would be a support for collective learning, guaranteeing a synergy between adaptation strategies to climate change and agroecological transitions. It would be a matter of increasing the technical substrate of the GI, by introducing into the specifications and the control plan the modalities of monitoring this evolving relationship of the product and its agroecosystem under climate constraint. Some options in this direction have already proposed the development of additional modules in the PDO specifications (PDO+), including elements of sustainability—sustainability of practices and maintenance of certain ecological values at the landscape scale—with financing mechanisms that would be partly linked to the Common Agricultural Policy (Flinzberger et al. 2022). More recently still, the new regulation no. 2024/1143 in its article 7, proposes to include in the specifications, "< *higher sustainability standards than those provided by Union law or national law on environmental matters* " and specifies a number of so-called sustainable practices in paragraphs a and b, which can be quoted here in their entirety: " *a) the mitigation of climate change and adaptation to it, the sustainable use and protection of landscapes, water and soil, the transition to a circular economy, including the reduction of food waste, the prevention and reduction of pollution, and the protection and restoration of biodiversity and ecosystems; b) the production of agricultural products using methods that reduce the use of pesticides and manage the risks resulting from such use, or reduce the risk of resistance to antimicrobials in agricultural production"* . It is then expected from the "producer group" to "provide *advice, organise training and disseminate guidance on good practices for current and future producers, including with regard to sustainable practices, particularly those provided for in Article 7, scientific and technical advances, the transition to digital, the integration of the gender dimension and equality between men and women, and consumer awareness* ". The actantial scheme of the GI would then be augmented with new "actors" of the GI. Not only the producers but also all the operators involved in the co-construction of adaptation choices to climate change in view of the state of the biophysical environment (agricultural advisers, certification bodies, local authorities, research operators, etc.). A terroir engineering informing a governance of local resources, would allow a co-construction of adaptation choices with the Producer Group. It would be a matter of informing along the way the new production practices so that they prove a certain degree of "fidelity" to the terroir in relation with the competent authorities and certification bodies. It would also be a matter of shaking up the stable definition of a quality/typicity of products, based on an objectification of stable organoleptic or physico-chemical characteristics and to obtain from consumers, a consent to pay for a product with evolving characteristics.

This third way designates a strategic possibility of evolution of European GIs that would affect all dimensions of the instrument. This perspective also calls for a reflection on the steering of this "strategic emergence" and on the form of learning processes (Miller 1996) to coordinate: an institutional type of learning at the level of national and European competent authorities, and an experimental learning at the level of production territories. However, if the institutional type of learning is mostly conducted from the development of standards (regulatory and legislative framework of GI) in a dynamic where choices are guided by values, and where the primary objective is to ensure overall coherence, experimental type learning follows a trial/error dynamic, in which action is central, choices are informed by feedback, and the primary objective is adaptation. These two types of learning do not have the same timing, nor the same tolerance for uncertainty or the same ability to push disruptive innovations. Their very different nature therefore raises the question of the means and modalities of their articulation.

26.5 Conclusion

These analytical elements on the process of appropriating European GIs allow us to define a (narrow) field of strategic emergence, in which the link-to-origin, based on the notion of terroir and currently undermined by climate change, can be reinvented. The delimitation of this strategic field takes into account determination effects linked to the different dimensions of the instrument (technical substrate, managerial philosophy and simplified organisational relations) and the way they have been mobilised, hybridised, transformed, during their adoption within the European Union and internationally. Thus, the interpretative flexibility of the European GIs model has resulted in the coexistence of two variants of a managerial philosophy of the link-to-origin, which is at the heart of the justification of this intellectual property right. These two interpretations (PDO and PGI) are solidly anchored by a technical substrate and a weakly divisible organisational structure, which despite their complexity, ensure a certain extensiveness of the GIs model, as an innovation during its diffusion. The majority adoption is to go for a broad interpretation of the link to origin, mainly based on product reputation (historicity of know-how, production practices). This avoids the thorny issue of terroir subject to changing climatic condition . This trend has been reinforced by the standardisation of controls, a nodal point in the organisation of the technical substrate of European GIs. These elements thus confirm the narrow path of a third way which would combine "perseverance and flexibility" and maintain the historical promise of the instrument. Adaptive terroir-based GIs would continue to base their justification   on a strong link-to-origin, but by reinterpreting the notion of terroir. It would then be a question of fundamentally re-inscribing the act of production at the heart of its biophysical environment. The terroir would then be a constantly updated (and verified) assertion of the co-evolution of production practices with their natural environment. This proposal goes against the dynamics of adoption of the GI over the last 20 years. However, it would open up synergistic paths between adaptation strategies to climate change and agroecological transition.

References

Allaire G, Sylvander B, Belletti G, Marescotti A, Barjolle D, Thévenod-Mottet E, Tregear A (2005) Les dispositifs français et européens de protection de la qualité et de l'origine dans le contexte de l'OMC: justifications générales et contextes nationaux. Paper presented at the Symposium international—Programme transversal de l'INRA "Pour et Sur le Développement Régional" PSDR, Lyon, France.

Ansari SM, Fiss PC, Zajac EJ (2010) Made to fit: how practices vary as they diffuse. Acad Manag Rev 35(1):67–92. https://doi.org/10.5465/amr.35.1.zok67

Barham E (2003) Translating terroir: the global challenge of French AOC labeling. J Rural Stud 19(1):127–138. https://doi.org/10.1016/S0743-0167(02)00052-9

Belmin R, Meynard J-M, Julhia L, Casabianca F (2018) Sociotechnical controversies as warning signs for niche governance. Agron Sustain Dev 38(5):44. https://doi.org/10.1007/s13593-018-0521-7

Bérard L, Marchenay P (2004) Du terroir révélé à l'indication géographique. In: Bérard L, Marchenay P (eds) Les produits de terroir : Entres cultures et règlements. CNRS Éditions, Paris

Bernard-Mongin C, Balouzat J, Chau E, Garnier A, Lequin S, Lerin F, Veliji A (2021) Geographical indication building process for Sharr Cheese (Kosovo): "Inside Insights" on sustainability. 13(10):5696. https://doi.org/10.3390/su13105696

Bienaymé M-H (1995) L'appellation d'origine contrôlée. Revue de Droit Rural 236:419–424

Clark LF, Kerr WA (2017) Climate change and terroir: the challenge of adapting geographical indications. 20(3–4):88–102. https://doi.org/10.1111/jwip.12078

David A (1996) Structure et dynamique des innovations managériales. Cinquième Conférence de l'Association Internationale de Management Stratégique, Lille, 12–15.

EUIPO (2016) Infringement of protected Geographical Indications for wine, spirits, agricultural products and foodstuffs in the European Union. Retrieved from Alicante, Spain. https://euipo.europa.eu/tunnel-web/secure/webdav/guest/document_library/observatory/documents/Geographical_indications_report/geographical_indications_report_en.pdf

European Commission (2021) Study on economic value of EU quality schemes, geographical indications (GIs) and traditional specialities guaranteed (TSGs), Final report, Directorate-General for Agriculture and Rural Development, Publications Office. https://data.europa.eu/doi/10.2762/396490

Flinzberger L, Cebrián-Piqueras MA, Peppler-Lisbach C, Zinngrebe Y (2022) Why geographical indications can support sustainable development in European Agri-Food landscapes 2. https://doi.org/10.3389/fcosc.2021.752377

Gangjee DS (2015) Proving provenance? Geographical indications certification and its ambiguities. World Dev. https://doi.org/10.1016/j.worlddev.2015.04.009

Gangjee DS (2017) From geography to history: Geographical indications and the reputational link. In: Calboli I, Ng-Loy W (eds) Geographical indications at the crossroads of trade, development, and culture: focus on Asia-Pacific. Cambridge University Press, Cambridge, pp 36–60. Available at SSRN: https://ssrn.com/abstract=2923892

Gersick CJG (1994) Pacing strategic change: the case of a new venture. Acad Manag J 37(1):9–45. https://doi.org/10.2307/256768

Grimand A (2012) L'appropriation des outils de gestion et ses effets sur les dynamiques organisationnelles: le cas du déploiement d'un référentiel des emplois et des compétences. Revue management et avenir 4:237–257. https://doi.org/10.3917/mav.054.0237

Hatchuel A, Weil B (1992) L'expert et le système. Gestion des savoirs et matamorphose des acteurs dans l'entreprise industrielle. Editions Economica, Paris

Le Masson P, Weil B, Hatchuel A (2006) Les processus d'innovation. Conception innovante et croissance des entreprises. Hermès-Lavoisier, Coll. « Science Publication », Paris

Marie-Vivien D, Bérard L, Boutonnet J-P, Casabianca F (2017) Are French geographical indications losing their soul? Analyzing recent developments in the governance of the link to the origin in France. World Dev 98:25–34. https://doi.org/10.1016/j.worlddev.2015.01.001

Miller D (1996) A preliminary typology of organizational learning: synthesizing the literature. J Manag 22(3):485–505. https://doi.org/10.1016/S0149-2063(96)90033-1

Millet M (2019) From Ossau and Iraty to PDO Ossau-Iraty: The long-term construction of a product based on two distinct places. Br Food J *121*(12):3062–3075. https://doi.org/10.1108/BFJ-10-2018-0719

Moisdon J-C (1997) Du mode d'existence des outils de gestion : les instruments de gestion à l'épreuve de l'organisation. Editions Seli Arslan, Paris, 286 p

Ollat N, Zito S, Richard Y, Aigrain P, Brugiere F, Duchêne E et al (2021) La diversité des vignobles français face au changement climatique: simulations climatiques et prospective participative. Climatologie *18*:3. https://doi.org/10.1108/BFJ-10-2018-0719

Quiñones-Ruiz XF, Penker M, Belletti G, Marescotti A, Scaramuzzi S, Barzini E et al (2016) Insights into the black box of collective efforts for the registration of Geographical Indications. Land Use Pol *57*:103–116. https://doi.org/10.1016/j.landusepol.2016.05.021

Réviron S, Chappuis J (2011) Geographical indications: collective organization and management. In: Barham E, Sylvander B (eds) Labels of origin for food: local development, global recognition. CAB International, On-Line, pp 45–62

Segrestin D (2004) Les chantiers du manager. Edition Armand Colin, Paris

Sylvander B, Allaire G, Barjolle D, Thévenod-Mottet E, Belletti G, Marescotti A, Tregear A (2006) Qualité, orgine et globalisation: Justifications générales et contextes nationaux, le cas des Indications Géographiques. Can J Reg Sci XXIX(1):43–54

Touzard JM, Ollat N (2022, 10/16) Les vins d'appellation vont-ils disparaître ou renaître avec le changement climatique? The Conversation.

Vandecandelaere E, Arfini F, Belletti G, Marescotti A, Allaire G, Cadilhon J-J et al (2009) Linking people, places and products. A guide for promoting quality linked to geographical origin and sustainable geographical indications. FAO-SINER-GI, Rome. 193 p

Zappalaglio A, Carls S, Gocci A, Guerrieri F, Knaak R, Kur A (2022) Study on the functioning of the EU GI system (March 18, 2022). Max Planck Institute for Innovation & Competition Research Papers, München. Available at SSRN https://ssrn.com/abstract=4061160

Chapter 27
The Issue of Geographical Indications in the Face of Climate Change in France

Gilles Flutet, Caroline Blot, Jacques Gautier, Laurent Mayoux, and Alexandra Ognov

Abbreviations

AOC	Controlled Designations of Origin (corresponding to the European term PDO)
PDO	Protected Designations of Origin
GI	geographical indication
PGI	Protected Geographical Indication
INAO	National Institute of Origin and Quality
ODG	defence and management organisation (association of producers of the geographical indication)

27.1 Introduction

Like all agricultural production systems in the world, the territories where products with geographical indications (GI) are produced are fully affected by climate change. The sectors must adapt to maintain their productions quantitatively and qualitatively and continue to benefit from the GI certification. But how can the conditions for the production of these products, meticulously codified in the specifications and resulting from local, often traditional, know-how, evolve without undermining their link to the origin, the very essence of the GI? In this context, maintaining the human activities developed around these GIs is both a major issue for the sustainability of the sectors and a real challenge as these changes question their fundamentals. It is therefore necessary to be able to propose adjustments to production rules

G. Flutet (✉) · C. Blot · J. Gautier · L. Mayoux · A. Ognov
French National Institute for Quality and Origin (INAO), Montreuil, France
e-mail: g.flutet@inao.gouv.fr

© The Author(s) 2025
E. Vandecandelaere et al. (eds.), *Worldwide Perspectives on Geographical Indications*, https://doi.org/10.1007/978-3-031-71641-6_27

367

without distorting the link that unites the product to its territory, based on a combination, sometimes complex, of natural and human factors. This is the challenge that must be met by GI production sectors and this chapter aims to illustrate the experience of GI actors in France and the actions that can be taken to support the adaptation of GIs to climate change.

27.2 Geographical Indications in France: Context

Set within a regulatory framework defined at the European level, geographical indications in France bring together 3 families of products: AOCs (Controlled Designations of Origin, corresponding to the European term PDO), PGIs (Protected Geographical Indications) and GI for spirits. These French GIs, numbering 754 in 2022 (of which 65% are linked to the vineyard) occupy a very significant place in the French agricultural and agri-food economy, and contribute to its qualitative image. They also contribute to creating value in disadvantaged rural areas and to maintaining a dynamic social fabric there.

Each GI is defined by a specification containing information on the boundaries of its production area, a definition of the products and precise rules regarding the method of obtaining (production and transformation).

A central element of the specification is the explanation of the link to the origin, allowing us to understand how human factors (uses, local know-how...) and natural factors (climate, geology, relief, soils, biodiversity...) of the delimited area interact to result in an original and specific product.

It is important to note that each specification is the result of a voluntary collective approach by economic operators gathered within a PDO (Defence and Management Organisation), producers, processors, downstream actors. It is approved by interministerial decree after approval by a decision-making body of the INAO, the state operator in France responsible for implementing public policy on official signs of quality and origin in the agricultural and agri-food sectors.

27.3 What Are the Impacts of Climate Change on GIs?

The production systems of GIs are, like all agro-systems, impacted in different ways by climate disruptions, and this has been the case for several years now.

The sensory and analytical characteristics of the products can undergo changes due to climate change. Wines, for example, have seen their alcohol levels increase in recent years in most wine-growing basins, while at the same time acidity levels have decreased. For many GI products, production volumes tend to be more fluctuating due to the increased frequency of exceptional climatic events such as spring frost linked to increasingly early bud break, drought (intensity, duration), hail and very intense rainy episodes. Thus, the severe frost that hit almost all vineyards in 2021, at the end of April, resulted in a global decrease in production of about 20% compared to the previous vintage. Fruit production under GI (*Mirabelles de Lorraine, Pruneaux d'Agen...*) is also regularly marked by drops in production linked to climatic hazards. Livestock sectors, on the other hand, are regularly faced with shortages of fodder, linked to the increasing irregularities of precipitation, which result in particular in an increased frequency of spring droughts. In 2023, water shortages and high summer temperatures impacted all sectors in France, further illustrating the urgency of the situation.

Climate change also induces significant changes in the calendars of vegetative cycles and cultural operations. Livestock sectors heavily reliant on grazing for animal feed are very exposed to these changes, with particularly early turnouts in the spring after winter, but also summer temperature peaks that can prevent animals from going out during the day or a summer drought forcing herds to be supplied with fodder outside of winter periods. Another illustration of the effect of global warming on crop cycles is provided by viticulture and the harvest, the date of which has been brought forward on average by 18 days between 1974 and 2019 in France and is approaching 30 days in some vineyards.

Finally, it can be observed that climate disruption tends to increase health risks in plant production (increased pressure from fungal diseases, increased damage from certain pests). A significant part of the vineyard was thus strongly impacted by mildew in 2023 in France. In dairy farms, periods of high heat can negatively affect the physiology of animals by reducing their milk production. Water shortages have also forced some farmers to reduce the size of their herd.

27.4 Adapting Without Undermining the Link to the Origin: A Difficult Equation for GIs

Various research programmes and studies conducted in recent years on climate change in the agricultural sector and its various industries have allowed us to precisely document its current and future effects, at increasingly fine scales, and to propose technical solutions for adaptation in the short, medium and longer term (see Box 27.1).

Box 27.1: Research Projects on Climate Change in the Agricultural Sector
In the viticulture sector the LACCAVE project, led by the French National Institute for Agronomic and Environmental Research (INRAe) and involving some twenty research teams, has, since 2012, provided the impetus and stimulus for multidisciplinary work on various topics with a forward-looking approach. By integrating the knowledge acquired in different disciplines (climatology, ecophysiology, genetics, agronomy, oenology, economics, sociology, geography, etc.) and with a systems approach, it has made it possible to

(continued)

Box 27.1 (continued)

develop modelling tools to simulate the impact of climate change on wine production and quality, but also to test the adaptive potential of certain levers and innovations. This programme has also allowed us to propose adaptation scenarios to wine professionals at different spatial scales and to assess their economic, sociological and environmental consequences. Finally, it should be noted that this large-scale project was the basis for the development of a national strategy for adaptation to climate change in the wine sector, which was approved by the public authorities. Another example is the AP3C (Adaptation of agricultural practices to climate change) programme launched in 2015 by agricultural development actors in the Massif Central, an area with a large number of GI cheeses. The project, which brought together climate experts from Météo France (the French national weather service), used local climate data to carry out a detailed analysis of the impact of climate change on the study area, with a view to adapting local production systems and raising awareness among all stakeholders. Thirty agro-climatic indicators were analysed in projections, with changes mapped over the period 2016 to 2050

In plant production, including forage, one of the major levers for adapting production conditions to climate change lies in the diversification of genetic resources and the development of more resilient varieties, capable in particular of producing in hotter and drier conditions. In outdoor livestock farming, adaptation recommendations are mainly based on a change in pasture management methods, a reduction in the load per hectare, redesigned livestock buildings and a diversification of forage crops. The introduction of new breeds into a herd can also be considered.

Even though they are based on models in tune with their natural environment not seeking intensification, IG productions may require changes in practices to maintain the economic viability of the farms concerned in the context of climate disruption.

However, the production rules written in the specifications (inherent to the link to the origin) limit the possibilities of easily modifying some of these practices.

Among the elements codified in the specifications that may limit rapid changes in practices, we can mention: genetic resources (limited list of varieties or breeds allowed, not necessarily the most rustic and best adapted to extreme climatic events), the prohibition or strict regulation of irrigation for IG of plant sectors, the minimum duration of outdoor grazing for IG cheese, imposed periods for certain cultural operations or herd management, the maximum yield per hectare and the choice of plots in AOP vitivinicoles (list of classified plots for grape production).

Note that in the vitivinicultural sector, there are mechanisms in France allowing operators of interested IGs to reserve a part of the harvest during years of generous production (linked to regular precipitation), these volumes (limited annually by the yield limits of the specifications) can be mobilised to supply the market during periods of deficit harvest. This system allows better coping with interannual variations in production reinforced by climate change.

27.5   Procedures and Tools Available to Evolve the Specifications

Even if it reflects traditional uses and know-how, the specifications of an IG must be able to evolve after the initial act of its official recognition. The basic condition for a modification to be approved is that it does not distort or weaken the product's link to its origin.

Any request for modification, necessarily issued by the ODG, is examined and analysed by the INAO following a procedure involving different actors. If validated, the revised specifications will be approved following the same formal scheme as a recognition, at national and European levels.

Regarding modifications induced by climate change, two main types can be distinguished:

- occasional requests, following an exceptional climatic event that does not allow certain technical requirements of the specifications to be met (temporary modifications)
- requests of a structural nature, aimed at adjusting certain recurring elements of temporary modification (addition of the possibility of irrigation for example) or facilitating adaptations to new constraints (by the introduction of varieties better adapted to dry conditions and high temperatures).

In the agri-food sectors and for several years, many modifications of specifications find their origin in taking into account the effects of climate change. This is why many specifications have been modified to remove fixed durations or calendar dates framing certain stages of production, to change the minimum size of certain fruits or vegetables, or to improve the food autonomy of livestock farms. Similarly, sectors have organised themselves to collectively manage climate hazards: fodder exchanges, increasing stocks to manage hazards, reducing densities or loads, etc. For example, in many French departments, fodder exchanges are being set up which allow, via a dedicated website, to share offers and demands for fodder and to connect farms. This allows partnerships to be created on farms but also between plain and mountain territories.

At the vineyard level, to anticipate the consequences of climate change, the INAO has endorsed a procedure allowing voluntary AOC winegrowers to plant varieties whose characteristics suggest an ability to adapt to future conditions, while maintaining a production that meets the characteristics of the AOC. This procedure allows observation on areas limited to 5% of the farm area of varieties of interest for adaptation (VIFA) and their wines over a period of 10 years among voluntary producers, in a participatory approach. At the end of the observation period, the ODG will then have to decide on the maintenance or not of each of these varieties in its specifications. In the same spirit, the INAO has also developed a device allowing the evaluation of innovations within the specifications themselves. This ability allows observation on a sufficient scale and in different farms of new practices which, if they are satisfactory and respect the link to the origin, can be integrated at the end of an observation period into the IG's specifications, before possibly being taken up by other appellations. This is for example the AOP *"Alsace"* which wishes to evaluate the impact of soil covering on the preservation of soil water reserves. Finally, the INAO and the various IG production sectors are developing links with research and development organisations working on measuring the impacts of climate change on agricultural production but also on solutions to mitigate them. These relationships contribute to better communicate on existing or explored solutions, but also allow the various sectors not to isolate themselves in the face of the challenges of climate change.

27.6   Conclusion

In conclusion, there are today various levers that can be activated to meet the challenges posed by climate change at the scale of an IG. But the solutions must guarantee a balance between preserving the link to the origin, speed of implementation and economic viability. The mobilisation of all actors (operators and ODG, researchers, administrations) is in this context essential to sustain the IG system. In addition to climate changes, other challenges must also be met by the IGs, such as the reduction of inputs (fertilisers, plant protection products, water) or animal welfare. Depending on the situations, the ODGs will have to prioritise the themes and the responses to be given to them.

Chapter 28
Geographical Indication Foods as Part of Healthy Diets; Their Contributions to Explore Further

Emilie Vandecandelaere and Florence Tartanac

Abbreviations

INFOODS   International Network of Food Data Systems
ICN2   Second International Conference on Nutrition
NCD   non-communicable diseases

28.1 Introduction

Global food systems has been changing rapidly due to a wide range of factors and are often recognized as a leading cause of unhealthy diets and environmental degradation (FAO/WHO 2018, 2019). Most of today's food systems need to be re-aligned from just supplying food to providing, in sustainable manner, high-quality foods that support healthy diets for all (FAO 2014). Healthy diets are linked to planet health: high attention needs to be given to the role of food systems in nurturing human health and supporting environmental sustainability, through producing a diversity of foods that enhance biodiversity rather than aiming for increased volume of a few crops (EAT-Lancet Commission on healthy diets 2019).

In 2021, the United Nations Food Systems Summit[1] and the Tokyo Nutrition for Growth Summit[2] worked together to advance solutions across agrifood systems,

---

[1] See Statement of Secretary-General's Chair Summary and Statement of Action on the UN Food Systems Summit, available at: https://www.un.org/en/food-systems-summit/news/making-food-systems-work-people-planet-and-prosperity

[2] See website: https://nutritionforgrowth.org/events/

E. Vandecandelaere (✉) · F. Tartanac
Food and Agriculture Organization of the United Nations (FAO), Rome, Italy
e-mail: emilie.vandecandelaere@fao.org

© The Author(s) 2025
E. Vandecandelaere et al. (eds.), *Worldwide Perspectives on Geographical Indications*, https://doi.org/10.1007/978-3-031-71641-6_28

based on the recognition that lack of availability and accessibility of healthy diets is one of the biggest challenges to ensure optimal health, resilience and prosperity for all, converged in concluding that food systems should promote healthy diets (UN 2021; N4G 2020, SOFI 2020).

For this transformation to happen, a territorial approach is of particular importance in rural areas where the economy is dominated by the agriculture and food sectors (SOFA 2017), and can take different forms. For example, from a consumption point of view, some territorial diets (sometimes called regional), have been shown as potentially capable of contributing positively to the health of people and the environment such as the Mediterranean Diet and the New Nordic Diet (Hachem et al. 2020). The recognition and promotion of traditional and indigenous foods are important to prevent from their possible disappearance, being under pressure in a modernizing and competitive food systems. Official registration of names of these products as geographical indications (GIs) and the associated process along the localized value-chain, represent therefore a territorial approach not only to contribute to maintain specific production practices for specific quality, but also to contribute to traditional diets and food system. By uniting local producers and other stakeholders in a territory to preserve and add value to some high quality local food product, GIs can not only stimulate social and economic development, enhance the preservation of natural and cultural environments but could also contribute to healthy diets (FAO 2018).

The objective of this chapter is to highlight some important aspects of the potential links between GI foods and healthy diets. It also initiates an advocacy for further research and knowledge development on these links. Further research could confirm the key mechanisms that we highlight here as well as best practices for stakeholders (in particular producers) to better preserve nutritional qualities and increase the contribution of GI systems to nutrition and health.

28.2   GI Foods as Part of Healthy Diets

Consuming a healthy diet throughout the life-course helps prevent malnutrition in all its forms as well as a range of noncommunicable diseases (NCDs) and conditions. According to the World Health Organization, a healthy diet contains fruits and vegetables, whole grains, fibers, nuts and seeds, and with limited free sugars, sugary snacks and beverages, processed meat and salt. In a healthy diet, saturated and industrial trans-fats are replaced with unsaturated fats (WHO 2018).

Food systems play an important role in shaping people's diets. Food systems across the globe are embedded in unique historical, religious, social, cultural and economic contexts, and are thus very diverse. Healthy diets are shaped by the way food is produced, procured, distributed, marketed, chosen, prepared and consumed. This is why social and cultural aspects of food systems must be taken into account in the dialogue on responses to improve diets (FAO/WHO 2019).

Since the 1990s, the consumption of energy-dense and high in fat, sugar and/or salt processed foods has increased relative to the consumption of nutritious foods as a result of rapid urbanization and changing lifestyles (FAO/WHO 2018). In this view, some traditional food systems, traditional diets and products could contribute to offer a possible answer to some of these challenges. They are recognized to have an important role in people's nutritional status by ICN2's Framework for Action and Rome Declaration on Nutrition (FAO/WHO 2014). Indeed increasing the adherence to traditional diets, or at least increasing the consumption of unprocessed and minimally processed foods better represented in traditional diets, has been proposed as an answer to the global health challenge (CIHEAM/FAO 2015; Sofi et al. 2008; Willett et al. 2019). Traditional diets (also called regional diets) are generally nutrient-rich and utilize locally available resources (Swanepoel and Raneri 2022). They play an important role in promoting and maintaining food security and nutrition, as it emerged from the analysis of regional diets such as the Mediterranean and new Nordic diets (Hachem et al. 2020) and of indigenous traditional food systems (Sidiq et al. 2022). Traditional food resources are essential for indigenous peoples because they provided source of essential micronutrients, while contributing to the preservation of cultures and the environments in which these are embedded (Sidiq et al. 2022). Traditional products are indeed integrated into regional diets which are "resulting from a complex, historical evolution reflecting the interaction between humans and their milieu. The products embody the knowledge and experience of past generations who optimized the local resources and ingredients to make palatable and healthy food" (Trichopoulou et al. 2007). From this viewpoint, GI food that use local raw material and relate to local food traditions in the way they are produced, processed and prepared, present an important potential to promote traditional diets and food systems.

GI products also normally concern either raw products (fruits, vegetables, meat and fish, honey, green coffee or cocoa, salt and mineral water) or minimally -processed products (including dairy products, vegetable oils, teas and herbal teas, processed cereals, meat fish or shellfish (e.g. smoked fish) (see the oriGIn worldwide compilation of GIs[3]). Actually the term "traditional" refers usually to *the time when populations used simple and time-honoured approaches to food production, before the introduction of technological innovations that substantially altered food production processes*" (FAO 2008), i.e. before the large scale industrialization after the second world war. Indeed, by reflecting the local traditions from a long time, GI foods may not be produced with modern food technologies or additives that would modify the cells and food texture and matrix as the ones used in industrial food ultra-processing. Despite the fact that traditions can evolve to integrate necessary or useful innovations such as mechanization that reproduce the human traditional gesture or preservation techniques, they need to maintain a meaningful traditional feature while the specific quality of the final product should remain intact.

---

[3] https://www.origin-gi.com/worldwide-gi-compilation

This aspect calls for further investigation and analysis of existing GI specifications and regulations on the level of processing of GI foods in general. More research should be also conducted on the alteration of specific and nutritional quality of GI foods when incorporating ultra-processing ingredients or procedures.

28.3 Nutritional Advantages of GI Foods Over Their Non-GI Equivalent

FAO published in 2021 a rather pioneer study on *"the nutrition and health potential of geographical indication foods"*(FAO 2021) which highlights important characteristics of some GI food products that contribute to healthy diets. This publication presents five case studies on the nutritional potential of registered GI foods: *Carnalentejana* (Portuguese beef), *furu* (Chinese fermented tofu), *Parmigiano Reggiano* and *Grana Padano* (Italian fermented cheese), *rooibos* (South African herbal tea) and indigenous rice varieties from the highlands of Borneo (Malaysia and Indonesia). The study explores the link between the production processes and the nutritional composition of the final products and shows that the nutritional characteristics (and bioactive compounds) of these foods can be largely attributed to their unique ingredients and production and processing methods, which are linked to their geographical origins.

Biodiversity in terms of animal breeds and plant varieties used for the GI product appears to be a crucial factor influencing the nutritional quality of the raw material and final food product (for example through the diversity in composition and profile of nutrients).

In the case of animal products, it is particularly interesting to see the influence of feed on the final product composition (milk, meat). Research shows that botanical composition of fodder ingested by animals, directly impacts nutritional quality of milk either directly (e.g. presence of carotine, terpenes...) or indirectly through the production of molecules by the animal (type of fatty acids such as oleic acid, plasmin, casein...) (Gregorini et al. 2022). The animal breed also influences the type of casein molecules in the milk (e.g. Tarine and Abondance dairy breeds produce milk rich in casein Kappa with B and C variants) (Macheboeuf et al. 1993). It is worth noting that local varieties and breeds have been usually developed over long periods of time through the interactions between the local environment and the human practices from the local communities, with the concept of the so-called "terroir"[4] which is at the heart of the GI concept. For example, as analyzed in the

---

[4] A delimited geographical space in which a human community has built up a collective intellectual or tacit production know-how in the course of history, based on a system of interactions between a physical and biological environment and a set of human factors, in which the sociotechnical trajectories brought into play reveal an originality, confer a typicity and engender a reputation for a product that originates in that terroir (FAO 2009).

FAO study (FAO 2021): the special features of Carnalentejana breed associated with breeding practices (e.g. free grazing on pastureland with oak trees) lead to a particular meat composition, including a much better nutrition profile compared to non GI meat: it contains significantly higher levels of several unsaturated FAs such as palmitoleic acid, oleic acid and conjugated linoleic acids (CLAs) (CLAs have been shown to protect against several non-communicable diseases (NCDs) in animal and in vitro studies (Yang et al. 2015) and it has a lower n-6/n-3 polyunsaturated fatty acids ratios than intensively fed beef (high levels of this ratio have been associated with depressive disorders and NCDs (Goodstine et al. 2003; Husted and Bouzinova 2016; Patterson et al. 2012; Wijendran and Hayes 2004).

Similarly, the specific quality of vegetal productions also depends on the variety characteristics and their interactions with the local environment (e.g. type of soils and the climatic conditions) together with the human practices (e.g. production with no or low use pesticides, and harvest at maturity) (FAO 2021). For example, in the case of the *Altopiano del Fucino carrots*, produced in the Italian province of L'Aquila, the combination of the soil characteristics (humus-rich soil of the fertile valley) and the harvest conditions as defined in the specifications to ensure good maturity leads to carrots with special organoleptic qualities and a higher level of vitamin C, beta carotene and saccharides compared to the non-GI carrots, as indicated in the GI specifications. This case as many others also illustrate the common correspondence between the special flavor and the nutritional quality.

The importance of the traditional methods on the nutritional quality of the GI foods are also noticeable at later stages, such as during processing (traditional processes are often slow and preserve better the food matrix and nutrients) and related to traditional modalities for food conservation. Such modalities can actually increase nutritional values not only in terms of quantity but also in relation with the availability of nutrients and bioactive compounds (Witkamp 2018). This is particularly the case of fermentation, a conservation method inherited from ancient generations. Many GI products are fermented food with a detailed description of the fermentation process in the specifications, as illustrated by the FAO study: dairy products, fermented soja *furu* and tea (FAO 2021). For example in the case of *furu*, a fermented tofu (soybean curd) produced in China, the GI *furu* use local materials (especially inoculating and maturing agents) and are fermented longer than the non-GI equivalents. For these reasons, the GI *furu* products contain higher levels of minerals and vitamins compared to the more industrial furu, and carry unique organoleptic properties. Such processes that enrich the food with more bioactive compounds may have a positive impact on human health through gut microbiota. The microbial diversity in foods plays a crucial role for nutrition and health protection in relation with the immune system and allergic responses (see research projects with cohort: PASTURE, GABRIELLA, MARTHA, Nicklaus et al. 2018). From this viewpoint, raw milk cheeses (often GI registered, especially in

South-European countries[5]) can represent providers of microbial diversity to contribute to human microbiota. The importance of food-processing microorganisms and of fermented foods in diets has been noticeably recognized by countries in the frame of FAO work on biodiversity and the need for strengthening research on traditional fermentation processes (FAO 2021). In this view, GI protection represents an interesting approach to preserve traditional fermented products and the use of local and biodiverse microorganisms.

28.4   The Crucial Role of GI Specifications

GI products are defined by their GI specifications, and once the GI is registered, these local rules are systematically applied throughout the production and certified as guaranteed by the GI labelled products. The fact that for example the varieties, the local environment, the production and processing methods linked to traditions are defined in the specifications ensure their continuous influence on the final product quality, including the intrinsic quality. From this perspective, the GI specifications could be considered as a good tool to ensure good levels of nutrients (guarantees of results) and/or enforce appropriate agriculture and processing practices that lead to a particular nutrition profile or influence certain health mechanism such as through the microbiota (guarantees of means).

When developing GI specifications, it would be valuable to consider the nutrition profile in the description of the product characteristics and the relations with traditional processes. For example, in the case of *Tushuri guda* cheese in Georgia, the producers, aware of the high level of salt in their cheese as a result of a tasting panel and chemical analysis organized by the project, have decided to improve the organoleptic and nutritional quality of their cheese by reducing the current level of salt content. With support of the Food Safety Agency that worked with them to elaborate specific food safety guidelines, producers could engage to improve their milk and cheese production methods and they could therefore introduce in the specifications a requirement for a maximum percentage of salt, which was lower than the average found in the traditional cheese samples. This effort allowed not only to improve the organoleptic quality of the cheese but also improved its nutrition profile. This result was possible thanks to the specifications that become compulsory for GI producers.

---

[5] In France, 76% of dairy appellation of origin are from raw milk and 70% of cheese with raw milk are PDO cheese.

28.5 Other Important Considerations of GI Processes to Contribute to Healthy Diets

Although GI registration and certification relate to specific quality and not food safety, going through the GI registration and then the GI products certification, usually lead GI producers to improve the food safety status of their production and to provide more food safety guarantees to consumers. This is particularly the case in traditional and small-scale production systems and informal marketing. In these cases small-scale producers are not able to comply with some requisites of the food safety legislations such as, for example, the number of product analysis or the interdiction of contact with some material (wood or cave which could be essential to the specific quality). The fact that GI specifications provide clear requirements based on traditional methods that have been used from generations and proved to be safe and that are then controls for certification is an important contribution to food safety. In addition, when food safety practices are not related to the specific quality and are then not mentioned in the specifications, guidelines for food safety and hygiene are always developed and applied by GI producers in order to always ensure the food safety of the GI products.

Certain regulations can facilitate food safety compliance in small-scale and traditional production systems and are particularly relevant for GI products. They are developed by public authorities with the objective to provide special rules for food safety adapted to the special conditions of small-scale or traditional production. Very interesting initiatives have been applied by countries in the framework of FAO projects (Serbia, Montenegro, Georgia), to design specific rules for food safety for traditional GI products (cheese and meat products), taking inspiration from EU regulation on food safety flexibility.

Another important aspect that may deserve more investigation, is the role of GI systems in contributing to dietary diversity. Many GI foods are using raw materials from neglected and underutilized species (breed and plant variety) or specific processing methods. This way, they make available a variety of final products on the market, including traditional foods that may have disappeared without the GI because they were not competitive compared to other food from a similar category or produced with lower production costs.

When considering healthy diet as a whole, GI economic impacts are essential to consider, not only to improve producers family access to quality and nutritious food products, but also to allow producers to reinvest in their local resources for a sustainable approach. By adding value to their products through the market and by improving value chain coordination and effectiveness, GI processes generate increased income and better livelihoods (FAO 2018). Also, many GI processes highlight women roles and enhance their entrepreneurship capacities. Increased income can represent an important factor in some communities and families to improve access to healthy diets.

Finally, there would be an important aspect to explore further, as research is rather scarce in this area: the link between traditional food products and preparation/eating practices. Actually, cultural preferences and reputation of the products may impact the way they are prepared and consumed, including frequency and combination with other types of foods. These combinations could contribute to a more balanced/diversified diets. It is also worth noting that many GIs are associated with specific diets recognized for their interest in terms of nutrition (e.g. Mediterranean diet). Several GI products are thus symbolic of such diversified diet (e.g. olive oil, legume, cheese, etc.)

28.6 Conclusion and Way Forward

In summary, the nutritional and health characteristics of at least some GI products can be attributed to the specific local conditions of production, the unique raw materials involved, including varieties/breeds, and the specific production and processing procedures, which are linked to the geographic origin and described in the specifications.

GI products can be part of healthy diets when they provide minimally or non-processed food from different categories, and food from specific local varieties with specific organoleptic characteristics and links to culture.

In these cases, GI specifications represent a tool to ensure and certify these contributions could include nutrition characteristics and inform consumers about the nutritional quality of the GI foods.

Building on these initial results—and mostly hypothesis to confirm at a larger scale-, a series of recommendations are provided to propose some milestones for the way forward.

28.6.1 Raising Awareness of Producers

As the specifications are crucial tool to preserve and increase GI nutritional quality and impacts of health, it is important to raise producers' awareness on these linkages, as well as awareness of practitioners and experts who support them in the elaboration of specifications. Strangely, nutrition is rarely a consideration as such when producers develop their product specifications (while health in relation with food safety is often at core of concerns). In this regard, public authorities could consider programmes to sensitize and provide support to help maintain and improve the nutritional value during production and processing when applicable.

28.6.2   Consumer Awareness

Research on consumers' willingness to pay for GI foods rarely include nutritional value. Such consideration can sometimes be seen in studies on willingness to pay for organic, but finer levels, like the content of micronutrients and other beneficial chemicals, haven't been explored. Building on the food composition data, it would be interesting to raise consumers awareness on the interest of some GI product as an unprocessed or minimally processed food, and other of its characteristics. This information should be always linked to the link to origin, culture and production processes, as to provide a narrative and information that makes sense for consumers in the perspective of a diversified and balanced diets. Indeed, consumers perceive quality as a combination of many aspects, and health and nutrition information alone seems to be not effective enough to stimulate behaviour change, possibly also because it is too complicated or abstract for the general population to understand (Guthrie et al. 2015).

Producers and value chain stakeholders should be encouraged to provide comprehensive information on their products and how they are linked to the origin, to feed the understanding of consumers on what is a food product and how to best prepare and eat them to contribute to a healthy diet.

28.6.3   Encourage Research and Data Publications on GI Nutrition Profile

Data on nutrition profile of GI products is of interest for consumers, business and policymakers. It would be important to encourage GI organizations and the researchers involved in the preparatory studies for developing specifications, to publish and disseminate the comparative data between GI product and its substitute on food composition, especially regarding certain nutrients and bioactive compounds.

Research programme should be encouraged to explore further the links between GI food products as part of healthy diets, including topics such as: impacts on nutrition and health of traditional practices and traditional material; food composition of GI food in various categories including bioactive compounds and microbial profiles of fermented foods, *in vivo* biological activities of food components, or robust clinical trials.; review of GI specifications in relation with their references or impacts on nutritional aspects. Results of specific food composition for GI products could inform the International Network of Food Data Systems (INFOODS).

Other important areas for research concern consumer perception, preparation and eating practices regarding GI products (and traditional products embedded in a specific culture in general), to identify the roles of social and cultural assets in healthy diets.

28.6.4 Regulations on Food Safety Flexibility

As mentioned above, national legislation on food safety flexibility have been developed in some countries to adapt to specific context in relation with small-scale production units and traditional methods and material used in some cases, including GI production. This legislation shows great benefits, as to provide further guarantee to consumers on food safety for products previously sold on informal market, while preserving a culinary heritage based on traditional methods, and a network of small scale production units,. Such approach could also be disseminated in other countries facing similar context and willing to preserve and promote their traditional food.

References

EAT-Lancet Commission on Healthy Diets (2019) From sustainable food systems. Summary report. Available at https://eatforum.org/content/uploads/2019/07/EAT-Lancet_Commission_Summary_Report.pdf. Accessed Apr 2024

FAO (2008) Promotion of traditional regional agricultural Products and food: a further step towards Sustainable rural development. Twenty-sixth FAO regional conference for Europe, Innsbruck, Austria, 26–27 June 2008. Agenda item 11

FAO (2017) The state of food and agriculture (SOFA). Leveraging food systems for inclusive rural transformation. Accessible at https://openknowledge.fao.org/handle/20.500.14283/i7658en

FAO (2018) Strengthening sustainable food systems through geographical indications: an analysis of economic impacts. FAO, Rome. https://openknowledge.fao.org/handle/20.500.14283/i8737en

FAO (2021) The nutrition and health potential of geographical indication foods, Rome. https://doi.org/10.4060/cb3913en

FAO/WHO (2014) Rome declaration on nutrition. Conference outcome document. Paper presented at Second International Conference on Nutrition, 2014, Rome, Italy. (also available at https://openknowledge.fao.org/handle/20.500.14283/ml542e)

FAO/WHO (2018) The nutrition challenge: food system solutions, p. 11. https://openknowledge.fao.org/handle/20.500.14283/ca2024en

FAO/WHO (2019) Sustainable healthy diets—guiding principles, Rome. https://doi.org/10.4060/CA6640EN

Goodstine SL, Zheng T, Holford TR, Ward BA, Carter D, Owens PH, Mayne ST (2003) Dietary (n-3)/(n-6) fatty acid ratio: possible relationship to premenopausal but not postmenopausal breast cancer risk in U.S. women. J Nutr 133(5):1409–1414. https://doi.org/10/gfzjs7

Gregorini P, Gordon IJ, Kerven C, Provenza F (eds) (2022) Grazing in future multi-scapes: from thoughtscapes to landscapes, creating health from the ground up. Front Media SA, Lausanne. https://doi.org/10.3389/978-2-88976-463-1

Guthrie J, Mancino L, Lin CTJ (2015) Nudging consumers toward better food choices: policy approaches to changing food consumption behaviors. Psychology & Marketing 32. https://doi.org/10.1002/mar.20795.

Hachem F, Vanham D, Moreno LA (2020) Territorial and sustainable healthy diets. Food Nutr Bull 41(2_Suppl):87S–103S. https://doi.org/10.1177/0379572120976253

Husted KS, Bouzinova EV (2016) The importance of n-6/n-3 fatty acids ratio in the major depressive disorder. Medicina 52(3):139–147. https://doi.org/10/gfzjtn

Macheboeuf D, Coulon JB, D'Hour B (1993) Effect of breed, protein genetic variants and feeding on cows' milk coagulation properties. J Dairy Res 60(01):43–54

Nicklaus S, Divaret-Chauveau A, Chardon ML, Roduit C, Kaulek V, Ksiazek E, Dalphin ML, Karvonen AM, Kirjavainen P, Pekkanen J, Lauener R, Schmausser-Hechfellner E, Renz H, Braun-Fahrländer C, Riedler J, Vuitton DA, Mutius EV, Dalphin JC (2018) Pasture study group. The protective effect of cheese consumption at 18 months on allergic diseases in the first 6 years. Allergy 74(4):788–798. https://doi.org/10.1111/all.13650. Epub 2018 Nov 19

Patterson E, Wall R, Fitzgerald GF, Ross RP, Stanton C (2012) Health implications of high dietary omega-6 polyunsaturated fatty acids. J Nutr Metabol. https://doi.org/10/f3m7fs

Sidiq FF, Coles D, Hubbard C, Clark B, Frewer LJ (2022) The role of traditional diets in promoting food security for indigenous peoples in low- and middle-income countries: a systematic review. IOP Conf Ser Earth Environ Sci 978 012001. https://doi.org/10.1088/1755-1315/978/1/012001

SOFI (2020) FAO, IFAD, UNICEF, WFP and WHO. 2020. The State of Food Security and Nutrition in the World 2020. Transforming food systems for affordable healthy diets. Rome, FAO. https://doi.org/10.4060/ca9692en

Swanepoel L, Raneri J (2022) Diet, human health and wellbeing in traditional food systems. A special issue of Sustainability (ISSN 2071–1050). This special issue belongs to the section "Health, Well-Being and Sustainability". Accessed at https://www.mdpi.com/journal/sustainability/special_issues/diet_small_island

Tokyo Nutrition for Growth Summit (N4G) (2020) Vision and roadmap. Available at https://nutritionforgrowth.org/wp-content/uploads/2020/12/N4G-Vision-and-Roadmap.pdf

Trichopoulou A, Soukara S, Vasilopoulou E (2007) Traditional foods: a science and society perspective. Trends Food Sci Technol 18(2007):420–427. https://doi.org/10.1016/j.tifs.2007.03.007

United Nations (UN) (2021) Science and innovations for food systems transformation and summit actions. In: von Braun J, Afsana K, Fresco LO, Hassan M (eds) Papers by the Scientific Group and its partners in support of the UN Food Systems Summit. Available at: https://sc-fss2021.org/wp-content/uploads/2021/09/ScGroup_Reader_UNFSS2021.pdf

WHO (2018) Healthy diets. Factsheet N° 394. Uploaded in August 2018. https://www.who.int/publications/m/item/healthy-diet-factsheet394

Wijendran V, Hayes KC (2004) Dietary n-6 and n-3 fatty acid balance and cardiovascular health. Annu Rev Nutr 24:597–615. https://doi.org/10/bzf7qt

Witkamp RF (2018) Bioactive components in traditional foods aimed at health promotion: a route to novel mechanistic insights and Lead molecules? Annu Rev Food Sci Technol 13:315–336. https://doi.org/10.1146/annurev-food-052720-092845. Epub 2022 Jan 18

Yang B, Chen H, Stanton C, Ross RP, Zhang H, Chen YQ, Chen W (2015) Review of the roles of conjugated linoleic acid in health and disease. J Funct Foods 15:314–325. https://doi.org/10/f7df3k

384 E. Vandecandelaere and F. Tartanac

Chapter 29
Everyday Food Practices: GI Products, Sustainable Consumption and Health

Gun Roos and Virginie Amilien

Acronyms

CAW	Cultural Adaptation Work
FAO	Food and Agriculture Organization of the United Nations
FQS	Food Quality Schemes
GI(s)	Geographical Indications
NNR2023	Nordic Nutrition Recommendations 2023
PDO	Protected Designation of Origin
PGI	Protected Geographical Indication
TSG	Traditional Speciality Guaranteed

29.1 Introduction

This chapter uses the prism of everyday food practices to take a closer look at Geographical Indications (GIs), sustainable consumption and health and wellbeing. Based on results from a qualitative consumer study, which was done in the context of the recent Strength2Food project (EU project in H2020 program aiming at Strengthening European Food Chain sustainability by Quality and Procurement Policy), we share insights on GI products, sustainability and health and reflect upon the way GI products can contribute to healthy diets. The qualitative consumer study aimed at providing an understanding of consumers' knowledge, perceptions, trust and

G. Roos (✉) · V. Amilien
Consumption Research Norway (SIFO), Oslo Metropolitan University, Oslo, Norway
e-mail: groos@oslomet.no

© The Author(s) 2025
E. Vandecandelaere et al. (eds.), *Worldwide Perspectives on Geographical Indications*, https://doi.org/10.1007/978-3-031-71641-6_29

valuation of EU, national and regional food quality labels, as well as everyday food practices including purchasing behavior with respect to food products promoted by those schemes (Amilien et al. 2018, 2022). An online ethnographic fieldwork gallery showcased photographs taken by households' family members and researchers of European households' practices around food consumption and how quality labels feature in their everyday food practices.

We and colleagues from seven European countries (France, Germany, Hungary, Italy, Norway, Serbia and the UK) did fieldwork among 41 households across the participating countries to get a better understanding of GI products as part of everyday food practices. The ethnographic approach gave us an opportunity to observe and understand the role of food quality labels, including GIs, and especially how and in which way labels were present or absent in everyday food practices. The chapter is mainly based on the fieldwork experiences in Norway (Haugrønning et al. 2018). In Norway we included two GI schemes, the one regulated by EU policies (PDO, PGI and TSG) and the other that is a Norwegian scheme "Beskyttede Betegnelser" (protected designations) modeled on the corresponding EU regulations and adopted in 2002 by the Norwegian government (https://www.beskyttedebetegnelser.no/protection-of-geographical-indications-in-norway/). As of November 2023, 32 Norwegian (including a variety of local fruits, vegetables, meat, fish and dairy products) and four Italian (*Prosciutto di Parma, Parmigiano Reggiano, Gorgonzola, Prosciutto di San Daniele*) have been granted the Norwegian protected designations. To better contextualize and understand the potential role of GIs, we will also refer to other relevant quality schemes used in Norway, as, for example, the organic label Debio and the Keyhole that reflects the healthiest choices in food product groups.

We will take a closer look at GIs and traditional food in everyday food practices through a conceptual framework "Cultural Adaptation Work" (CAW) (Hegnes 2023), which aims at describing, understanding and explaining socio-cultural adaptation of GIs. The socio-cultural adaptation of GIs means tailoring to specific food-cultural contexts and includes translations of meaning, reorganizations of social relations and transformations of things in the tension between the global and the local and between tradition and innovation.

The chapter is structured as follows. We first give some background and present main concepts before we take a closer look at GI products and traditional food in everyday food practices, with focus on sustainability and health. We will have a special focus on Norway, trying to create a dialogue between the new Nordic Nutrition Recommendations NNR2023 and the doings and sayings of our Norwegian informants in the Strength2Food project. We will end the chapter with a discussion of how to promote GI products, sustainability and health and reflect on possible contribution of GI products and traditional products to healthy sustainable diets.

29.2 GI Products and Sustainability and Health— Background and Main Concepts

The European Geographical Indications (GIs) include Protected Designation of Origin (PDO), Protected Geographical Indication (PGI), and traditional products are certified under the Traditional Specialty Guaranteed (TSG) schemes. The aims of GI products are to support local agriculture, regional traditions and know how, preserve local identities and enhance cultural heritage, protect the intellectual property and rights of local products, and protect traditional food from imitation and unfair competition.

Sustainable consumption combines social, cultural, environmental factors and economic aspects. The perspective of sustainable healthy diet (FAO and WHO 2019) that guides this chapter builds on the human right to food and a holistic view of diet, food security, health and wellbeing. The human right to food security is realized when all people, at all times, have physical and economic access to sufficient, safe and nutritious food to meet their dietary needs and food preferences for an active and healthy life. FAO has described four key dimensions that should be fulfilled simultaneously: physical availability of food, economic and physical access to food, adequacy of diet, and sustainability as food should be accessible for both present and future generations (FAO et.al 2023).

The perspectives emerging from the new Nordic Nutrition Recommendations NNR2023 are also important to have in mind. The NNR2023 report represents the scientific basis for national dietary guidelines and recommendations in the Nordic and Baltic countries (Blomhoff et al. 2023). The overall recommendation is: "a predominantly plant-based diet rich in vegetables, fruits, berries, pulses, potatoes and whole grains, ample amounts of fish and nuts, moderate intake of low-fat dairy products, limited intake of red meat, white meat, processed meat, alcohol, and processed foods containing high amounts of added fats, salt and sugar." The role of processed and ultra-processed (energy dense products, high in added or free sugars, salt and total fat/saturated fat, and low in fiber and micronutrients foods) are discussed in the recommendations and the recommendations include a focus on whole foods and limiting processed foods.

Using the Cultural Adaptation Work (CAW) will shed light on the combination of the several approaches proposed here, including GIs and their juridical frame, informants' perceptions and wordings, FAO's definition of sustainable food and official Nutrition Recommendations. CAW is a conceptual framework (Hegnes 2023) aiming to better understand how people continually adapt their understanding, their interaction forms and their material environment relationally, in time and space through translations of meaning, reorganization *of social relations* and the *transformation of things*. Understanding food consumption as transformative practices permits us to identify, describe and understand the construction, power relations, adaptations, justifications, consequences and strategies to cope with food, in an over-civilized risk and health focused society. Exploring sustainable and healthy food consumption through perception and understanding of GIs by our informants, will *expand our* knowledge base of the *cultural conditions* underlying the potential use of GIs towards a more healthy and sustainable consumption.

29.3   GIs, Consumers and Everyday Food Practices

The researchers in the Strength2Food project did fieldwork and gathered data on household everyday food practices to shed light on the participants' interest for and use of Food Quality Schemes (FQS), including GIs. In Norway, we visited six households (named H1 to H6) three times and participated in the different phases of consumption (planning, purchasing, using/cooking/eating and disposal practices) together with the households. During the first visit we brought a basket with FQS food products to stimulate semi-structured interviews focusing on FQS. The interviews were transcribed, anonymised and analysed.

In the Norwegian households, GIs were seldom present neither in dialogues and justifications nor in everyday food practices. In general, the households' practices related to cooking and eating were, as with their purchasing practices, much routinized. The trip to the grocery store was in general quick and ruled by routines and habits, and we observed participants efficiently assembling their food staples in the store and they seemed to ignore labels. In the interviews they said that they may check the appearance of fruits and vegetables, and that price, taste and trust based on earlier experiences are decisive for what they choose. Also, when the households at the first visit were presented with food baskets including products with these labels they didn't seem to pay attention to or mention the labels. Participants focused more on the type of product, appearance, design, quality and brand. When evoking a discussion with the participants about GIs, most of them expressed their uncertainty, lack of understanding, lack of visibility, limited interest and knowledge about these labels (Box 29.1).

Box 29.1: Quotes from Interviews with Norwegian Households H1, H3 and H5 (Names Are Pseudonyms)

It's a disadvantage that I didn't reflect on this or that label. I don't have a relation to what it is. Now it's been in front of my nose. We have talked about labelling, you have asked questions about labels, and I still don't see them. (H1—Arne)

We don't look although we are very interested in food, we buy what we want and we look for if it's organic and choose the organic if they have it. Then we start looking and if it's not [organic], like ham, then we take the one that the kids think tastes the best. (H3—Dagny)

It's like Chardonnay and Chablis, that it's the same but made in different places. But the whole country [Norway]. A little bit too wide for it to be interesting. (H5—Sofie).

The households appeared to stick with their routines and the products they had become familiar with. One of the participants pointed to what many of the participants found difficult with food provisioning and labelling schemes: *"I would like to make real choices without spending too much time on it"* (H4—Elisabeth).

During our kitchen tours in the households, we didn't observe any food products with the Norwegian protected designations labels. However, several households had olive oil with a PDO or PGI, but the label was not the reason why the products had been bought. GI products were bought, used, or found in homes because of their taste, quality or because of habit, not because of the label itself. However, labelling and other information that the consumer can trust were recognized as relevant (Box 29.2).

Box 29.2: From Interview with Norwegian Household H4 (Name Is Pseudonym)

Elisabeth: I think we should have more labelling that we can trust, I thought maybe organic was a solution, but it turns out it is not, there is always a dark side.

Interviewer: Then what can we do? You say we need more labelling, but we see that we already have all these different labels.

Elisabeth: I think maybe more visibility would be nice. And I think [the existing labels] are not enough because right now animal welfare and ethical issues are not communicated.

One of the participants expressed several flaws with the Norwegian protected designation labels mostly because they do not show enough traceability and the history of the product (Box 29.3).

Box 29.3: From Interview with Norwegian Household H5 (Name Is Pseudonym)

Sven: Once upon a time, you had a bag of potato chips that said that the potatoes were from a place down the street. You know, traceability. When it says that it comes from this place, with this producer, it has more of a story line. That it is from a place in Lillehammer. This is the kind of thing I find interesting.

Interviewer: But why do you find it interesting?

Sven: It does something to the product if it has a history. Many times, you get a packet of minced meat which is one out of 40.000 animals at Tulip [Danish brand], then I don't want to buy it.

29.4   GIs and Healthy Diet

Health and healthy diet have definitions that vary from culture to culture. GIs are also different from product to product and country to country. Thus, we need to explore the meaning of consumer understanding, knowledge and food practices linked to GIs through cultural adaptation work.

The fieldwork revealed that consumers do not know much about the GI specifications that producers follow and respect, and GIs did not shape consumers' everyday food practices, but concerns about health and nutrition were observed during all phases of consumption. For example, the participants talked about healthy meals, having fish at least once a week for dinner, limiting minced meat and checking ingredients. Some consumers associated GIs with "better", "greener" and "healthier", but some GIs that are based on older conservation methods were also associated with high contents of salt (for example, cured leg of lamb). Fresh GIs and traditional products were often perceived as closer to organic production that uses less chemicals and pesticides. This was, for example, the most important thing for one of the participants: *"To me the most important thing is that the food is not sprayed [. . .] I think about the poison, I do not want that, and I do not want to expose my children to chemicals"* (H3—Dagny).

Another participant told us that she doesn't look for labels, but she reads the list of ingredients to check for additives. She also said that she avoids some imported vegetables and fruit because she thinks they are heavily sprayed: *"It is a lot of preservatives with the exception of what is in jam that is a natural substance, but it's a lot of Es and coloring and other things, then I'm skeptical"* (H6—Linda).

Several participants voiced concern that labels have been misused and thus consumers do not trust these. A typical example was the Keyhole, the Nordic label for healthier food (Box 29.4).

Box 29.4:   From Interviews with Norwegian Households H2, H3 and H5 (Names Are Pseudonyms)

The Keyhole, it has been on many products that are not typical healthy products. It's not negative that there is the Keyhole on products that we buy, but it's not decisive for if we buy or not. (H2—Mona)

David: The Keyhole it is actually a fake logo in one way, because it means that it is a little bit healthier than the product next to it. It doesn't mean that it is healthy even if it is a Keyhole.

Dagny: and I think it's wrong, I think it's deception.

David: When Grandiosa pizza has a Keyhole, it is many who don't understand, they buy because it's a Keyhole. I never ever follow the Keyhole.

(continued)

Box 29.4 (continued)

Dagny: For our sake they could have not done it, this fake logo, the breakfast cereal shelf is the worst shelf, it has many Keyholes. (H3—David and Dagny).

Sofie: You may risk getting the Keyhole on pizza Grandiosa, it's better than other pizza. . . . so meaningless.

Sven: They have spoiled the label themselves. (H5—Sofie and Svein).

The Nordic Nutrition Recommendations NNR2023 include recommendations to limit intake of processed foods and meat (Blomhoff et al. 2023), but the exclusion of guidelines for ultra-processed foods in NNR2023 is discussed in Norway (Kolby 2023). The group "processed foods" includes a wide spectrum of products and the amounts of added fats, salt and sugar vary. Ultra-processed products are energy dense products, high in added or free sugars, salt and total fat/saturated fat, and low in fiber and micronutrients. Some of the participants were concerned about ingredients, processed and ultra-processed food (Box 29.5).

Box 29.5: From Interviews with Norwegian Households H1 and H3 (Names Are Pseudonyms)

I would like to say that it is important that I know what's in it, but if I really meant it. I should maybe have been better at reading the list of ingredients on the pre-prepared foods that I buy, because I don't do it very much. (H1—Arne)

David: I'm very interested in good food and to get good food you have to have proper raw materials.

Dagny: I think you should buy what is sustainable and stop buying the other, so that they stop making it, I think there is a lot we could manage without.

Daniel: Like what for example?

Dagny: All the pre-prepared food; learn to make food from scratch. All sweets, it's not very sustainable. Now is chocolate not sustainable anymore, there is soon no cocoa left, isn't that true? It's not sustainable to eat as much chocolate as we Norwegians do.

Daniel: But with sustainability, it is to also to use the food that you buy, many fill the garbage can with leftover food and all kinds." (H3—David, Dagny, Daniel).

Local food, place and sustainability were important to the households although these values did not always seem to be manifested in their food practices. As Arne (H1) said *"we more often choose based on what we want or think is good rather than where it's from."* When talking about sustainability the participants mostly referred to sustainability as environmental sustainability and few discussed the connection to the ethical implications of consumption. Especially food packaging and plastic wrapping was mentioned: "*I think some of the most important to do is to cut down on packaging. . . I get frustrated when we, for example, buy taco. Then it's plastic on the avocado and plastic and cardboard on the tomatoes. In some places you can bring your spice jars and fill them up and I think it's cool, but I think it's a too big step for many. But to cut unnecessary plastic wrapping*" (H1—Anne).

Some of the participants suggested that the Government should take actions enhancing the bargaining power of the producers in dealing with the large distribution channels and should support and promote small local producers. Some were critical about the use of the word sustainability because it's been misused and all wish to be sustainable. Svein said that he wanted to decide himself based on information if it's local or not and that they prefer to buy local food: "*we prefer to buy organic, we prefer to buy Norwegian, we prefer to buy local, if these three meet, it's the best, but if we can choose it's local that is the first*" (H5—Svein). He was not the only participant that said that he prefers to buy local food (Box 29.6).

Box 29.6: From Interviews with Norwegian Households H4 and H6 (Names Are Pseudonyms)

I try to buy food that is local, with organic it's not necessarily good because it comes from the other side of the world. (H4—Elisabeth)

I'm not fanatic, but I want quality and I want it to be local as much as possible. (H6—Linda)

The consumer study points to that there is a potential for sharing more information about how GI specifications may contribute to a more diversified and healthy diet. Small scale produced GIs can emphasize the closeness between local place and local people who consumers meet. Large scale produced GIs can find a similar relationship, but this has to be virtual—hyperreal.

29.5   Conclusions

This chapter underlines that the six Norwegian households were quite similar to the 35 other Strength2Food households in France, Germany, Hungary, Italy, Serbia and the UK (Amilien et al. 2018, 2022): consumer awareness of GIs was generally low, and GIs were not a visible part of everyday food practices. Quantitative consumer surveys in the same Strength2Food project also showed that GIs were not recognized by consumers (Hartmann et al. 2019).

The ethnographic approach gave us an opportunity to observe and try to understand the role of GIs in everyday food consumption. GIs (PDO, PGI) and TSG were not perceived as interesting by our participants, but rather invisible and difficult to understand. However, some other labels such as the organic label, fair trade and national quality labels seemed to be well known and integrated into participants' practices. GIs were perceived as quite complex, especially because products are very different from each other. GIs did not play a central role in planning nor performing everyday routinised food purchasing. Participants were seldom aware of the presence or meaning of the GI labels on the staple foods in their kitchen and they didn't pay attention to the labels. GI products were mainly used as part of special occasions with the exception of specific GI products such as *Parmigiano Reggiano* that was well known and integrated in food practices. *Parmigiano Reggiano* has also been granted the Norwegian protected designations.

On the one hand, the GI labelling scheme is a system for protecting the name and preserving the characteristics of a product because of the place it is coming from. But specifications of GIs also often include better quality and more sustainable production. On the other hand, the consumer approach shows that there was a great interest in health. In a cultural adaptation perspective this emphasizes not only that consumers knew very little about GIs but they seldom associated them with sustainability and health. Transforming the original aim of GIs would permit to translate it to a new meaning, where the link between GIs and health is more important.

Our informants focused on the importance of organic food and pesticides and chemicals in food products for them. A better knowledge and recognition of GIs would also promote their potential roles in everyday food consumption. This would require a reorganization of social relations, not only within producers, but also between producers and consumers.

Moreover, consumers participating in this study brought up that the Government should increase controls of production practices and promote educational programs addressing consumer awareness not only of what to eat, but also about the dynamics behind the food production and provision systems. Several participants noticed that labelling, information flyers and traditional marketing methods have had limited effect, thus maybe better with other types of measures like educational programs, workshops and interactivity with food shoppers and eaters.

The new Nordic Nutrition Recommendations NNR2023 emphasize the need to decrease meat consumption, have less processed food and be more aware of food quality. The media debate around the recommendations has been intense and tense in Norway, and we suggest that GIs would be an opportunity to highlight many of these dimensions to consumers.

Acknowledgements The work has been carried out as part of the Strength2Food project that has received funding from the European Union's Horizon 2020 research and innovation programme under grant agreement No. 678024. We would like to thank the participating households and the Strength2Food partners who participated in the ethnographic fieldwork.

References

Amilien V, Roos G, Haugrønning V et al (2018) Deliverable 8-2 ethnographic study: qualitative research findings on European consumers' food practices linked to sustainable food chains and food quality schemes. Strength2Food Project

Amilien V, Discetti R, Lecoeur J-L et al (2022) European food quality schemes in everyday food consumption: an exploration of sayings and doings through pragmatic regimes and engagement. J Rural Stud 95:336–349

Blomhoff R, Andersen R, Arnesen EK et al (2023) Nordic nutrition recommendations 2023. Nordic Council of Ministers, Copenhagen

FAO, WHO (2019) Sustainable healthy diets—guiding principles. FAO, Rome. https://www.fao.org/3/ca6640en/ca6640en.pdf. Accessed 20 Nov 2023

FAO, IFAD, UNICEF et al (2023) The state of food security and nutrition in the world 2023. Urbanization, agrifood systems transformation and healthy diets across the rural–urban continuum. FAO, Rome. https://doi.org/10.4060/cc3017en. Accessed 20 Nov 2023

Hartmann M, Yeh C-H, Amilien V et al (2019) Deliverable 8-1 report on quantitative research findings on European consumers' perception and valuation of EU food quality schemes as well as their confidence in such measures. Strength2Food Project

Haugrønning V, Amilien V, Roos G (2018) Quality labels lost in everyday consumption: an ethnographic study of six Norwegian households food practices linked to food quality schemes and sustainable food chains, SIFO project note no. 8. OsloMet, Oslo

Hegnes AW (2023) Food cultures and geographical indications in Norway. Routledge, London

Kolby M (2023) Mer redelighet i diskusjonen om ultraprosessert mat etterlyses. Tidsskr Nor Legeforen. https://doi.org/10.4045/tidsskr.23.0588